Infrared Tools for Solar Astrophysics: What's Next?

Proceedings of the Fifteenth
National Solar Observatory/Sacramento Peak Summer Workshop

Infrared Tools for Solar Astrophysics: What's Next?

Sunspot, New Mexico, USA 19 –22 September 1994

Edited by

J. R. Kuhn & M. J. Penn

National Solar Observatory, USA

Sponsored by

The National Solar Observatory
The National Science Foundation
The Air Force Office of Science Research
The National Aeronautical and Space Administration
The National Optical Astronomy Observatories

Published by

World Scientific Publishing Co. Pte. Ltd.

P O Box 128, Farrer Road, Singapore 9128

USA office: Suite 1B, 1060 Main Street, River Edge, NJ 07661

UK office: 57 Shelton Street, Covent Garden, London WC2H 9HE

INFRARED TOOLS FOR SOLAR ASTROPHYSICS: WHAT'S NEXT?

ISBN 981-02-2188-6

This book is printed on acid-free paper.

Printed in Singapore by Uto-Print

Preface

The papers in this volume were presented during the 15th National Solar Observatory Workshop at Sacramento Peak during September 19-23, 1994. This meeting was organised with two objectives: 1) to sample the new results from extended-object infrared observations, and 2) to explore the potential capabilities of large-aperture, low-scattered light instrumentation. The steady percolation of new infrared array detectors into astronomical instrumentation has stimulated interest in a broad range of problems that have resisted previous attempts at solution (for example "true-field" magnetometry and coronal imaging). Although much of our discussion centered on problems in solar astrophysics, we also explored some important issues in planetary and extragalactic astrophysics that bear on the workshop's second objective.

The meeting had an unusually diverse group of participants, with a significant fraction of astronomers interested in nighttime problems along with a majority of solar astronomers. Several commercial and industrial participants brought the latest news on infrared detectors and instruments to the discussions.

Much of the discussion at the meeting centered on the characteristics and potential for a unique new coronagraphic telescope. Many of us were excited by the possibility that such a novel instrument could be used for both day- and night-time astronomical problems. The engineering constraints that minimized scattered light in the telescope also lead to an instrument with the smallest possible IR telescope emissivity. Some of the comments that were generated during a panel discussion of this project are recorded in this volume. We also organized a gedanken experiment whereby each of the workshop participants was asked to submit an "observing proposal" for this coronagraphic telescope. A summary of those proposals is also presented here.

The organizing committee for this meeting consisted of Jeff Kuhn, Serge Koutchmy, Don Neidig, Matt Penn, Doug Rabin, Ray Smartt, and Jack Zirker. This truly was a group effort and would not have succeeded without the help of these individuals. Financial support for the workshop came from several sources. Don Neidig worked hard to obtain support from the Air Force Office of Scientific Research and its European (London) and Asian (Tokyo) offices. David Sime at the NSF division of Atmospheric Science, Bill Wagner from NASA, and the NOAO Director's office also provided critically needed funding. And finally, we would like to thank Vickie Fowler for the enormous amount of work she did on the workshop proceedings.

Jeff Kuhn and Matt Penn
National Solar Observatory at Sunspot, NM
January 1995

Contents

ADVANCED CORONAGRAPHY

SERGE KOUTCHMY
Institut d'Astrophysique-CNRS
98bis Bd Arago F-75014 Paris

ABSTRACT

In this introductory lecture we only try to cover a few selected topics which seem to justify our excitement about the possibility of having, in the not too distant future, a Large Mirror-objective Coronagraph (LMC) available. This could be the main IR tool for Solar Astrophysics and coronagraphy but would also be an ideal instrument for several other important applications. Specifically it could also be used in fields such as space debris detection, IR low-telescope-emissivity astronomy in general and night-sky coronagraphy. Selected topics of this presentation are: extra-solar astronomy, optical detection of small orbital debris, solar-disc and solar limb observations, off-limb observations in both low-excitation emission lines and forbidden coronal lines and finally, spectro-polarimetry to measure the magnetic field in the corona (and the chromosphere, the transition region and in prominences). The latter appears to be one of the major challenges in solar physics, requiring new advances in instrumental methods, as well as greatly improved theoretical analysis.

1. Preliminary Remarks

In this 45 min talk, it is possible to cover only a very few highly selected topics which, in no case, reflect the full potential of new possibilities for the observational astrophysics opened by the construction of a Large Mirror Coronagraph – LMC (see the parameters in J. Becker's paper of this workshop). We believe this is the main IR tool for future Solar Research, but also for a whole series of additional observational topics in Astrophysics and the whole workshop will indeed deal and more deeply develop the subject. However, a first remark should be made. Considering the development of an entirely new instrument like the LMC, based on several new concepts and technological advances like the super-polishing of mirrors, thermal control and very low-emissivity and scattering, IR capability and zero-parasitic polarization, etc., there is no need to find any scientific justification: undoubtedly such instrument will lead to several discoveries we cannot at present anticipate, as it is usual each time technological progress is introduced into scientific research.

An other remark concerns the need for a truly new solar facility to open new observational windows, like the measurement of the coronal magnetic field and the sophisticated analysis of faint and subtle surface phenomena of prime importance for solar-terrestrial physics. In addition to the optical windows already used in conventional telescopes, we mean the UV accessible on the ground, and especially the IR windows. Such an instrument does not exist at present and has never been planned; accordingly we deal with a fully novel domain and speculations are inevitably permitted. So, unconfirmed or even provocative statements will intentionally be formulated in this contribution; I believe this is the best way to proceed, at least at the beginning, when dealing with new things. Hopefully, these statements will be corrected and/or substantiated by the presentation to be given during this workshop by specialists in the different fields, including theorists and instrumentalists, as well as scientists dealing with the data and their interpretation.

In that sense, we will hear with great interest what scientists studying the solar atmosphere, including the corona, have to say, as well as those who want to use the LMC to analyze space debris, or to look around stars, or even around very far extragalactic objects. It will be equally interesting to hear about technological progress made in the last few years, especially in the IR, which is well suited for a very efficient use of a LMC which has a totally unobstructed aperture of very low emissivity and where there is great promise for the whole of astrophysics.

2. Doing Exciting Optical, IR and Spectro-photo-polarimetry of Relatively Faint and Extended Fields close to a Very Bright Source: Advanced Coronagraphy

Let us first try to define what we mean by "advanced coronagraphy". A true coronagraph was invented more than half a century ago by B. Lyot and it is still the source of inspiration of today's design of coronagraphs, which should include low scattered-light optical components, apodization methods, very low instrumental polarization, etc.. However, astrophysics today needs in additional properties which were not present in the Lyot design, namely access to IR and a large aperture. Accordingly, we should reconsider what is a modern coronagraph:

It is a low-scattered-light large reflecting telescope capable of measuring objects of the order of 10^7 times fainter than the solar disc at 1 arcmin from the limb (see Figure 1 and Koutchmy, et al.[1]). We notice that a large aperture coronagraph was never designed, up to now, mainly due to technological difficulties with mirrors; these difficulties were recently overcome.[2] Obviously true stellar coronagraphs does not exist, because the chromatism in Lyot-type instruments drastically limits the detection of passband. The result of this situation is:

1. The absence of suitable large aperture coronagraphs necessary to reach sub-arcsec spatial resolution in coronal researches, keeping enough signal to noise ratio to permit useful diagnostics of densities, velocities and temperatures simultaneously, which is not possible even in space with SXT's, due to the lack of photons!
2. The inability to incorporate in a coronagraph modern techniques like imaging (CCD) polarimetry, active optics, IR technology, etc., keeping an apparently fair spatial resolution.

3. The complete lack of observational results about fundamental and well-predicted phenomena in solar physics, like the identification and analysis of fast mode waves or the measurement of the true coronal magnetic field and its variation, even without considering small scales! Note that the pioneering attempts made up to now (e.g. J. Harvey Ph.D. thesis) were done with instruments of small apertures by the today standards for photospheric magnetic field measurements, with which a signal of an amplitude 10^6 larger is available and array detectors are used!!

4. The lack of results on stellar coronagraphy, circumstellar imaging and so forth, although post-focus stellar coronagraphs were used to circumvent the disadvantage of using a conventional telescope with its large amount of spurious scattered light.

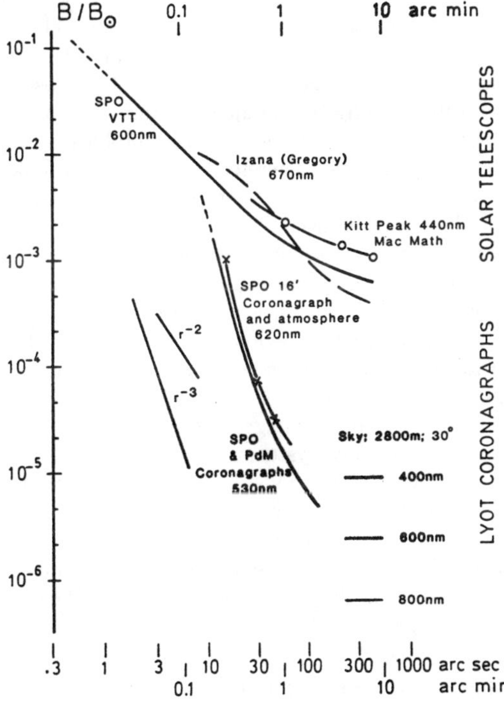

Figure 1: Levels of scattered light measured outside the solar limb using different selected solar telescopes and using conventional Lyot-coronagraphs on good coronal skies. The sky level far enough from the solar limb is indicated at the bottom-right, for different wavelengths and a high altitude site at h=2800 m.

Today, the availability of new mirror technology makes feasible the goals of precursors, such as R. Bonnet in France,[3] and H. Zirin and G. Newkirk in USA.[4] Fast progress in IR technology makes the project even more attractive in opening new observational windows and making the idea of a 4m aperture LMC a very exciting opportunity.

3. Advanced Coronagraphy in Extra–Solar Astronomy

There are many obvious "night-sky" objects to be observed with a coronagraph, starting from the analysis of highly energetic circumstellar phenomena: binary interacting systems, accretion disks, non-thermal emission and jets in nebulae around symbiotic stars the prototype of which is R Aquarii.[5] A quite similar situation, from the pure observational point of view, is encountered in radio galaxies, active galactic nuclei, etc.; many extra-galactic more or less compact objects would greatly benefit by being analyzed with the LMC, to better evaluate their close neighborhood which could certainly give also a new insight into the problem of the so-called missing mass. However, the field which could undoubtedly benefit the most is the search for protoplanetary disks around young main-sequence stars and other stars. It is possible that the Astrophysics of the XXIst century will emphasize the search for evidence of extra-solar planetary systems. Therefore, it is time to look with a suitable instrument around but close to the nearest stars! Only one example of a dust disk is known: β Pictoris which was discovered a decade ago thanks to the use of a post-focus "coronagraph" attached to a two meter aperture telescope. Recently[6] the disk was observed in the IR at the 3.6m aperture ESO telescope and the observed asymmetry of the disk was interpreted as evidence of a gravitational drag effect produced by a giant planet, forming a dust-free region in the dust-disk. This would be the first indirect evidence for an extra-solar planet, although the inferred distance from the central star seems too large. It would be extremely interesting to look more closely at this system with more advanced coronagraphic systems. Even more interesting would be to find other examples of such systems, taking advantage of a larger aperture and especially, a far lower level of spurious scattered light (see again Figure 1). This was recently rather clearly pointed out in a review paper[7] considering the perspective of "Science with the VLT of ESO"; in its conclusion about the problem of the detection of planets, the author confessed "...direct detection of planets in the next decades is thinkable but needs dedicated coronagraphs". We certainly agree with this statement!

4. Optical Detection of Space Debris

This field is of prime importance for those who design spacecraft, but perhaps most important is the question of the evolution of the orbital debris populations and the search for evidence of collisions among debris, which would produce still more debris.

Thanks to the large amount of scattered light due to forward scattering of the solar light by debris particles, it was shown that it is possible with a two-meter-aperture coronagraph to see debris of millimeter size.[8] This is by far the best method to analyze small debris from the ground, see D. Neidig paper of this workshop.

Note that orbital debris will be seen during coronal observations "in focus". Earth atmospheric aerosols up to an altitude of 80 km will be out of focus. Both populations produce a noise background in fine coronal observations, but this noise can be extracted using a suitable algorithm. Using the Mac II coronagraph[2], we obtained movies at a video-rate which beautifully illustrates the method described by Koutchmy, et al.[1], but more extended work can be done in the future to advance further in the analysis of space debris recorded with the LMC.

5. Solar Disc and Solar Extreme-limb Observations

a) A high spatial resolution low-scattered-light and polarization-free solar telescope as the LMC will bring fresh and fundamental insights in several fields of photospheric studies. This will be a direct consequence of the increased contrast of all measurements resulting from the vastly decreased magnitude of the wings of the instrumental smearing function (to keep this advantage, earth-atmospheric induced effects should also be limited.[1]) Let us mention some relevant measurements: intergranular contrasts; penumbral filaments; pores, bright rings; magnetic elements: tubes, sheets, ropes, etc.. Obvious progress will be made in problems of magneto-convection, because the best sites to analyze are, with present instruments, full of spurious scattered light; they are the cores of sunspots where umbral dots (u.d.) and chains of dots are painfully observed with the best instruments. Are u.d. the remnants of some magneto-convective effects occuring in the inhomogeneous core, or are they as suggested by the apparent connections with definitely chromospheric umbral structures, the result of hot gas streaming downward toward the core, a kind of inverse Evershed-effect? Needless to say, observations near 1.6 μ in the opacity minimum region are important, as well as line-profile analysis in polarized light, where too often the poor concept of a filling factor systematically replaced the search for more adequate direct measurements when small-scale phenomena are concerned. Finally, on the disc and in chromospheric lines, we should mention the analysis of the very bright flare kernels that could be resolved during the whole flaring phase and which are often indistinguishable from the brightening of parts of small loops.

b) However, extreme-limb observations including chromospheric structures are probably the most perspicacious observations to be performed with the LMC First, looking at facular elements and sunspots at the photospheric limb, the temperature minimum region is directly measured, which means the extension of the inferred geometric depression or hillock above magnetic regions.

Further, in photospheric and especially chromospheric emission lines, the transition region (T.R.) toward the corona will be at last analyzed from the ground. This starts with the "shell" phenomena analyzed with the D3 He lines in the fifties by

Athay, Menzel, White and later by Russians in 1083 nm of He I. The He emission lines are due to the coronal ionizing back-radiation[9] which permits the T.R. to be "illuminated". These layers are clearly highly inhomogeneous (see Figure 2) and this is one of the keys to understanding the whole problem of chromospheric heating and of solar mass loss. Without even trying to enter into the details, we mention some of the outstanding problems of T.R. physics: turbulent and oscillatory motions, propagation of waves including fast modes; flows, explosive and/or impulsive events, spicules and macro-spicules; deep sites of magnetic energy release; down-flows, bottom "face" of the corona especially in coronal holes; shocks and mechanical heating; the feet of coronal flux loops; connections with the cool "chromospheric" area where CO-emissions are observed; etc..

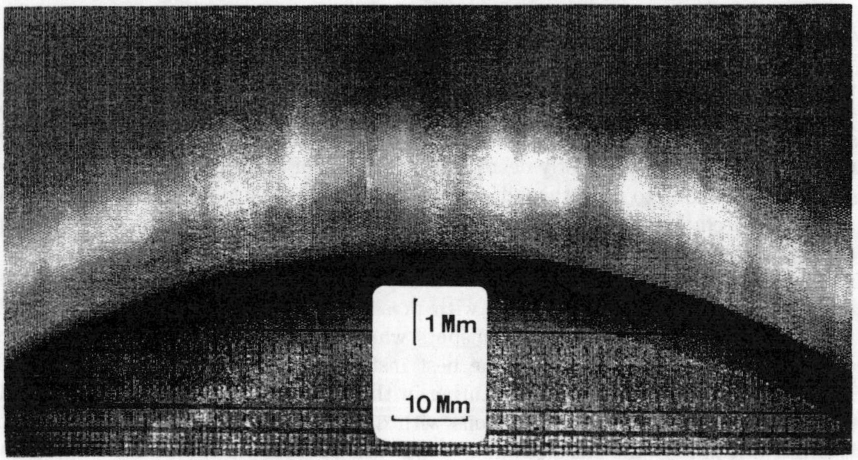

Figure 2: A cut through the T.R. He-shell as seen in the D3 emission-line near the solar equator; intensity variations in the radial directions were stretched by a factor × 6 to better show inhomogeneities.

6. Out of the Limb Observations in both Low-excitation Emission Lines and Forbidden Coronal Lines

Going further in the radial direction, at distances larger than 5 to 10 Mm above the photospheric limb, regions where the hot corona dominates are measured, but in low-excitation lines where the "cool" component can be analyzed. This component is highly dynamical, thus the large throughput of the LMC will be very much appreciated in simultaneously determining thermo-dynamical parameters, densities of neutral, ions and electrons, their temperatures and microturbulent velocities, their Doppler and proper motion velocities and the surrounding magnetic field. This is possibly the location where rapidly propagating waves heating the chromosphere could best be identified, because radiative transfer effects present in disc observations [10] are avoided and fast modes should show up essentially as transverse motions (see Figure 3). In the case of more impulsive and possibly explosive events, multi-line analysis of H, He and Ca II IR lines could already enable us to disentangle parameters like the temperature, the turbulence and the transverse motions which are completely mixed up in T.R. spaceborne UV measurements, without mentioning the temporal resolution which would be vastly improved with the LMC! This could be fundamental to identify shocks and/or highly non-linear effects.

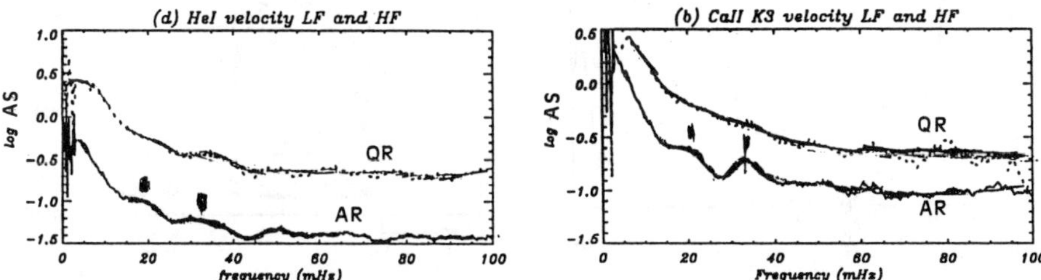

Figure 3: Amplitude spectra of velocity variations measured [10] in chromospheric lines to show the possible occurrence of waves with short periods.

Jumping now to more specific questions related to the physics of the corona, again we start with the search for resonance oscillations in loops and the identification of waves (fast-modes), especially in magnetically dominated and/or confined regions, see Figure 4. First attempts to observe these oscillations[11] as well theoretical expectations[12] require in addition to spectral and to spatial resolution, a high temporal resolution, which using coronal lines, can be reached only with the LMC We feel that in the very first runs of observations with the LMC, coronal oscillations and waves measuring phase velocities will be resolved in Doppler velocity (instead of speculating about the tedious Doppler width of coronal lines) and will permit a giant step in coronal physics.

Doing fast imaging in coronal lines would also permit progress. Let us mention exciting topics like i) loop-loop interactions and/or encounters considered by R. Smartt and V. Airapetian at this workshop; ii) dark loop dynamics (are they high beta empty magnetic flux tubes embedded in the corona?) and iii) very small plasmoids, rapidly evolving at T.R. temperatures and coronal heights (indeed above the coronal height-scale of 65 Mm), which were recently discovered during the 1991 total-eclipse using the largest optical telescope ever pointed on the Sun.[13] These last phenomena are probably the key to understand both coronal mass loss and coronal heating by magnetic energy release in the shell of the plasmoid vortices; surprisingly enough we do not see at present, besides numerical experiments on computers, any diagnostics which could be attempted except observations to be made with the LMC!! The explanation is once again the need to reach a high photon throughput in coronal radiation and a simultaneous high temporal resolution, which is impossible to envisage at the moment in spaceborne experiments.

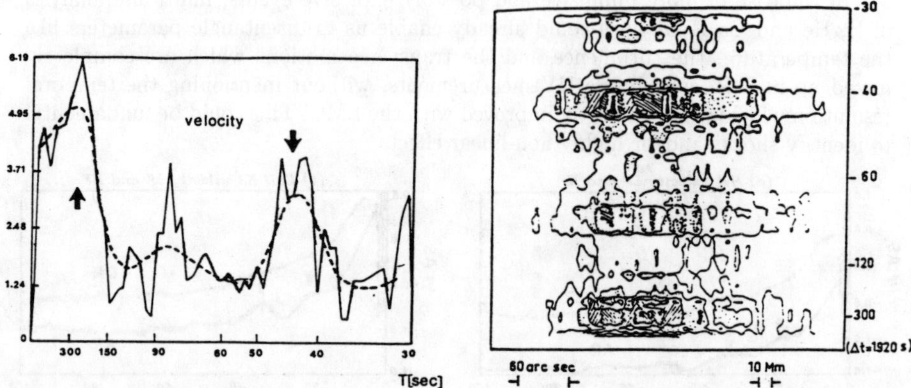

Figure 4: Power spectra of coronal transverse oscillations showing high frequency waves.[11]

7. Spectro-Polarimetry to Measure the Magnetic Field

This is the field where the LMC possesses the greatest potential of discoveries for Astrophysics and Solar Physics. Thus, let us try to go into some details, starting from a very short review of the existing knowledge about coronal magnetic fields.

A. Inferred values from different methods

The solar atmosphere is extremely structured starting from the bottom of the chromosphere, the main agent responsible for that being the magnetic field. A first method to guess what is going on is to extrapolate the measured (with the Zeeman effect) photospheric field; many photospheric magnetographs are in operation everywhere, but controversies still exist on their accuracy. However, the models developed to extrapolate the field through the chromosphere into the corona are even more controversial, although extended theoretical studies, including sophisticated computer codes, are performed. Both potential or current-free fields and force free fields are considered, with mixed success. Computations are based on the use of highly averaged maps of the photospheric fields where, e.g., close opposite polarities would "disappear", etc. Generally speaking, highly concentrated fields are ignored, including those of the core of sunspots. Note that the corona seems "empty" of gas above sunspot umbra, which is not true at their edges where a confinement of the plasma could occur, with loops and filaments confined in twisted and sheared fields, etc. Magnetic surfaces called separatrices are also predicted.

The field inside the corona was also inferred from pressure balance equilibrium at the edge of discontinuities like the well observed tangential discontinuities in the intermediate corona at eclipses.[13] Rather low values are deduced. This is not the case with the magnetic fields deduced from the interpretation of radio emissions of non-thermal origin, including gyromagnetic emission.[14] In figure 5 we assembled all values we could find in the literature, as a function of the radial distances. A large range of values seems to co-exist and there is no doubt that a real dispersion exists. However this analysis convincingly shows that the plasma has a low beta in the very inner corona but not everywhere in the intermediate corona.

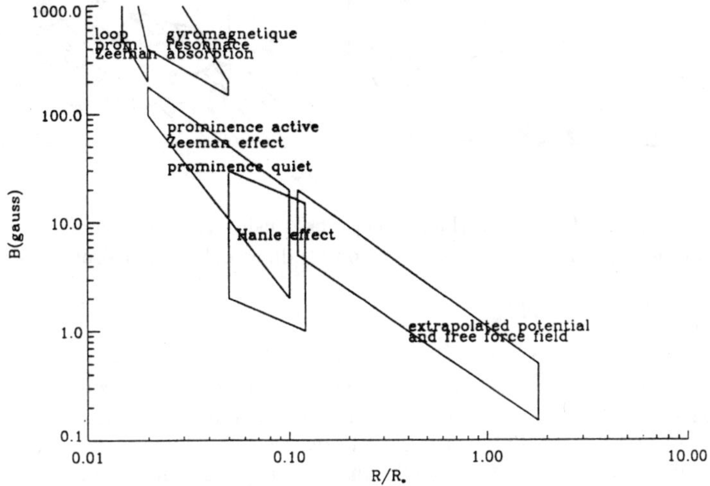

Figure 5: Log-log display of the strength of the coronal magnetic field as a function of radial distances deduced using various methods."

B. Observations made up to now

1. Prominence magnetic field using emission lines with $\tau \lesssim 1$

In figure 5, we include measurements performed using both the Zeeman effect and the Hanle effect[15] over prominences because they are embedded in the corona and, accordingly, give a good diagnostic of the coronal field. For quiescent prominences, the latest values coming from the interpretation of the Hanle effect correspond to rather low values. The questions which emerge from the statistical analysis of more than 10 years of accumulated data on the Hanle effect on prominences are: absence of vertical fields, although vertical threads are observed; inverse polarities with respect to the polarity predicted by the potential fields computed from solar disc observations; no increase of the strength of the field when the resolution is improved, although the field seems non-uniform at least in moderately active prominences (see Figure 6). Much progress could be made by improving the resolution of those measurements using a LMC; hopefully the magnetic field could be measured at the sub-arcsec level needed to reach the scale of threads in prominences. Using a similar technique, the field in spicules could also be observed; this is fundamental for understanding the mechanism responsible for driving the gas impulsively in spicules.

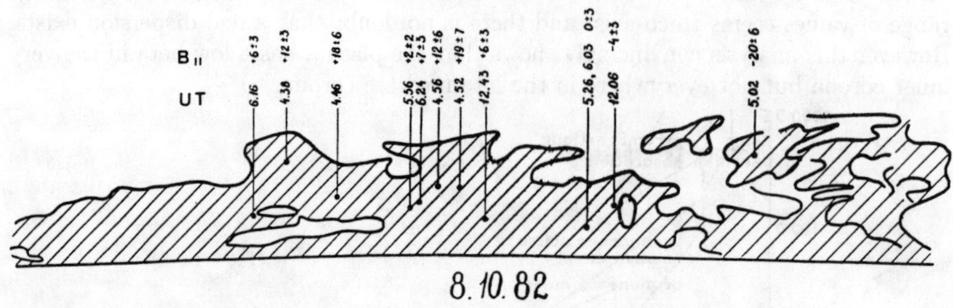

Figure 6: Example of values of the magnetic field using the Zeeman effect [16] to show variable amplitudes and polarities of the longitudinal field in Hα prominences.

2. Coronal Magnetic Field

Only very few attempts have been made (the first attempts were reported in J. Harvey's thesis) using the Zeeman effect because with existing coronagraphs, a very long integration time is needed to extract the circular polarization from a coronal line in the visible (the green line of the Fe XIV was used). The Hanle effect has been also measured and more systematically analyzed e.g. Arnaud[16]. However, only the direction of the field can be extracted from these painful measurements with a very crude spatial resolution, made with small aperture Lyot-coronagraphs. We are clearly in a "state of infancy" at this point; for example, no attempt has been made up to now to resolve structures along the l.o.s., although good perspectives exist.[17]

C. <u>What is needed ?</u>

To measure the coronal magnetic field it is clearly more advantageous to use IR lines. For the moment, the most promising lines are the strong Fe XIII lines near 1075 and 1079 nm which are quite close to the well known HeI line at 1083 nm. A whole line profile analysis in polarized light is needed. However, as a powerful first step, the analysis could be limited to circular polarization and intensity profiles (longitudinal Zeeman effect) and, simultaneously, to the measurement of the amount and direction of the linear polarization integrated over the whole profile to take advantage of the Hanle effect. New methods like Integral Field Spectroscopy should be introduced.

Let us now estimate the requirements to adequately perform a measurement using the Zeeman effect. We take the Fe XIII emission at 1074.7 nm with a Lande factor of g=1.5[18] and consider a rather bright coronal loop with a surface brightness of

50×10^{-6} per 0.1 nm passband of the solar disc brightness. With the usual formula to estimate the amplitude of the Zeeman signal,[19] we find that 2×10^8 photons are needed to measure a 10 gauss field with a signal-noise ratio of 10, assuming the signal could be spatially integrated over an extension of 1×10 arcsec^2 (or 3.3×3.3 arcsec^2) which could represent a part of the loop; we also assume an overall efficiency of the polarization analysis of 0.1; with a 1m-aperture telescope we find that an integration time of the order of 50 sec is needed to perform the Zeeman effect measurement. Our number is possibly optimistic, but the important point is to show that the aim of the experiment is well achievable with the LMC.

To give a last really optimistic prediction, remember that compared with the attempts made in the past, when a 40 to 50 cm aperture Lyot-coronagraph was used on the Fe XIV-green line scanning the whole profile with one single photo-multiplier, we have to-day at our disposal a gain of several orders of magnitude, because of the use of multiplexed detectors of better quantum efficiency and the access to the IR-lines where the Zeeman effect is more efficient. As noted before, there is no way to measure, in the near future, the coronal magnetic field with spaceborn X-UV telescopes. This makes the development of the LMC a well-identified priority for coronal researches.

8. References

1. S. Koutchmy et al., *SPIE Conf.*, **1235** (1990), 849-857.
2. R. Smartt et al., *SPIE Conf.*, **1236** (1990), 206; R. Schwenn et al., these Proceedings.
3. R. Bonnet, *L'Astronomie*, **80** (1966), 192.
4. H. Zirin and G. Newkirk, *Appl.Optics*, **2** (1963), 977.
5. R. Viotti and M. Hack in *Cataclysmic Variables and related objects*, NASA SP-507 (1993).
6. P.O. Lagage and E. Pantin, *Nature*, **369** (1994), 628.
7. A.M. Lagrange, in *"Science with the VLT"* in ESO-Pub. (1994), The Messager.
8. S. Koutchmy and Ch. Nitschelm, *Astrophys. Space Sc.*, **143** (1988), 45 ; D. Neidug et al., these Proceedings.
9. E.H. Avrett, J.M. Fontanla and R. Loeser, *IAU Symp.54 on IR Solar Phys.* (Kluwer), (1994) **35**.
10. K. Bocchialini et al., *Ap.J.L.*, **428**, (1993) L67 and *Sp.Sc.Rev.*, **70**, (1994) 57.
11. S. Koutchmy et al. *Astron.Astrophys.*, **120** (1983), 185.
12. P.J. Cargill, J. Chen and D.A. Garren, *Ap.J.*, **423**, (1994), 854-870.
13. S. Koutchmy *Adv.Space Res.*, **Vol.14**, No 4 (1994), 29-39 (COSPAR92).
14. C.E. Alissandrakis, M.R. Kundu and P. Lantos, *Astron. Astrophys.*, **82** (1980), 30; G.A. Dulk and D.J. McLean, *Solar Phys.*, **57**, (1978) 279.
15. J.L. Leroy in *Physics of Solar Prominences*, IAU Coll. 44 (1978), Jensen et al. Eds, 56; I. Kim in *Lectures Notes in Physics*, **363**, 49 (IAU Coll. 117).

16. J. Arnaud in *ESA SP-348, 285* and *Astron. Astrophys*, **178** (1987) 263.
17. R.N. Smartt and C.W. Querfeld in *Solar Polarimetry*, NSO/SPO Workshop 11, L. November Ed. (1991) p.326.
18. L.B. Demkina and G.M. Nikolsky, *Soln.Dannye*, **12** (1981), 105.
19. S. Koutchmy and R.N. Smartt in *High Spatial Resolution Solar Obs.*, NSO/SPO Workshop 10, O. von der Lue Ed. (1989), p.560.

MECHANISMS OF CORONAL HEATING

JACK B. ZIRKER
National Solar Observatory/Sacramento Peak*
Sunspot, NM 88349, USA

1. Introduction

This paper will be a short summary of my talk at the workshop. For a more extended discussion and references, please refer to my review, "Coronal Heating" in Solar Physics[12]. The excellent paper by P. Cargill in these proceedings updates that review and suggests observational tests of different mechanisms.

The origin of the high coronal temperature has puzzled solar astronomers since its discover in the 1930's. Currently, two mechanisms are viable: wave heating and heating by electric currents. In addition, an alternative scheme "velocity filtration" which bypasses the whole issue of heating, has recently been suggested.

Withbroe and Noyes[11] estimated the power requirements any mechanism must satisfy:

- Quiet corona 3 x 10^5 erg s cm^2
- Active corona 10^7
- Coronal hole 10^6

L. Biermann and M. Schwarzschild (ca. 1948) independently proposed heating by sound waves, that originate below the photosphere and shock in the upper atmosphere. A long train of theoretical work showed, however, that shock dissipation would occur in the chromosphere not the corona. Athay and White[1] effectively disproved this idea with an argument based on line widths of Si II lines observed by OSO 8.

With the recognition that the corona is permeated by magnetic field, with its roots in the photosphere, attention turned to a variety of magnetohydrodynamic waves. Recent simulations of convection in the presence of magnetic field[10] have been useful in sorting the possibilities. They stress the importance of mode coupling, particularly conversion of acoustic power into magnetic waves. Convective motions, with diverging upflow and filamentary downflows, produce ample acoustic power, plus fluxtube waves with torsional, kink or sausage deformations. About a third of the acoustic flux converts to magnetosonic waves, the rest is reflected downward.

Magnetosonic waves are compressible and could deliver power to the corona when they shock. Moreover they propagate across magnetic field lines and so can distribute

*Operated by the Association of Universities for Research in Astronomy, Inc., under cooperative agreement with the National Science Foundation.

power evenly. However, they are strongly refracted by the height gradient of the Alfven velocity and hence have difficulty in reaching the corona.

Thus the most promising type of wave appears to be some form of tube wave, with transverse displacements. Alfven waves have received the most attention from theorists. To a first approximation, they are incompressible, so the principle problem has been how to allow them to dissipate.

Resonance absorption[2] appears to be a viable dissipation mechanism. At the interface between a filled coronal loop and its rarefied environment, a resonance layer is formed. An Alfven wave that propagates along the axis of the loop excites another wave within this surface layer. The surface wave is resonant if its phase speed matches the local Alfven speed at the layer. The surface wave will refract into the loop, creating strong cross-field gradients of velocity. Joule and viscous dissipation can then occur.

To date the evidence of coronal Alfven waves has been scanty. Recent work by Hassler and Moran[3] showing a height increase of "turbulent" line width, does support the idea of Alfven wave heating, however.

2. Heating By Electric Currents

If a flux tube is twisted by photospheric motions a field-aligned electric current is produced. Or if magnetic fields with orthogonal components are pressed together by external motions, a cross-field (neutral) electric current is produced.

Spicer[9] has shown that field-aligned currents must be highly filamentary in order that the current density be sufficient to dissipate in the weak resistivity of the corona. To match the typical radiative losses, lateral dimensions of the filaments must be at most a few tens of meters. Then to provide sufficient power to heat a typical coronal loop, some 10^4 filaments must exist. Such a configuration would be vulnerable to the coalescence instability. The details of this proposal have not been worked out in detail.

Parker, in a series of papers,[12] has proposed that coronal heating takes place in neutral current sheets, that separate misaligned field configurations. He pictures a random walk of loop footpoints in the photosphere that tends to braid the loops. Such a braid has a non-uniform twist along its length and according to a theorem proved by Parker[4], cannot be in static equilibrium. A current sheet must form, in which Joule dissipation converts the free energy of the current to heat. Parker pictures this release of energy as part of an impulsive reconnection process "nanoflare" that simplifies the braided field lines. He estimated that nanoflares have energies of 10^{24} ergs or less and must appear in a loop at least several times per second.

Several unsuccessful attempts have been made by the author and collaborators to detect nanoflares in optical forbidden coronal lines or at radio wavelengths. Smartt et al[8] have reported a type of event in which coronal loops interact, with the release of some energy. These events may be evidence for reconnection. Shimizu et al[7] has counted microflares (energy $< 10^{26}$ ergs) in Yohkoh data and finds too few to heat the corona (see Cargill in these proceedings for a discussion). Thus, while plausible, the nanoflare mechanism remains unconfirmed.

3. Velocity Filtration

Scudder[5,6] has proposed a scheme for generating a hot corona without a conventional heating mechanism. He postulates the formation of non-Maxwellian tails to the electron velocity distribution functions in the corona, because of the rapid decrease of the Coulomb collision cross-section with increasing electron speed. The majority of chromospheric electrons have insufficient energy to overcome the gravitational potential barrier to the corona, so only these hot electrons arrive there.

Scudder works out the consequences of his elegant proposal in considerable detail. His mechanism can explain, for example, the observed height dependence of coronal line widths but has some serious conflicts with other data (see Cargill).

Ultimately the power to maintain the corona against its radiative losses must derive from the hypothetical mechanism that produces large non-Maxwellian tails. Scudder has not explained as yet the origin of these tails, which remain a serious question for his proposal.

4. References

1. R.G. Athay and O.R. White, *Ap. J.* **226** 1135 (1978).
2. J. Davila in "Mechanisms of Chromospheric and Coronal Heating", P. Ulmschneider, E.R. Priest, and R. Rosner, (eds.), p. 464. (Berlin: Springer-Verlag) 1991.
3. D. Hassler and T.G. Moran, *Bull. Am. Astron. Soc.* **25** 1200 (1993).
4. E.N. Parker *Ap. J.* **174** 499 (1972).
5. J. Scudder *Ap. J.* **398** 299 (1993a).
6. J. Scudder *Ap. J.* **398** 319 (1993b).
7. T. Shimizu, S. Tsuneta, L.W. Acton, J. Lemen, and Y. Uchida *Pub. Ast. Soc. Japan* **44** L147 (1992).
8. R.N. Smartt, Z. Zhang, and M.F. Smutko *Sol. Phy.* **148** 139 (1993).
9. D. Spicer "Mechanisms of Chromospheric and Coronal Heating", P. Ulmschneider, E.R. Priest, and R. Rosner, (eds.), p. 547. (Berlin: Springer-Verlag) 1993.
10. R. Stein and A. Nordlund: "Mechanisms of Chromospheric and Coronal Heating", P. Ulmschneider, E.R. Priest, and R. Rosner, (eds.), p. 386. (Berlin: Springer-Verlag) 1993.
11. G.L. Withbroe and R.W. Noyes *Ann. Rev. Astron. Ap.* **15** 63 (1977).
12. J.B. Zirker *Sol. Phy.* **148** 43 (1993).

DIAGNOSTICS OF CORONAL HEATING

PETER J. CARGILL
Beam Physics Branch, Plasma Physics Division, Code 6790
Naval Research Laboratory, Washington, DC 20375-5320

ABSTRACT

At present, theories of coronal heating can be split into three general types: dissipation of MHD waves, small scale magnetic reconnection (also sometimes referred to as nanoflares) and velocity filtration. There are a number of observable features in the corona that should be compared to the predictions of theoretical models. These include the shape of the emission measure vs. temperature curve, the coronal filling factor, hot plasma components, the distribution of the size of X-ray brightenings, non-thermal line profiles and the production of energetic particles. A comparison of the predictions of the nanoflare model leads to some agreements with observations, some disagreements and some predictions. Neither of the other two classes of heating mechanisms satisfies all of the existing observational tests.

1. Introduction

The fact that the solar corona is at a temperature in excess of 10^6K has been known for over half a century[20]. It was rapidly realized that such a temperature could only be the result of in situ deposition of energy (sometimes referred to as mechanical heating), and, with one important exception to be discussed later, this has remained the paradigm to the present day. The basic requirements for coronal heating have been tabulated by Withbroe and Noyes[72]. They divided the corona into three separate components: active regions, quiet regions and coronal holes. We will not be concerned with coronal holes in this paper, and since active regions have more severe energy requirements than quiet regions, we will concentrate on them. The energy loss from active regions is approximately 10^7 ergs cm^{-2} s^{-1}.

It is convenient to split coronal heating mechanisms up into three distinct classes. Two rely on the free energy in the coronal magnetic field and associated plasma motions, and the third relies on chromospheric electric fields and turbulence. The rest of this Section contains a brief review of these mechanisms. Space limitations preclude us from providing a comprehensive review and an exhaustive reference list. Readers are referred to the many reviews that have appeared over the past decade: the "review of the reviews" compiled by Zirker[73] is a good place to start.

Magnetic heating mechanisms can be divided into two, depending on the frequency of the photospheric motions that injects energy into the corona. The photosphere has a broad spectrum of fluctuations with periods of order a few minutes, which generates waves that may be transmitted to the corona. Thus ***wave heating*** has long been considered as a possible heating mechanism. Of the various MHD wave modes, acoustic waves have been discarded for active region heating on the grounds that their flux observed at transition region temperatures is too small[3]. Fast mode (magnetosonic) waves undergo strong refraction due to the steep density gradient in the transition region[25], but may be important if generated in the corona itself[21,54]. Alfven waves are generally reflected in the transition region, but can access the corona if a resonance condition ($P \approx L/V_A$) holds[26] (P is the wave period, L the typical length of a coronal field line, and V_A the coronal Alfven speed). If we look at short active region loops ($L \approx 2 \times 10^9$ cm), a coronal density of 5×10^9, and magnetic field strengths between 50 and 500 G, then waves with periods between 1.3 and 13 secs can access the corona. It is unclear whether sufficient wave power is generated in the photosphere in this range. An alternative would be if the Alfven waves are generated in the upper chromosphere or transition region[17], perhaps due to the mode-conversion of upward propagating non-linear waves[28] or by impulsive events such as spicules or high speed jets[8].

While fast and slow waves can dissipate readily in a uniform medium[54], Alfven waves dissipate very slowly, on timescales not of interest in the corona. In structured atmospheres, this is no longer the case since the waves can undergo resonance absorption and, in principle, be dissipated effectively[15,23,27,56]. However, the precise dissipation mechanism is still a matter of some controversy. Steinolfson and Davila[66] and Ofman et al.[45] examined the linear theory of dissipation by resistivity and viscosity respectively and found that the velocities in the resonance layer were unrealistically large. More recently Ofman et al.[46] have noted recently that nonlinear effects associated with shear flow instabilities can limit the size of these velocities to more reasonable values.

Heating by small scale currents involves different types of photospheric motions. As the photospheric plasma moves around at highly subsonic speeds, the footpoints of coronal field lines tend to random walk, so that the coronal field becomes very tangled and non-potential. Parker[49] suggested that such a magnetic field was in a state of permanent non-equilibrium, with regions of intense current being spontaneously formed and dissipated. It was argued that the dissipation of these currents heated the corona to X-ray temperatures. The approximations made in Parker's analysis have been criticized[69,70], so that it is unclear (and still controversial) whether current sheets can form spontaneously[7]. However, there seems to be little doubt that small scale current sheets can form due to the very contorted photospheric motions[70] - in other words, equilibria have a chance of existing (contrary to Parker's assertion), but probably don't. The picture one then has is of the corona being heated by a series of small, discrete events, which presumably occur when the localized currents reach a certain intensity.

Parker[51] has taken this concept a step further to suggest that the size of these discrete events is such that $\approx 10^{24}$ ergs is released in each one. He called these "nanoflares" since the energy released is about 9 orders of magnitude smaller than the very largest flares. The experimental evidence for the existence of nanoflares at this time is slender. Parker based his conjecture on brightenings seen in CIV[53] that could be interpreted as elemental energy releases. The value of 10^{24} ergs should not be taken too literally, with the important point being that energy is released in small, discrete bursts. Unlike Alfven wave heating, there is little difficulty in accounting for the dissipation of the intense current that will exist in nanoflares. The theory of magnetic reconnection is sufficiently advanced that we understand rather well how such dissipation takes place on short timescales[55].

A second form of direct electric current has been investigated, and differs in significant ways from the nanoflare idea. A number of workers[1,6,24,67] noted that when coronal electric currents become highly filamented, their strength is such that plasma microinstabilities can be excited. For typical solar parameters, the scale of these currents needs to be of order a few meters, and so instrumentally unresolvable. The energy release in each of these filaments is also small - Cargill[11] named such events "femtaflares" - and so many such events are required to account for the observed energy losses. A serious problem with this picture is the following. The threshold for instability in most cases depends on T to some positive power. Thus as the current is dissipated, the temperature increases due to plasma heating, and the instability shuts off rapidly. One then has to continually drive the system harder and harder, and the conditions on the current required for instability soon become extreme[33].

The third mechanism is based on a different physical premise. Scudder[60,61,62] has argued that if the distribution functions in the chromosphere are modified by the addition of significant non-Maxwellian tails (a κ distribution), then the particles in the tails can escape into the corona, and give rise to the required hot population. This mechanism is often referred to as ***velocity filtration***, a term we shall use in this paper. The question that has generally been asked by other workers is: how are these tails generated? Scudder[63] has answered this question with two suggestions. (1) In the presence of a gravitational force, a weak field-aligned electric field is required to satisfy quasi-neutrality. This electric field can draw electrons from the distribution function into tails (the well-known electron runaway effect). (2) the observed non-thermal motions (turbulence) observed in the chromosphere can accelerate ions (and possibly electrons, although difficulties exist here[43]) by a second order Fermi process, thus producing enhanced tails.

2. Observational Evidence for Heating

The wealth of observations obtained over the past three decades has given us much useful information on the state of the corona in active regions. The forthcoming SOHO and TRACE missions will refine what we already know and give important new information on potential diagnostics that are currently uncertain. In this Section we review the anticipated state of observational knowledge once the SOHO data becomes available, and discuss other potential diagnostics that have not yet been attempted.

Emission Measure vs Temperature Curve The original detection of a hot corona was from observations of highly excited states of heavy ions such as iron. Since then, a wide variety of observations in UV and X-rays have been carried out, that gives not only a clear picture of the temperature structure of the corona, but also of how much material is radiating at any given temperature. This is conveniently shown as a plot of how the line-of-sight emission measure (defined as $E_l = \int n_e^2 dh$, where dh is the line of sight) scales as as a function of temperature[31,47]. Between $\approx 10^5$ and

10^6K, $E_l \propto T^{3/2}$ and below 10^5K $E_l \propto T^{-2}$. An important issue is whether the former scaling is determined by the competing effects of conductive and radiative losses as they redistribute deposited energy or by the energy deposition itself. It is unlikely that the low temperature behavior of E_l is direct consequence of the coronal heating mechanism. One alternative suggestion is that a population of cool loops exists[4,58], that are presumably heated by some unspecified mechanism.

High Temperature Coronal Plasma A question of increasing interest is how much material is at higher temperatures than the commonly accepted "coronal" temperature. It has been known since Skylab that a significant part of the corona is at $2 - 3 \times 10^6$K[47], but recent work suggests that there may be plasma that is much hotter than this. The SXV channel on the BCS instrument on Yohkoh has detected a significant number of counts over a three year period that are not related to flares[41]. SXV has a peak formation temperature of $10^{7.2}$, but is formed over a quite broad temperature range, so that it is unclear whether the emission comes from a little plasma at $10^{7.2}$ or a lot at $10^{6.6}$. The SXT on Yohkoh has also presented evidence for hot plasma in active regions. Klimchuk and Gary[34] have noted that significant emission measures exist at at least $10^{6.5}$: indeed it may possibly be the "coronal" temperature.

Non-thermal Line Broadenings A third observable that has been well documented is the non-thermal broadening of spectral lines. This is generally ascribed to unresolved turbulence of an unknown type. The Skylab mission produced evidence of non-thermal broadenings of 10 - 30 km/s in the transition region and low corona[32]. The XRP experiment on SMM observed the broadening of coronal lines in excess of 60 km/s[59]. To our knowledge, clear Doppler shifts have never been observed in these data, unlike in flares where they are common[42].

Coronal Filling Factor An observable that may be of critical importance is the coronal filling factor. The scales of dissipation in some heating mechanisms are small, so that at any given time not all of the coronal volume may be filled with hot plasma. The filling factor can be defined as the ratio of the radiating volume to the total volume. It can be readily determined if the density can be evaluated in two independent ways. This is possible if the coronal volume emission measure (n^2V) and the (unresolved) radiating volume are known, as well as an absolute density from pairs of spectral lines[22]. Note that if the radiating source is resolved, a filling factor of unity should be obtained. Evaluations of the coronal filling factor for active regions have been few. Dere[18] estimated a filling factor of order unity, whereas Thomas[68] obtained $\approx 10^{-2}$. SOHO will hopefully resolve this controversy, but the detection of density-sensitive IR pairs would also be most useful.

Distribution of Coronal X-ray Brightenings An issue that has become of increasing interest in recent years is the connection of coronal heating by small, discrete events such as nanoflares to larger events such as microflares[36] and flares. Hudson[30] pointed out that, if he were to extrapolate the number of events at a given energy ($N(E)$) from observed energies ($> 10^{28}$ ergs) to nanoflare energies, the total amount of energy released was about an order of magnitude too small to heat the corona. Further analysis by Lee et al.[35] confirmed this result. At high energies,

Hudson noted that $N(E) \sim E^{-1.8}$, and pointed out that the exponent had to be < -2 (and probably quite a bit steeper than -2) to get enough low energy events. He also pointed out that if observations at lower energies confirmed the $E^{-1.8}$ scaling, coronal heating based on nanoflares would be untenable. The SXT Yohkoh data has provided an opportunity to extend this analysis to less energetic events. Shimizu[65] has identified brightenings in soft X-rays and has extended Hudson's curve of $N(E)$ down to events of $\approx 10^{26}$ ergs. Rather than the curve steepening as is required to heat the corona, it turns over the other way, so that the problem becomes even worse than before.

Energetic Particle Signatures Finally in this Section, we wish to suggest a future observational test of heating mechanisms. As will be discussed below, an important difference between some models may lie in their production of energetic electrons (by energetic we mean > 10 keV or so). Flares are well known to produce copious numbers of energetic electrons, which can be detected both from their hard X-ray and radio emissions[43]. It is unlikely that hard X-ray signatures of coronal heating can be seen at this time, but radio signatures may be detectable. A first effort was made by S. White and J. Zirker at Aricebo, but no definitive result has been obtained at this time. Future efforts along this line should be encouraged.

3. Signatures of Nanoflare Heating

Thus far, we have discussed a number of diagnostics of coronal heating processes. The next question that arises is how a comparison between theory and observations can be carried out. In this Section we present results that have been obtained for nanoflare heating and in the next Section discuss wave heating and velocity filtration. Any comparison of nanoflare heating with observations requires some sort of statistical treatment. In the nanoflare model, small heating events are occurring randomly in the corona, presumably with individual energies distributed randomly between some upper and lower limit. It is difficult to see how the dynamics and energetics of such a complex system can be dealt with in a way that does not deal with averaged properties. In this Section we discuss two such statistical treatments.

The first model[10,11,12] assumes that the corona in an active region comprises a large "megaloop" that is broken up into many hundred narrow, elemental flux tubes. Each of these elemental flux tubes is moved by the photospheric flow, and, when neighboring flux tubes are misaligned by a given amount, rapid reconnection takes place[50,51]. Cargill[10,11,12] assumed that an energy Q was released in each of these reconnection events and that a cross-sectional area A_h of the megaloop was heated. If the length of the dissipation volume is $\propto A_h^{1/2}$, then we have:

$$Q \approx \frac{B_t^2}{8\pi} A_h^{3/2} . \quad (1)$$

Here B_t is the reconnecting field component and should not be confused with the total coronal field strength. Estimates place B_t at 10 - 25% of the coronal field strength[50,51].

Cargill[11] studied the thermal response of such a coronal model to nanoflare heating. Since we need to follow the thermal evolution of possibly many thousand elemental loops, an analytic treatment was used in preference to the more usual numerical studies of loop evolution[40]. It was argued that the cooling took place in two distinct phases, with conductive cooling dominating initially, and radiative cooling being important at later times. This transition from conductive to radiative cooling is to be expected, since as a loop cools, its temperature falls and the density first rises and then falls. A study of the characteristic times for a loop to cool by conduction (τ_c) and radiation (τ_r):

$$\tau_c = 4 \times 10^{-10} \frac{nL^2}{T^{5/2}}\ s, \quad \tau_r = \frac{3kT^{1-\alpha}}{n\chi}\ s \tag{2}$$

shows that there is a temperature below which $\tau_r < \tau_c$. Here L is the loop half-length and the radiative loss function is assumed to be of the form $n^2 \chi T^\alpha$, generally parameterized as a piecewise-continuous function[14,57]. We use the function of Cook et al.[14] which has more contemporary coronal abundances than that of Rosner et al.[57]. To model the cooling, the evaporative cooling model of Antiochos and Sturrock[5] and a radiative model[13,64] that assumes $T \propto n^2$ are used.

Examples of the results using this approach have been presented elsewhere[11,12] and need not be repeated in detail here. In one case our megaloop had a cross-section of 4×10^{16} cm^2, a length of 10^9 cm and was made up of 500 small elemental loops. Assuming nanoflare energies between 5×10^{23} and 2×10^{24} ergs and an active region energy loss of 10^7 ergs cm^{-2} s^{-1}, each elemental loop is heated on average every 1500 s. The elemental loops were found to have a broad distribution of temperatures and densities, with the maximum being at reasonable coronal values ($10^{6.3}$ and 10^{10} respectively). Up to 10^6, the emission measure scales as $T^{1.5}$, in agreement with the observations discussed in the previous Section. Above 10^6, the curve steepens.

These results are encouraging in that they satisfy the most basic requirements of a heating model. In addition it is straightforward to compute the filling factor, which turns out to be ≈ 0.5. This is closer to Dere's[18] than Thomas'[68] value, but when the number of elemental loops is increased, the filling factor falls. For example, when 5000 loops are used, we get a filling factor < 0.1. Cargill[9] suggested that this result could be inverted to determine details of the energy release volume if the filling factor (ϕ) could be determined. He showed that

$$\phi \approx 10^{-4} \frac{\Lambda L A_h^{7/6}}{Q^{7/6}}, \tag{3}$$

where Λ is the coronal energy requirement in units of ergs cm^{-2} s^{-1}. Perhaps SOHO data can be used to attempt this inversion.

More recently, we have used the output of the nanoflare model to compute a predicted disc-wide SXV count rate (the Yohkoh BCS has no spatial resolution). The observations[41] detect $\approx 100 - 300$ counts/s. If we assume that 15% of the solar disc is covered by active regions, then we get reasonable agreement with observations, for a case where the filling factor is 0.5. Smaller filling factors give significantly greater count rates. An interesting feature is the temperature of the plasma that the

SXV counts are coming from. We have shown that the peak number of SXV counts are coming from plasma near $10^{6.6}$K, which is well away from the peak formation temperature of SXV so that the SXV counts are coming from the former of our two options discussed above: a dense, relatively cool plasma.

An earlier effort to compute observables[9,12] involved calculation of the non-thermal line broadening of a nanoflare heated corona. It is well known from the basic theory of magnetic reconnection that a pair of plasma jets with velocities of V_{At} are produced at each reconnection site. Here V_{At} is the Alfven speed based on the strength of the **reconnecting** field, **not** the total field. This is an important point, since V_{At} is usually much smaller than V_A. It was again assumed that the corona was heated by many nanoflares, and line broadenings well in excess of those observed are generated. Note that line broadenings rather than Doppler shifts arise, since the reconnection sites will be randomly oriented with respect to the line of sight.

At first sight, this result would appear to pose a problem for the nanoflare model. However, it has been pointed out that it is likely that the flows will be either rapidly thermalize[73] or may just fizzle out after a short lifetime[75]. The latter objection was made on the basis of calculations of flows along a magnetic field driven by an impulsive pressure pulse. However, this is not relevant in the case of reconnection jets since these flows are driven by field line tension and will persist as long as the reconnection does. The first criticism is more relevant, since in a complex magnetic topology like our megaloop or Parker's tangled corona, the flows will not be able to travel indefinitely but will run into neighboring loops. The thermalization process itself is unclear. The flows are highly sub-Alfvenic with respect to the magnetosonic speed in the neighboring loop, so cannot be thermalized by a standing magnetosonic shock. It is possible that interaction of the flows with neighboring loops sends a pressure pulse back to the reconnection site, so that the flows are then shut off.

A final point about this model is its energetic particle production properties. At the reconnection site, there is a field-aligned electric field satisfying $E_{\|} = \eta J_{\|}$, that is directed perpendicular to the reconnecting field component. Such DC electric fields are well known to be excellent potential sites for electron acceleration[19,48]. The properties of such DC fields have not yet been worked out for coronal heating applications, but in the case of solar flares, they have been shown to be effective accelerators[29,44].

Another approach to nanoflare heating addresses the problem in a different way. A number of workers[38,39,74] argued that energy release in the corona could be described in terms of a self organized critical system. They developed a simple model of the coronal field in which it relaxed with respect to its nearest neighbors, with an elemental amount of energy being released in each relaxation. The number of events as a function of energy ($N(E)$) scaled as $E^{-1.8}$, in good agreement with observations in the flare and microflare energy range[30]. More recently, Vlahos et al.[71] have extended this model to lower energies by modifying the relaxation criteria to take into account the field at more distant points. They find that at lower energies $N(E) \propto E^{-3.5}$ - the steepening being what is required to account for the energy release in active regions. This is a most encouraging development, even though it does not agree with the work of Shimizu[65] discussed above. In concluding, it should be noted that the physical basis (i.e. their interpretation in terms of the basic equations of MHD) of these models is non-existent.

4. Can Observations Really Help? - a Comparison of Predictions

In the previous Section, we discussed some of the experimental predictions of the nanoflare model. It is useful at this point to see how the other models discussed in the Introduction (wave heating and velocity filtration) compare. Table 1 provides a useful framework within which to organize the comparison. The left column lists the seven observables along with one other factor that we discuss in a moment. The observables are the emission measure (E_l) vs. T curve (1) above and (2) below 10^5K, (3) the filling factor, (4) the non-thermal line broadening, (5) the distribution of the size of X-ray events, (6) SXV emission (i.e. a hot component) and (7) energetic particle production. We have added a eighth factor, namely the need for cross-field heat transport. For most values of resistivity and viscosity, the actual regions of dissipation are rather small, perhaps a few tens or hundreds of meters. This has led to investigations of the efficiency of cross-field heat transport mechanisms[37].

Table 1: A summary of the comparative signatures of coronal heating mechanisms

Test	Waves	Nanoflares	Velocity Filtration
$E_l(T > 10^5)$	Yes	Yes	No
$E_l(T < 10^5)$	No	No	Yes
Filling Factor	$\ll 1$ to $O(1)$	$\ll 1$ to $O(1)$	1
Line Profiles	Perhaps O.K.	Possible Problems	O.K.
Brightness Distributions	Unlikely	Makes Prediction	None
SXV	Yes	Yes	Unknown
Energetic Particles	Maybe	Almost Certainly	Unlikely
Cross-field Transport	Possible Problems	O.K.	O.K.

The center column summarizes the results of the previous Section for the nanoflare model. It successfully reproduces the $T^{1.5}$ slope of the emission measure curve, but does not model the low temperature part[11]. It predicts a filling factor of between 0.5 and 0.1. The computed spectral line widths appear to be too big: some resolutions have been discussed above. The recent work of Vlahos et al., (1995) makes predictions about the slope of the distribution of X-ray brightenings, but the result is in contradiction to the preliminary Yohkoh analysis of such events. We can get reasonable count rates for SXV. Finally, heating by small scale reconnection imposes no special requirements on cross-field heat transport. Although the mechanism does heat the plasma over very small scales, the hot plasma is continually removed from the heating site by the jets, with new plasma being swept in. Thus, new plasma is continually being heated, and so a significant volume of plasma can be heated in a nanoflare.

Turning to wave heating, it is likely that any mechanism that deposits energy directly into the corona and relies on conduction and radiation to determine the spatial structure of the coronal temperature will reproduce the $T^{3/2}$ scaling with the emission measure. Therefore we believe wave heating will satisfy both this and the SXV requirements. For the case of Alfven wave heating, the filling factor will, in part, depend on the thickness of the resonant layer. If coronal values of the dissipation coefficients are used, then very small filling factors are expected. For fast and slow mode waves, filling factors close to unity are likely. An apparent advantage of wave heating is its seeming ability to account for the non-thermal line widths. The flux in Alfven waves given by a 40 km/s broadening is comfortably enough to account for the corona power losses, so it seems natural to explain the line broadenings as coronal waves. There is one problem with this explanation which is that if the Alfven waves in the ambient corona have amplitude of 40 km/s, then those in the resonance layer may have a much larger amplitude. Wave heating makes no predictions about the distribution of X-ray brightenings, but there is no reason to believe that it results in a continuum of events all the way up to microflares and flares. In this regard, it is in agreement with the preliminary Yohkoh results. Finally, the issue of whether wave heating should produce energetic electrons has not been examined. There are two ways that energetic electrons could be produced. If the waves are dissipated by parallel resistivity, then a parallel electric field exists, and electrons can runaway, with the runaway number depending on the properties of the electric field. However, viscosity is another possible dissipation mechanism, and should it be more effective than resistivity, then $E_{\|} \approx 0$ and no electrons will be accelerated. The second alternative is that electrons can be energized by the waves themselves, through a second order Fermi process. This requires that a) either the electrons be energized initially by some other means so that scattering can commence (the injection problem) or b) the waves cascade to very short wavelengths so that they can interact with the thermal electrons[43].

A major potential problem with wave heating by resonance absorption is the narrowness of the absorption layer for coronal diffusion coefficients. This leads to a need for cross-field transport. No plausible mechanism has yet been proposed for such transport. However, Davila[16] has estimated that the resonance layer could be broadened to a significant fraction of a coronal loop if the coronal resistivity is enhanced to give a magnetic Reynolds number of 10^3. The method of production of such an eddy resistivity is unclear at present. Another advantage of such an enhanced resistivity is that the magnitude of the wave velocities in the absorption layer is in agreement with the X-ray observations of line broadening.

Finally, we turn to velocity filtration. Unlike the other mechanisms discussed above, this is presumably a steady state process. Electron tails are continually being created in the chromosphere, move into the corona, and loose their energy there to account for the observed losses. Thus, the method should be operating along all magnetic field lines, so that there are no cross-field heat transport problems and the filling factor should be of order unity. Recently Anderson[2] has calculated the E_l vs. T dependence of this model and finds that the emission measure decreases continually from low to high temperatures. This result can thus perhaps account for the observed distribution of emission measure at low temperatures, but fails for the high temperature region. Of course, this is the exact opposite to the coronal models discussed above. It is also unclear at present whether this model can produce a SXV component. One of the early successes of velocity filtration was the generation of "non-thermal" line profiles. By incorporating the κ distribution functions into the computation of the line widths, Scudder[61] was able to reproduce broadly similar line profiles to those observed without the need to invoke random, small-scale bulk motions. Finally, one would not expect the model to produce any power law distribution of soft X-ray brightenings, nor should it be associated with hard X-ray emission.

5. Conclusions

Despite being the topic of intensive research, we still appear to be a long way from pinning down the identity of the mechanism responsible for producing the hot solar corona. Table 1 shows that all of the three principal mechanisms being discussed at this time have failings when compared with existing observations. Wave heating does well on many counts, yet there is a worrying problem associated with cross-field heat transport. Nanoflares also do quite well, with the major exception of the predicted non-thermal line profiles and the concern about the preliminary results of the Yohkoh X-ray brightenings. Velocity filtration has the major failing that it does not appear to be able to reproduce the coronal part of the emission measure curve. It should be stated that more work may resolve these problems. For example, the way the Alfven wave resonance absorption model fits into the global corona has never really been addressed. It may be that the resonance layer can move around, heating different plasma volumes, or that there are many layers, all close to each other. The problems with the line profiles in the nanoflare model may be refined by further work on reconnection in highly structured medium, but the problem with the brightness spectra may be tougher to resolve. Finally, computing intensities of coronal lines with a non-Maxwellian distribution function requires the discarding of many assumptions about spectral line formation that workers have become used to. Thus, velocity filtration may yet give the correct coronal emission measure.

The need for new observations is still pressing. SOHO will provide much improved diagnostics of the filling factor and the non-thermal line broadenings that may resolve some of the issues discussed above. Perhaps future radio observations can see fluctuations that can be attributed unambiguously to energetic particles. Turning to the infrared, current thinking is directed toward a large reflecting coronagraph. There are a number of areas where IR observations could be of interest. The presence of pairs of Iron lines at 1074.6 and 1079.8 nm raises the possibility of doing density diagnostics, similar to those possible with UV and X-ray lines. This could provide an independent estimate of the coronal filling factor, which, given the present discrepancies in the current data, would be invaluable. Perhaps the most interesting prospect with an IR coronagraph could be the ability to measure coronal magnetic fields. This has long been regarded as one of the major goals of coronal physics, since at present estimates of the coronal field are based on a number of extrapolations from the data and/or theory. While it may not be possible to determine the detailed structure of the field that is so important in the nanoflare model, even a rough estimate of its magnitude would be welcome. Thirdly, it would seem to be possible to get reasonably accurate estimates of non-thermal velocities in coronal lines (a few km/s). As noted in the previous Sections, this may be a most important discriminator between mechanisms. Finally, a most important issue is the region of the Sun that the coronagraph will observe. Estimates of the location of the occulting disc go as low as 10 arcseconds above the limb[52]. Thus, it would be possible to observe the low corona at the limb at rather high resolution. Another interesting issue would be to determine some of these key coronal heating parameters in large scale structures such as streamers. Most of the work discussed in this paper has been concerned with active regions that are usually observed on the disc. Yet streamers are observed to be in excess of 10^6K as well, and have been difficult to observe by traditional means (UV and X-rays). Perhaps such an investigation can shed some light on the heating of streamers: - are they heated throughout their volume, as coronal loops probably are, or is the heating going on at the base, with the energy being conducted to the rest of the volume? This issue is not understood at present. In summary, IR observations of the corona can help advance our understanding of coronal heating by both complementing future missions such as SOHO and observing regimes that other observing methods generally ignore.

6. Acknowledgments

This work was sponsored by ONR. I am grateful to Jeff Kuhn, Matt Penn and Jack Zirker for inviting me to attend a most stimulating workshop and to the staff at Sac Peak for their hospitality. I thank Jim Klimchuk, John Mariska and Jack Zirker from many useful discussions, Steve Anderson, Joe Davila and Jack Scudder for information about their unpublished work, and Matt Penn for answering my questions about coronal IR observation.

7. References

1. P. Amendt, and G. Benford, G., *Ap. J.*, **341**, (1982), 1082.
2. S. W. Anderson, *EOS, Trans AGU*, **75(16)**, (1994), 264.
3. R. G. Athay and O. R. White, *Ap. J.*, **229**, (1978), 1147.
4. S. K. Antiochos and G. Noci, *Ap. J.*, **301**, (1986), 440.
5. S. K. Antiochos and P. A. Sturrock, *Ap. J.*, **220**, (1978), 1137.
6. G. Benford, *Ap. J.*, **269**, (1983), 690.
7. P. K. Browning, *Plasma Phys. and Controlled Fusion*, **33**, (1991), 539.
8. G. E. Brueckner and J. D. F. Bartoe, *Ap. J.*, **272**, (1983), 329.
9. P. J. Cargill, in *ESA SP-348*, (1992), 245.

10. P. J. Cargill, *Solar Phys.*, **147**, (1993), 263.
11. P. J. Cargill, *Astrophys. J.*, **422**, (1994a), 381.
12. P. J. Cargill, in *Solar System Plasmas in Space and Time*, ed. J. L. Burch and J. H. Waite, Jr., Geophysical Monograph 84, (AGU, 1994b) 31.
13. P. J. Cargill, J. T. Mariska and S. K. Antiochos, *Ap. J.*, (Feb 1, 1995), in press.
14. J. W. Cook, C. C. Cheng, V. L. Jacobs, and S. K. Antiochos, *Ap. J.*, **338**, (1989), 1176.
15. J. M. Davila, *Ap. J.*, **317**, (1987), 514.
16. J. M. Davila, (1994), private communication.
17. J. M. Davila and S. M. Chitre, *Ap. J. Lett.*, **381**, (1991), L31.
18. K. P. Dere, *Solar Phys.*, **75**, (1982), 189.
19. H. Dreicer, *Phys. Rev.*, **115**,(1959), 238.
20. B. Edlen, *Ark. Mat. Astr. Och. Phys.*, **28B** (1941), No.1.
21. T. Fla, S. R. Habbal, T. E. Holzer, and E. Leer, *Ap. J.*, **280**, (1984), 382.
22. A. H. Gabriel and C. Jordan, *MNRAS*, **145**, (1969), 241.
23. A. Hasegawa and L. Chen, *Phys. Fluids,* **19**, (1976), 1924.
24. S. Hinata, *Ap. J.*, **232**, (1979), 915.
25. J V Hollweg, *Geophys. Res. Lett.*, **5**, (1978), 731.
26. J. V Hollweg, *Ap. J.*, **277**, (1984), 392.
27. J. V. Hollweg, *Ap. J.*, **306**, (1986), 730.
28. J. V. Hollweg, *Ap. J.*, **389**, (1992). 731.
29. G. D. Holman, *Ap. J.*, **293**, (1985), 584.
30. H. S. Hudson, *Solar Phys.*, **133**, (1991), 357.
31. C. Jordan, *Astron. and Ap.*, **86**, (1980), 355.
32. O. Kjeldseth Moe and K. R. Nicholas, *Ap. J.*, **211**, (1977), 579.
33. J. A. Klimchuk, private communication, (1993).
34. J. A. Klimchuk and D. E. Gary, *Ap. J.*, (1994), submitted.
35. T. T. Lee, V. Petrosian and J. M. McTiernan, *Ap. J.*, **412**, (1993), 401.
36. R. P. Lin, R. A. Schwartz, S. R. Kane, R. M. Pelling and K. C. Hurley, *Ap. J.*, **283**, (1984), 421.
37. C. Litwin and R. Rosner, *Ap. J.*, **412**, (1993), 375.
38. E. T. Lu and R. J. Hamilton, *Ap. J. Lett.*, **380**, (1991), L89.
39. E. T. Lu, R. J. Hamilton, J. M. McTiernan and K. R. Bromund, *Ap. J.*, **412**, (1993), 841.
40. J. T. Mariska, *Ap. J.*, **319**, (1987), 465.
41. J. T. Mariska, private communication, (1994).
42. J. T. Mariska, G. A. Doschek and R. D. Bentley, *Ap. J.*, **419**, (1993), 418.
43. J. A. Miller et al, *J. Geophys. Res.*, (1995), to be submitted.
44. E. Moghaddam-Taaheri and C. K. Goertz, *Ap. J.*, **352**, (1990), 361.
45. L. Ofman, J. M. Davila and R. S. Steinolfson, *Ap. J.*, **421**, (1994a), 360.
46. L. Ofman, J. M. Davila and R. S. Steinolfson, *Geophys. Res. Lett.*, **21**, (1994b), 2259.
47. R. Pallavicini, G. Peres, S. Serio, G. S. Vaiana, L. Golub, L. and R. Rosner,

Ap. J., **247**, (1981), 692.
48. K. Papadopoulos, *Revs. Geophys.*, **15**, (1977), 113.
49. E. N. Parker, *Ap. J.*, **174**, (1972), 499.
50. E. N. Parker, *Ap. J.*, **264**, (1983), 642.
51. E. N. Parker, *Ap. J.*, **330**, (1988), 474.
52. M. Penn, private communication, (1994).
53. J. G. Porter, J. Toomre and K. B. Gebbie, *Ap. J.*, **283**, (1984), 879.
54. L. J. Porter, J. A. Klimchuk and P. A. Sturrock, *ApJ*, **435**, (1994), 482.
55. E. R. Priest, *Rep. Prog. Phys.*, **48**, (1985), 955.
56. I. C. Rae and B. Roberts, *Geophys. Astrophys. Fluid Dyn.*, **18**, (1981), 197.
57. R. Rosner, W. H. Tucker and G. S. Vaiana, *Ap. J.*, **220**, (1978), 643.
58. G. Roumeliotis, *Ap. J.*, **379**, (1992), 392.
59. J. L. R. Saba and K. T. Strong, *Ap. J.*, **375**, (1991),789.
60. J. D. Scudder, *Ap. J.*, **398**, (1992a), 299.
61. J. D. Scudder, *Ap. J.*, **398**, (1992b), 319.
62. J. D. Scudder, *Ap. J.*, **427**, (1994a), 446.
63. J. D. Scudder, private communication, (1994b).
64. S. Serio, F. Reale, J. Jakimiec, B. Sylwester and J. Sylwester, *Astron. and Ap.*, **241**, (1991), 197.
65. T. Shimizu, *Proc. Kofu Meeting*, (1994), in press.
66. R. S. Steinolfson and J. M. Davila, *Ap. J.*, **415**, (1993), 354.
67. P. A. Sturrock, *Solar Phys.*, **121**, (1989), 387.
68. R. J. Thomas, *BAAS*, **25**, (1993), 1210.
69. A. A. Van Ballegooijen, *Ap. J.*, **298**, (1985), 421.
70. A. A. Van Ballegooijen, *Ap. J.*, **311**, (1986), 1001.
71. L. Vlahos, M. Georgoulis, R. Kluiving and P. Paschos, *Astron. and Ap.*, (1995), submitted.
72. G. L. Withbroe and R. W. Noyes, *Ann. Rev. Astron. Ap.*, **15**, (1977), 363.
73. J. B. Zirker, *Solar Phys.*, **148**, (1993), 43.
74. J. B. Zirker and F. M. Cleveland, *Solar Phys.*, **145**, (1993), 119.
75. J. B. Zirker and F. M. Cleveland, *Solar Phys.*, (1994), in press.

EMISSION-LINE SIGNATURES OF CORONAL LOOP INTERACTIONS

VLADIMIR S. AIRAPETIAN
Visiting Scientist from the Los Alamos National Laboratory, Los Alamos, NM
National Solar Observatory/Sacramento Peak, Sunspot, NM 88349 USA*

and

RAYMOND N. SMARTT
National Solar Observatory/Sacramento Peak, Sunspot, NM 88349 USA*

ABSTRACT

Coronal loop interactions (CLI) are observed in visible and soft X-ray coronal emission lines as occasional transient brightenings at sites where adjacent loops come into contact. The detailed morphology of these events as observed in the 5303 Å (Fe XIV), 6374.5 Å (Fe X) and Hα lines are discussed. A scenario for CLI events, based on their morphology as observed in the data, is discussed. Advantages of using the Fe XIII IR and Ca XV yellow lines to try to determine directly densities and temperatures at the site of an interaction are considered.

1. Introduction

Coronal loop interactions (CLI) are evident in images of post-flare loop systems as recorded with a 20-cm aperture, emission-line coronagraph in the 5303 Å line of Fe XIV (T $\sim$ 2 x 10^6 K) and the 6374.5 Å line of FeX (T $\sim$ 10^6 K), and also in Hα. Interactions occur between two or more concentrated magnetic flux systems driven by persistent mesogranular flows. A coronal loop interaction is seen as a localized transient enhancement at the site of interaction.[13] Such an enhancement appears first as a gradual brightening at the projected intersection of two loops in the green-line image, reaching a maximum on average $\sim$ 15-20 minutes after the first evidence of the start of the event. Following the maximum, the brightness gradually fades on the same time scale as the rise time. The red-line image shows the same behavior, but its maximum is delayed on average 8.6 minutes (based on 90 events) after that of the green line, while Hα reaches a maximum at the interaction site on average 9.3 minutes later. Based on the assumption that the plasma is cooling during the observed period in these events, we derive a characteristic maximum temperature in CLI's of $T_e \sim (3-5)$ x 10^6 K, and a density of $N_e \sim (1-3)$ x 10^{10} cm^{-3}, which is $\sim$ 10 times larger than the average plasma density in coronal loops.[1] The characteristic energy release in an elementary CLI event is found to be $\sim$ 3 x 10^{26} erg. Similar enhancements have also been observed in Yohkoh soft X-ray (SXR) data, including clear examples of two-loop interactions.[3,5] It is found that CLI events can make a significant contribution to coronal heating.[1]

---

*Operated by the Association of Universities for Research in Astronomy, Inc. (AURA) under cooperative agreement with the National Science Foundation.

Here we extend earlier studies to discuss briefly the efficiency of observations of CLI's in visible and IR coronal lines. This leads to key diagnostics for mechanism responsible for a CLI formation and provides some predictions about observational possibilities in other lines.

2. Morphology of CLI's

Brightenings that occur at an early stage in the development of a post-flare loop system, and hence at a relatively small height, tend to be brighter than those that occur at a relatively large height. Measurements of two enhancements that occurred relatively high in the corona ($\sim 5 \times 10^9$ cm) showed increases in brightness of three to five times. However, it should be pointed out that these values are likely to be underestimates because of the small size of the recorded images of the enhancements. By eye, there is the appearance of a central bright core that is probably below the resolution limit of the microdensitometer used to scan the images. When a brightening occurs at an early stage in a post-flare loop development, the specific loop system geometry is usually not obvious because of the complexity of the total system. Especially in the case of enhancements that occur at a much later stage (after several hours) and where the "plane" of the interacting loops are not highly inclined to the line-of-site, it is evident that enhancements occur when two loops come into contact at some localized region. It is surmised that all enhancements are due to such loop interactions, where portions of two (or more) loops come into contact, or even just portions of the same loop where helical structure brings such portions into close proximity. It has been noted that post-flare loops often display enhanced tops. Detailed examination of these enhanced tops reveals the same characteristics as localized enhancements - small knots of plasma in the tops become bright independently, and then fade, in a manner similar to that of other individual loop enhancements. It is suggested that such brightenings in loop tops could be due to localized interaction of a compact arcade system of loops, or loop segments in close proximity, of a helical loop top.

Since the loop configurations are changing continuously as they grow in height, the time history of an event involves the direct effect of the interaction as well as the changing loop geometry. An enhancement typically shows some brightness extending along the loops, away from the region of an overlap of two loops, forming an X-type configuration. A further common characteristic is some filling-in of the inner region of the 'X', such regions typically having a relatively sharp boundary in the form of an arc. These occur predominantly in the lower part of the 'X'.

In their early phase of development, post-flare coronal systems are associated with complex Hα loop structures . Hence, for events that occur in an early phase, associated changes in the detailed Hα structure are difficult to interpret, but the development of a bright knot of material at the interaction site is usually evident. For events that occur much later in the development of a post-flare loop system, Hα emission is absent initially. A typical scenario is the gradual appearance of Hα material that becomes concentrated at the site, followed by a gradual connection forming between the region of interaction and the solar surface, with material predominantly moving down from the site to the surface. In some cases, Hα simply appears at the interaction site with only a partial connection between this material and the solar surface.

Since red-line emission is not observed at an interaction site prior to an enhancement appearing in the green-line image, it appears that the plasma must be heated very rapidly at the onset of coalescence of two loops. If initial heating raises the plasma temperature higher than the characteristic temperature of the green-line, appropriate for high-temperature coronal lines, it is expected that an enhancement will be observed in such lines prior to a maximum occurring in the green-line enhancement, following plasma cooling.

Finally, very faint enhancements have been detected in nominally quiescent coronal loop structures. Therefore, we conclude that CLI events are observed commonly in post-flare loop systems only because the dynamical character of these systems greatly increases the probability of an event, as compared with the occurrence of CLI's in normal coronal loop systems. With this perspective, we now consider mechanisms that might describe the actual plasma processes involved in the interactions, while testing these ideas against the constraints set by the observational data.

3. Mechanisms

The basic mechanism of energy release in a CLI event according to Sakai[9] is due to the loop coalescence instability. The physical process depends significantly on the geometry of the interacting magnetic field structures, or magnetic loops, namely on the ratio of the characteristic scale along the loop (involved in the interaction) to the cross-sectional loop radius. If this ratio is much larger than 1, we have the case of magnetic reconnection with the formation of a one- or two-dimensional current sheet, similar to that considered to explain long-lived X-ray bright points.[7] A distinctly different physical situation occurs when the ratio is close to 1. This is represented by a three-dimensional X type current loop coalescence where the interaction takes place over the scale of the cross-sectional radius of the intersecting loops. This case describes what is observed in coronal emission-line interaction events, at least in clear cases where there are few loops in the immediate vicinity of CLI[11,8] and in the SXT data of Yohkoh.[1,3]. Depending on the relative direction of the longitudinal magnetic field in the interacting current loops, we can have partial reconnection (when the energy of the azimuthal component of the field is relaxing) or complete reconnection (with the release of the energy of the longitudinal field also). Since the loops appear to have the same general form after an event as before, we assume that partial reconnection does in fact occur as a basic mechanism in the formation of a CLI.

It has been shown that the pinch effect initiates a fast plasma compression[2,9], increasing the plasma density and resulting in a relatively hot (5×10^6 K) and dense (3×10^{11} cm^{-3}) blob at the place of a loop interaction. Our data indicate that the initiation of an interaction is rapid with an increase in density and temperature higher than the characteristic temperature of the green line; to this extent this model appears to fit our data. Radiative plasma cooling can result in the production of a plasma inflow in the region of a blob (similar to what occurs in a filament due to the thermal radiative instability). This can explain the long-lived (on a scale of $\sim$ 10 min) brightening (or density enhancement) at the place of a loop interaction seen in coronal emission lines, and also the formation of dense and cool Hα-emitting plasma several tens of minutes later[8]. We conclude that loop coalescence involving the partial reconnection and the magnetic pinch effect appears to be a plausible scenario to describe these events. Certainly it is not inconsistent with details of the events as seen in the optical data.

4. Spectroscopic Plasma Diagnostics

We can check the above scenario by determining the time evolution of the plasma density and temperature in CLI events. Because a CLI represents a transient impulsive process, we need to investigate the conditions under which the non-equilibrium effects (non-steady state ionization of the plasma and the deviation from Maxwellian distribution) can be neglected when defining coronal line intensities. The formation of a spectral line under equilibrium conditions implies that the characteristic Maxwellization time and the time for establishing ionization equilibrium for Fe X and Fe XIV ions should both be shorter than the characteristic cooling and/or conductive times. From Spitzer[12], we estimate the self-collision time for electrons, protons and Fe ions. Calculations show that $t_c(e) \sim 0.004$ s, $t_c(p) \sim 0.1$ s, and, t_e(Fe ions) ~ 2 s, values which are close to the characteristic Maxwellization times ($t_{Maxwell} = t_c/1.14$). From the above mentioned parameters, the radiative cooling time is $\sim$ 25 s, while the conductive time is several times greater. Therefore, the cooling time is larger than the Maxwelization and ionization times.

The two Ca XV yellow lines (5694.5 Å and 5446.4 Å) have a characteristic temperature $\sim 3.3 \times 10^6$ K. Their relative intensities are very sensitive to density[13]. Hence, they could provide useful diagnostics for CLI events. These two lines give a unique opportunity to observe CLI's in the temperature range overlapping that of the SXT of Yohkoh[1]. Therefore, simultaneous observations in the yellow line (by a coronagraph with a high angular resolution (≤ 1 arcsec)) and in SXR (by Yohkoh) can be carried out in the case of limb CLI events. For the particular event of Dec 7, 1981, $T_{max} \leq 5 \times 10^6$ K and and $EM_{enh} \sim 10^{28}$ cm^{-5}, so that we can estimate fluxes, J_y, for these lines[11]:

$$J_y(5694.5\text{Å}) \sim 1.2 \times 10^2 \text{ erg cm}^{-2} \text{ s}^{-1},$$

$$J_y(5446.4\text{Å}) \sim 10^2 \text{ erg cm}^{-2} \text{ s}^{-1}.$$

At the characteristic temperature of these lines ($\sim 3.6 \times 10^6$ K),

$$J_y(5694.5\text{Å}) \sim 7 \times 10^2 \text{ erg cm}^{-2} \text{ s}^{-1},$$

$$J_y(5446.4\text{Å}) \sim 6 \times 10^2 \text{ erg cm}^{-2} \text{ s}^{-1}.$$

Hence, such an enhancement should be observable in both these lines provided the estimates for N_e and T_e are realistic.

The three Fe XIII lines, 10747 Å, 10798 Å, and 3388 Å, can also be used to determine plasma parameters of CLI's, the characteristic temperature is $\sim 1.8 \times 10^6$ K. Both of the IR lines should have stronger fluxes than the green line in relatively dense regions, such as in a CLI. Further, their intensity ratio is also very sensitive to the plasma density[4], and their intrinsic polarization is high[13]. First, for typical densities in CLI's, the ratio of the intensities of the infra-red lines I(10747 Å)/I(10798 Å) approaches 1, while for a density, $N_e \sim 10^8$ cm^{-3},[12] it is $\sim$ 2.5 - 3 for a temperature, $T_e \sim 2 \times 10^6$ K. Because of the increased role of collisional excitation from the ground state, the ratio of these two lines for $N_e \geq 3 \times 10^9$ cm^{-3} is relatively insensitive to density variations. By contrast the ratio I(3388 Å)/I(10747 Å) becomes increasingly sensitive to an increase in the plasma density in the range from 10^9 to 10^{10} cm^{-3}. Since these lines arise from the same ion, their emission arises from the same plasma volume. We can derive the emission measure, $N_e^2 V$ (where V is the volume of the emitting plasma) directly from the 3388 Å line, and the plasma density, N_e, using the ratio I(3388 Å)/I(10747 Å), of the IR lines. Thus, it is possible to find the time evolution of the density and volume of the emitting plasma and, therefore, to distinquish between several possible mechanisms by obtaining high-resolution observations of limb CLI events from ground-based coronagraphs (IR lines) and by the SOHO satellite in the 3388 Å line.

Another interesting feature of Fe XIII lines is their high intrinsic polarization. House[5] analyzed the polarized radiation in 10747 Å as a function of coronal height. If the plasma density is high enough ($\geq 10^9$ cm^{-3}), the rate of collisional excitations affecting the depolarization in 10747 Å emission line is comparable to the radiative excitation rate. House[5] shows that the expected polarization for a such dense plasma should be very small, 0.1 - 0.01 %, which should be measurable with advanced polarimetry techniques. Such observations should be valuable to determine the geometry of the magnetic field in the vicinity of a loop interaction site.

5. ACKNOWLEDGMENTS

V.S. Airapetian was supported by Los Alamos National Laboratory under grant # 1081 L0014-9Q, the Scientific Office of USAF and partially by NSO/Sac Peak.

6. References

1. V. S. Airapetian & R. N. Smartt, , ApJ May 10, 1995
2. B. Chargeishvili, J. Zhao, & J.-I. Sakai , Sol. Phys. **145**, 295 (1993)
3. C. De Jager, M. India-Koide, S. Koide, & J.-I Sakai in "Symposium on Current Loop Interaction in Solar Flares", ed. J.-I. Sakai (Toyama: Toyama Univ.), 149 (1993)

4. D.R. Flower, & G. Pineau des Forets , A&A **24**, 181 (1973)
5. L.L. House , ApJ **214**, 632 (1977)
6. H.E. Mason , MNRAS **170** 0, (1975)
7. C.E. Parnell, E.R. Priest, & L. Golub , Sol. Phys. **151**, 57 (1994)
8. G. Roumeliotis, private communication
9. J.-I. Sakai in "Symposium on Current Loop Interaction in Solar Flares", ed. J.-I. Sakai (Toyama: Toyama Univ.), 13 (1993)
10. T. Shimizu, S. Tsuneta, L.W. Acton, J. Lemen, & Y. Uchida , PASJ **44**, L147 (1992)
11. R.N. Smartt, Z. Zhang, & M.F. Smutko , Sol. Phys. **148**, 139 (1993)
12. L. Spitzer, Jr. 1967, "Physics of Fully Ionized Gases", 2d ed. (New York: Interscience) (1967)
13. H. Zirin, "Astrophysics of the Sun" (Cambridge: Cambridge University), 237 (1988)

WHITE-LIGHT CORONAL FINE STRUCTURE

LAURENCE J. NOVEMBER
National Solar Observatory/Sacramento Peak
National Optical Astronomy Observatories
*
Sunspot NM 88349 USA
and

SERGE KOUTCHMY

Institut d'Astrophysique, CNRS, 98 bis Bd Arago, F-75014 Paris, France

ABSTRACT

White-light coronal observations are reported from the eclipse of July 11 1991 taken with the 3.6 meter Canada-France-Hawaii Telescope on Mauna Kea. The observations provide the highest resolution observations ever achieved because of the unique opportunity of observing the totally eclipsed sun with a large aperture telescope. We study the time series obtained with a 70mm acme photograph camera in a 10×10 arcmin field of view just above the solar limb. Spatial analysis using unsharp masking and mad-max analysis, a form of second derivative spatial filter, reveals subarcsecond dark and bright loops. Using computed line-of-sight path integrals through the corona based upon our measured radial decrease of electron density, we estimate that the electron density depletion and enhancement in the dark and bright loops, respectively, is about 10%. The contrast of the dark loops can be explained by a gas pressure decrease of 10% indicating a magnetic field of 0.7 gauss there, and the bright loops by a temperature decrease of 20% compared to the surrounding corona.

1. Introduction

The total eclipse of July 11 1991 presented a unique opportunity for studying the white-light solar corona, as the eclipse path passed directly over some of the largest telescopes in the world in Mauna Kea in Hawaii. We report analysis of the high resolution broad-band visible white-light observations obtained with the 3.6 m Canada/France/Hawaii Telescope.[4]

2. The observations

Observations were obtained with 2 ccd cameras and a photographic acme system in two pointing positions during totality. We report on the 70 mm photographic acme observations from one position above the west limb. About 380 photographic images were obtained on 2415 film with a 1/48 sec exposure at a rate of 2 frames per second

*NOAO is operated by the Association of Universities for Research in Astronomy, Inc. (AURA) under cooperative agreement with the National Science Foundation.

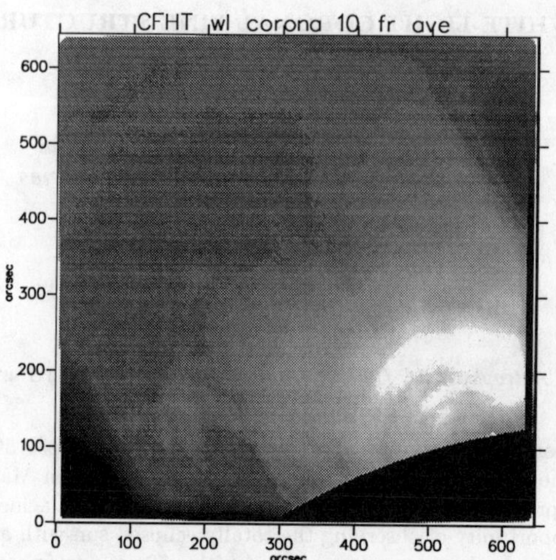

Figure 1: The photographic density from a single 70 mm frame with the azimuthally averaged radially smoothed density subtracted. The plotting range was 1.5 photographic density units.

through a Wratten 69 filter. The filter admits broad-band red spectral light including the Hα line. The film was developed in HC-110D with 16 minute developing time giving a photographic gamma of about 2.82. The region was a 10×10 arcmin region lying at the base of two large streamers. The main feature in the field of view is an 'arcade' of loops.

The large coronal radial intensity gradient was not balanced with a neutral density filter and the resulting density variation across the field of view is very large, almost 5 density units. The exposure in the inner corona was almost saturated recording on the reduced-slope upper part of the photographic characteristic curve, while the most radially distant portions were almost clear. The National Solar Observatory/Sacramento Peak (NSO/SP) Fast Microphotometer (FM)[2,5] is a laser spot scanner well suited for wide-latitude images. Light scattering from adjacent portions of the image is not a problem with this system as the illumination is from one laser spot, and the high intensity source and analog log-ratio module give a good signal over 6 full decades. The FM records the digitized log signal with a 14 bit A/D, and exhibits a noise of less than 0.5 millidensity units.

Figure 1 shows the photographic density from a single white-light image with the azimuthally averaged radially smoothed density subtracted. The main structure in the field of the view, the coronal arcade between 350-650 arcsecs in the horizontal and 100-300 arcsecs in the vertical shows an excess of about 0.5 density units above the average, substantially smaller than the mean background radial variation. Also some finer contrast structure is apparent as relatively bright and dark loops in the arcade.

When we subtract the locally smoothed image, fine structure in the coronal ar-

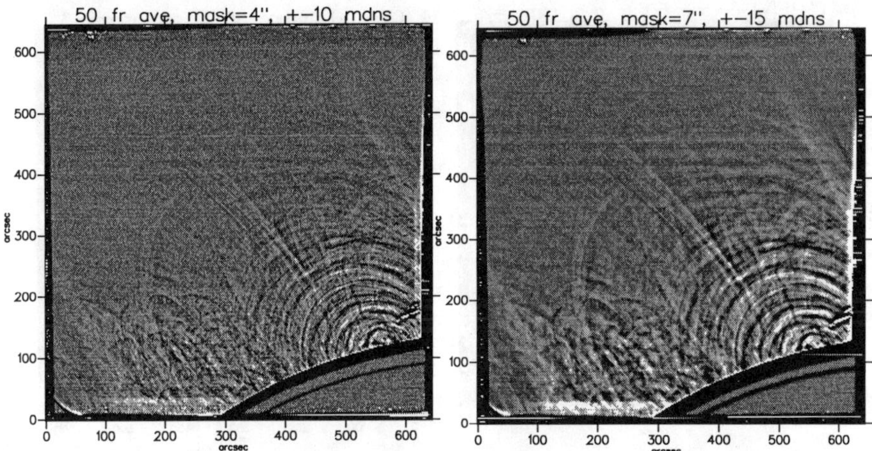

Figure 2: Unsharped masked image. An average of 50 images like the one shown in Figure 3 with a 4 arcsec Gaussian convolution subtracted on the left and a 7 arcsec Gaussian convolution subtracted on the right. The plotting range for the left image is ±.010 density units and for the right image is ±.03 density units.

cade becomes evident as low contrast fluctuations smaller than 0.05 density units in amplitude. An average of 50 frames from the time series is shown in Figure 2 spatially filtered in this way. The unsharp masked image from a single frame shows substantial photographic film noise and a residual fringe pattern in the more distant corona. The NSO/SP FM introduces low contrast interference fringes below a density of about 0.6 due to the laser illumination. The time average greatly reduces the noise and appears to eliminate the residual fringe pattern.

Substantial fine structure is evident. The arcade consists of many loops that are both brighter and darker than their surroundings. The total contrast of the loops is small, less than 0.03 photographic density units.

A superior form of spatial filtering is obtained using the mad-max algorithm, like that described by Koutchmy and Koutchmy.[3] The 'mad max' is the second derivative in the direction for which the absolute value of the second derivative is a maximum. The behavior of the mad max qualitatively resembles a second-derivative filter, but by selection of the direction of the maximum variation it tends to enhance linear structures substantially. We have modified the mad-max filter for this work using the convolution with the second derivative of a Gaussian. The process defines the maximum value of the convolution with the second derivative of a Gaussian in the direction that maximizes its absolute value.

Figure 3 shows the mad-max images using a Gaussian of half width at $1/e$ of 2 and 3 arcsecs applied to the average of 50 images. The same loop structures can be

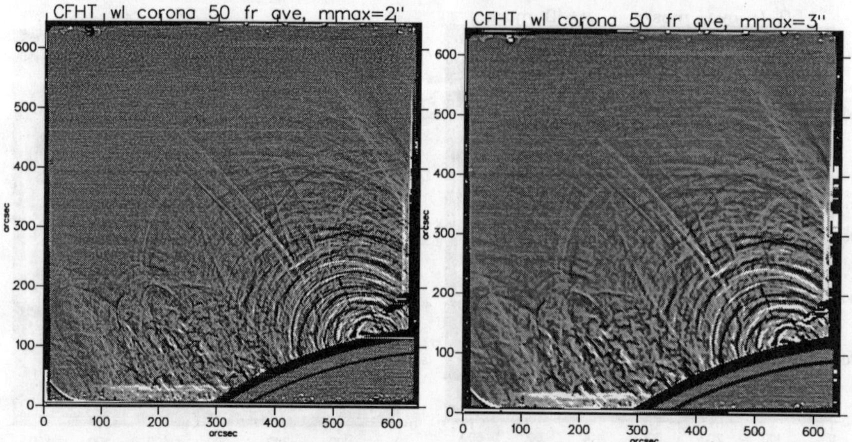

Figure 3: Mad-max image for 2 and 3 arcsec spatial scales in left and right panels respectively from 50 frame average.

found in the unsharp masked images, Figure 2, but better defined.

Loop structures smaller than 1 arcsec are apparent in the images. Figure 4 shows the enlargement of a portion of the images in the vicinity of the loops. (The labeling on the axes remains consistent with the rest of the figures for locating the subfield.) The features in the coronal loop structure exhibit a distribution of density units in the range of roughly ± 20 millidensities for the 7 arcsec unsharp mask, as is evident in the histogram, Figure 5. The histogram for the 7 arcsec unsharp mask may be slightly skewed, favoring a larger density excursion for the dark loops.

3. Physical interpretation of dark and bright loops

The images presented in Figures 2, 3, and 4 show that there is coronal loop structure finer than a few arcsecs. Presumably these indicate a fine magnetic morphology with corresponding density inhomogeneities in the corona.

The measured photographic density fluctuations must correspond to relative white-light intensity fluctuations due to Thomson scattering. Estimates of the mean free path for photon scattering based upon coronal densities indicate that the medium is optically thin, and therefore the intensity is proportional to the integrated electron density along the line of sight. The darkest loops must indicate partially or fully evacuated structures. The maximum density fluctuations for the dark loops must be limited by their thickness which limits their contribution to the total emission in the integral along the line of sight.

The photographic density D is generally expressed in terms of the intensity I by

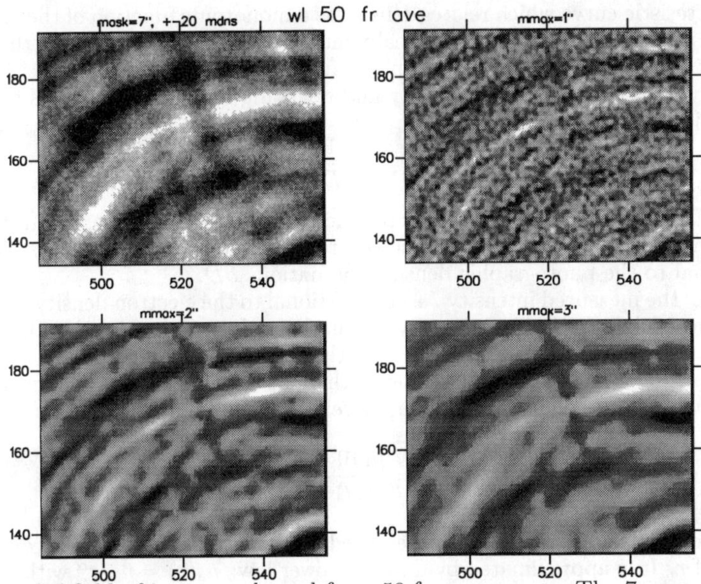

Figure 4: Subfield of images enlarged from 50 frame average. The 7 arcsec unsharp-masked field plotted in the range ±.015 density units is shown in the upper left corner. The mad-max-filtered images for the spatial scales of 1, 2, and 3 arcsecs are shown in the top right, bottom left and right images, respectively. It is easy to identify common features in the images.

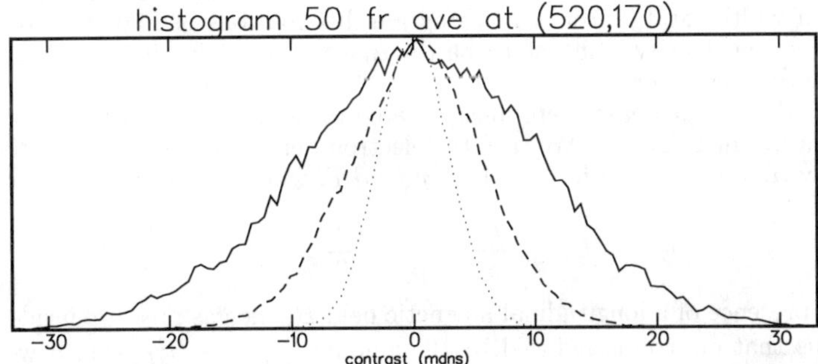

Figure 5: Histogram of photographic density distribution from the unsharp masked images for the region shown in Figure 6. The three histograms are for the 2 arcsec (dotted), 4 arcsec (dashed), and 7 arcsec (solid) unsharp masks. The horizontal scale represents the feature contrast in millidensities on the film.

the characteristic curve which relates density as a monotonic function of the logarithm of the intensity: $D(\log_{10}(I))$. We are really interested in the differential of the inverse relation, $I(D)$, which defines I as a unique function of D. Taking the derivative of the density with respect to the intensity and rewriting for the inverse gives the useful relationship:

$$\frac{\partial I}{\partial D} = \frac{\ln 10}{\gamma(D)} I, \tag{1}$$

where $\gamma(D)$ is the slope of the characteristic curve, which in general is a function of the density D. Thus we have that the relative intensity fluctuations $\partial I/I$ are proportional to the photographic density fluctuations ∂D.

We take the measured intensity I as proportional to the electron density integrated over the line of sight, $n_e L$. The intensity fluctuation, ∂I, is proportional to the electron density fluctuation in the line of sight from the thickness of that volume: $\partial n_e d$. The small distance d is the effective diameter of the loop structure along the line of sight. Substituting into the photographic density relation Equation (1) we have:

$$\frac{\partial n_e}{n_e} = \frac{\ln 10}{\gamma(D)} \frac{L}{d} \partial D, \tag{2}$$

The mean solar radial density n_e measured from these data is a rapidly decreasing function of r. It is approximated by a radial power law: $n_e(r) = n_0/r^\alpha$ with $\alpha = 14.2$ at small radii, for r in units of solar radii. Our measured slope is consistent with the standard coronal radial electron density variation near 1.1 solar radii, the location of the subfield in Figure 4.[1]) Because the radial dependence of n_e is so steep, only a small effective length significantly contributes to the path integral $n_e L$. We have estimated the path integrals by numerical integration, and find that $L \approx 0.12r$ for our measured radial slope, or $L \approx 127$ arcsec for the subfield. We take the diameter of the loops d to be twice the horizontal scale used for the unsharp masking filter, since the scale used for the filter is the half width at $1/e$ for the Gaussian filter. Using the measured density contrast ∂D from the images in the subfields from Figure 4 given by the half widths at $1/e$ from the histogram distributions of Figure 5, we find the relative electron density contrast to be $\partial n_e/n_e \approx -0.105$ in the dark loops for all three unsharp mask sizes.

With a density decrease there must be a corresponding gas pressure decrease ∂p or temperature increase ∂T. We take the electron density n_e as representative of the gas density n. From the perfect gas law: $p = nkT$, where k is Boltzmann's constant, we have:

$$\frac{\partial n}{n} = \frac{\partial p}{p} - \frac{\partial T}{T} \tag{3}$$

In the presence of a longitudinal magnetic field $B_{\|}$ the gas pressure inside the flux tube minus that outside is reduced by the amount: $\partial p = -B_{\|}^2/(8\pi)$. If we assume there to be no temperature change, then $\partial p \approx -0.105p \approx -1.74 \cdot 10^{-2}$ erg/cm^3, giving $B_{\|} \approx 0.7$ gauss.

The bright loops must indicate regions of enhanced electron density. The histograms, Figure 5, show that the bright loops are comparable in amplitude to the dark loops, meaning: $\partial n \approx 0.105n$ there. If we assume that the bright loops exhibit the same magnetic field strength as the dark loops, then the temperature there must be reduced by just twice the electron density fluctuation: $\partial T \approx -0.21T$ compared to what is found in the dark loops. But without additional constraints it is not possible to define the thermodynamic quantities uniquely.

A variety of coronal structures and evolution can be studied with this data set. We expect to be able to address issues on the size distribution of magnetic loop structures and their temporal evolution from further analysis.

4. Acknowledgments

It is a pleasure to acknowledge the efforts of many people who helped to obtain these observations: M. Belmahdi, R. L. Coulter, P. Démoulin, G. Monnet, J. Mouette, J. C. Noéns, J. Sovka, and J. P. Zimmermann.

5. References

1. , C. W. Allen, *Astrophysical Quantities*, third edition, Athlone Press, University of London (1973), pp. 176.
2. M. R. Arrambide, R. B. Dunn, A. W. Healy, R. Porter, A. L. Widener and L. J. November, and G. E. Spence, "The Sacramento Peak Fast Microphotometer", *Proc. Microdensitometry Conf. Astronomy*, ed. D. Klinglesmith, NASA Publication 2317, Goddard Space Flight Center, Goddard Maryland (1983), pp. 243-254.
3. O. Koutchmy and S. Koutchmy, "Optimum Filter and Frame Integration: Application to Granulation Pictures", *Proceedings of the 10th NSO/SPO summer workshop*, ed. O. von der Lühe (1988), pp. 217-231.
4. S. Koutchmy, M. Belmahdi, R. L. Coulter, P. Démoulin, V. Gaizauskas, R. M. MacQueen, G. Monnet, J. Mouette, J. C. Noéns, L. J. November, R. W. Noyes, D. G. Sime, R. N. Smartt, J. Sovka, J. C. Vial, J. P. Zimmermann, and J. B. Zirker, "CFHT eclipse observation of the very fine-scale solar corona", *Astron. Astrophys.* **281** (1994), pp. 249-257.
5. L. J. November 1993, "FITS library: FITS interactive task and shell-script library", *Software for Solar Image Processing, Proceedings for the LEST Mini Workshop*, ed. Z. Yi, T. Darvann, and R. Molowny Horas, Inst. of Theoretical Physics, Univ. of Oslo, POB 1029 Blindern, N-0315 Oslo Norway (1993), pp. 111-122.

Association of Solar Coronal Temperature and Structure from Ground-Based Emission-Line Data with Global Magnetic Field Models and Yohkoh SXT Data

R. C. Altrock
Phillips Lab./Geophysics
National Optical Astronomy Observatories†
PO Box 62, Sunspot, NM 88349, USA

P. Hick and B. V. Jackson
UCSD, CASS 0111, La Jolla, CA 92093-0111

J. T. Hoeksema and X. P. Zhao
CSSA ERL 328, Stanford, CA 94305

G. Slater
Lockheed Palo Alto Res. Lab., 3251 Hanover St., Palo Alto, CA 94304

T. W. Henry
National Solar Observatory/Sacramento Peak
National Optical Astronomy Observatories†
PO Box 62, Sunspot, NM 88349, USA

ABSTRACT

The large-scale structure of the solar corona is investigated through use of synoptic maps produced from Fe XIV 530.3nm, Fe X 637.4nm and Ca XV 569.4nm data from NSO/SP, Yohkoh SXT CMP data and Wilcox Solar Observatory (WSO) Source Surface Field maps. Temperatures have been calculated with the Fe X and Fe XIV data using an analytic approximation to the technique developed by Guhathakurta et al. (Ap. J. 388, 633, 1992). We find that the 1.15 Ro Fe XIV data are an excellent proxy for the degraded Yohkoh SXT data. Both isolated emission features and the large-scale structure are nearly identical throughout the entire period studied. In addition, coronal holes and other low-emission regions are very similar. Synoptic maps of coronal temperature calculated as above show a tendency for the highest temperatures to occur where the large-scale solar magnetic fields change polarity at high latitudes (cf. Guhathakurta, Fisher and Altrock, Ap. J. 414, L145, 1993). Temperatures over lower-latitude features, including active regions, appear to be significantly lower than over the high-latitude polarity reversal. In addition, we find that the regions of enhanced temperature generally follow the heliospheric current sheet as defined by the neutral line in the WSO maps, particularly when the temperature is calculated at higher altitudes. However, we observe that emission in Ca XV, which is formed at approximately 3MK, generally occurs only in low-latitude regions that are bright in Yohkoh SXT, Fe X and Fe XIV. Thus, the Guhathakurta et al. technique may fail here due to the presence of fine structure.

† Operated by the Association of Universities for Research in Astronomy, Inc. (AURA) under cooperative agreement with the National Science Foundation

IR Coronal Lines from EUV Data

Edward S. Chang
Department of Physics and Astronomy, University of Massachusetts
Amherst, MA 01003, USA

and

Drake Deming
Code 693, NASA/Goddard Space Flight Center, Greenbelt, MD 20771, USA

ABSTRACT

Coronal IR line intensities can be deduced from the abundant UV spectral lines when they share common upper levels and the relevant Einstein A coefficients are known. We have examined the intensities of a numbers of IR lines in Fe XIV, both forbidden and allowed, including the analogue of the photospheric 12 μm emission lines. Other IR lines from the excited states of ground configurations appear to be more promising. We discuss deduced intensities for the 1.561 and the 2.218 μm lines of Fe XII and other weaker lines.

1. Introduction

Great strides have been taken in understanding the solar magnetic fields as is evident in the recent review by Solanki.[20] Yet all that knowledge is confined primarily to the photosphere, a little to the chromosphere, and virtually none to the corona. From the review, it is clear that the most powerful method is to measure the Zeeman-split Stoke parameters of infrared lines.

Recently, rapid advances in infrared detector technology make it possible to study coronal lines in the infrared. These lines, like the underlying continuum, are a million times weaker than lines in the photospheric spectrum. Surprisingly, there have been no surveys of IR coronal lines since the 1970 eclipse observation of Olsen *et al.*[18] Of the 13 lines observed between 1 and 3 μm, only 7 were identified as coronal. Even then, the identity of some have been questioned. Indeed Kastner[14] has even suggested that one line is actually a chromospheric neutral helium line rather than a highly ionized coronal line. On the other hand, hundreds of lines have been measured in high resolution ultraviolet (UV) and extreme ultraviolet (EUV) spectra, e.g. the SERTS data of Thomas and Neupert[21] (referred as TN). It is possible to infer IR line intensities from these lines if the IR and UV lines share the same upper level. In that event, the relative line intensities are given simply by the (usually known) Einstein A coefficient ratio.

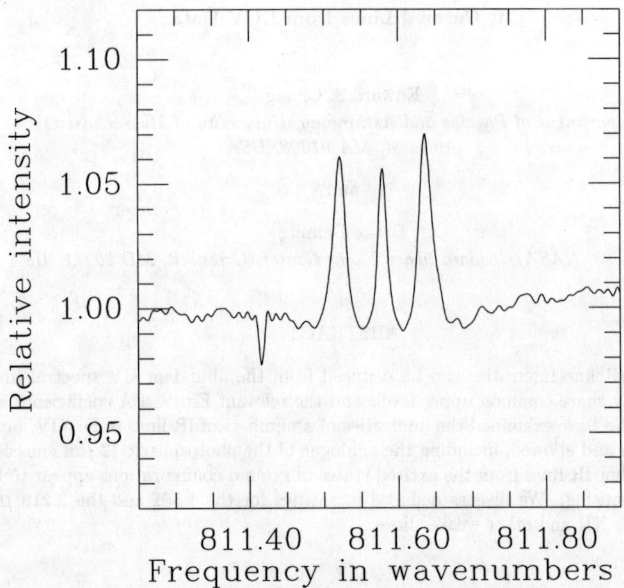

Figure 1: Spectrum of the 7i-6h emission from a sunspot penumbra, taken using the McMath-Pierce Fourier transform spectrometer. The Zeeman splitting into three resolved components is very obvious. The absorption line near 811.42 cm-1 is telluric.

2. High-ℓ Rydberg Allowed transitions

The 12 μm emission lines are the most magnetically-sensitive lines in the solar spectrum (Deming 1994). Extensive measurements of the upper photospheric sunspot magnetic fields have been carried out using the 7i-6h transition in Mg I.[13] The very prominent Zeeman splitting seen in this upper photospheric emission is illustrated in Figure 1.

These Rydberg levels of high n and ℓ are only a few tenths of 1 eV below the ionization limit, and are strongly populated by recombination and cascading from yet higher levels. In contrast to the low-excitation absorption lines formed in the deep photosphere, these highly excited emission lines are formed in the upper photosphere near the temperature minimum.[5,3] Nevertheless, their line intensities are comparable because the low populations of the highly excited levels are compensated by their large oscillator strengths, which vary as n^4. [Summary of magnetic field measurement results.]

Naturally, one wants to know if similar or analogous (in whatever sense) lines can form in the corona. Since neutral atoms cannot exist in the hot corona, we investigate abundant coronal ions in high n and ℓ states, which have all the same atomic characteristics of the 12 μm lines. Thus their oscillator strengths are accurately and readily calculated in the hydrogenic approximation. Again their populations are expected to be much lower than the those belonging to the ground configurations, but their oscillator strengths are much larger, especially when compared with the forbidden lines observed in ground configurations. We now discuss the theory for these high n and ℓ ionic lines.

As explained in an earlier review[4], the energy level structure of an atomic system is determined by its core, specifically the quadrupole moment Q. When the core is an S state, its Q vanishes, and we have simply one level for each set of quantum numbers n and ℓ. However, when the core is not an S state, coupling with Q causes each n, ℓ level to split into several levels (as many as 50 levels in the case of Fe I.[14] Correspondingly, a single line in the former case becomes many lines in the latter, thereby diluting its oscillator strength. Since we are only interested in strong lines, we will restrict our discussion to the Q=0 case.

For the purpose of this discussion, it is of sufficient accuracy to represent the Rydberg energy level by[8],

$$E_{n\ell} = -\frac{Z^2 R}{n^2} - Z^4 \alpha P(n, \ell). \tag{1}$$

The first term is simply the Rydberg formula, where Z is the net charge (atomic number minus the number of core electrons) and R the Rydberg constant. The second term represents the main correction, due to the polarizability of the ionic core α. Values for α for most stages of ionization can be found in Fraga *et al.*[11], but they must be multiplied by the conversion factor of 6.75 (to change from Angstrom cubed to Bohr radii cubed). P is a well known function of n and ℓ and has been tabulated by Edlén.[8] Since the tabulated values for α are not accurate to better than 10 % , we will ignore the next order correction term which typically contributes even less to Eq.(1).

Consider Fe XIV which produces the famous green coronal line identified by Edlén.[7] The core is magnesium-like ^1S, which satisfies the above condition with Z=14. By the Rydberg formula, the n=7 to 6 lines are now shifted from 12 μm to 600 Å, and the n=5 to 6 lines from 7 μm to 370 Å. The latter falls into the spectral region covered by the high resolution TN data. The polarizability of the magnesium-like core is only given to one significant place by Fraga, but fortunately it has been calculated in the more accurate coupled Hartree-Fock method to be 0.3518 a_o^3.[17] Comparison with other calculated values for lower members of the iso-electronic sequence suggests that the above value is accurate to better than 1%. From Eq.(1). we calculate the 6h-5g level difference to be 268,940 $\pm$ 70 cm^{-1}, which correspond to a wavelength of 371.830 $\pm$ 0.097 Å. [Actually, the uncertainty in wavelength is probably greater because we have not accounted for higher order terms neglected in Eq.(1).] In the TN unidentified line lists, there is a line at 371.834 $\pm$ 0.012 Å, with an observed intensity of 10 $\pm$ 4 ergs cm^{-2} s^{-1} sr^{-1}, including re-calibration of the intensity.[22] We assume that the identification is correct and raise the question of what IR lines would originate from the same upper level 6h.

An obvious allowed transition in the IR is 6h-6g, whose frequency is found from Eq.(1) to be 3303 ± 48 cm^{-1} corresponding to a wavelength of 3.028 ± 0.044 μm . Actually, there are relativistic (spin-orbit) corrections which split the line into two main components at about a standard deviation apart. For the most favorable scenario, we ignore these effects in calculating the predicted IR line intensity from the EUV line. The Einstein A coefficients for both the EUV and the IR lines are hydrogenic and are calculated[11] to be $A_{6h-5g} = 2.0 \times 10^{11}$ s^{-1} and $A_{6h-6g} = 4.7 \times 10^{5}$ s^{-1}. Since they share the same upper level, we infer an IR line intensity of 2.4×10^{-5} erg, cm^{-1} s^{-1} sr^{-1}. From the only observation of coronal IR lines in this spectral region was the airborne eclipse FTS data of Olsen [18], the weakest observed line had an intensity of about 3 ergs cm^{-2} s^{-1} sr^{-1}. Thus we conclude that the proposed IR line in Fe XIV, which is weaker by 5 orders of magnitude, is not likely to be observable.

To gain a better understanding of the physics of high-ℓ Rydberg states in highly stripped ions as in Fe XIV, we have begun modeling these levels in a collaboration with Bhatia. His previous calculations of Fe XIV[1,2] had included all n=3 states, totaling 5 configurations and 40 levels, and had succeeded in predicting and identifying solar XUV lines. Our present model does not yet include the 6h level, but only the 5f as the highest level. For an electron density of 10^{10} cm^{-3} and a temperature of $2 \times 10^{6} K$, the relative population of the 5f levels is calculated to be less than 10^{-12}. Therefore we can surmise that the relative population of the higher 6h level is no more than 10^{-13}, while the upper level of the green line $^{2}P_{\frac{3}{2}}$ is calculated to be 1.8×10^{-1} . The loss of 12 orders of magnitude is insufficiently compensated by the gain of 4 orders of magnitude in the Einstein A coefficients, leaving the IR 6h-6g line weaker by 8 orders of magnitudes in the photon counting rate.

Unfortunately the above considerations for the 6h Rydberg level for Fe XIV appear to be quite general for other high-ℓ Rydberg levels as well as for all coronal ions. Undoubtedly, 12 μm -like lines exist for many coronal ions, but they are at present all too weak to be observable by at least 3 orders of magnitude. For example, the ratio of the Einstein A coefficients for the IR/UV lines originating from the same upper level will always be about 10^{-6}, from the frequency ratio of 10^{-2} cubed. The dipole matrix element plays no significant role since it is roughly $n^{2}a_{o}$ for both IR and UV transitions. Since no strong unidentified UV lines have been observed, any Rydberg UV lines have to be weak, and the corresponding IR lines would be 6 orders of magnitude weaker in the photon counting rate. Alternatively, all Rydberg levels are by definition near the ionization limit. Therefore they have comparable high excitation temperatures and populations too low to produce observable emission lines.

The prior discussion suggests that strong coronal lines must originate from low excitation levels. However in highly stripped ions, the low levels are usually far apart in energy, typically corresponding to the XUV range. Therefore IR lines are largely confined to forbidden transitions between low-lying multiplet sublevels, separated by the small fine structure splittings.

3. Low Excitation Forbidden Transitions

Fortunately a list of forbidden lines in the ground configuration, including Einstein A coefficients, has been compiled by Kaufman and Sugar[16] from laboratory and calculated data. The nearly 300 lines reduce to a couple of dozen if we confine ourselves to the most abundant elements at a sufficiently high stage of ionization (say, VIII or greater). Of these, most belong to the ground spectroscopic term, like the FE XIV green line, for which predictions of line intensity have to come from detailed atomic modeling, but cannot be inferred from the abundant UV data. Among these are the near-IR lines (Fe XIII), observed in the 1965 eclipse by Jefferies *et al.*[13] Coronal observations with those Fe XIII lines and another ground spectroscopic term line (Si X) are reported by Penn[19] at this workshop. However, a few lines in the ground configuration belong to excited spectroscopic terms. We will show that their line intensities can be inferred from the appropriate UV data.

A promising example is provided by Fe XII, whose ground configuration is $3p^3$. In ascending order, the 3 spectroscopic terms are ^4S, ^2D, and ^2P. UV transitions between these have been extensively studied by Feldman *et al.*[9] including blending of certain Ni II lines. The IR lines of interest arise from fine structure transitions within the two excited spectroscopic terms, $^2D_{\frac{5}{2}} -^2 D_{\frac{3}{2}}$ at 4509 cm^{-1}or 2.217 $\pm$0.003 μm and $^2P_{\frac{3}{2}} -^2 P_{\frac{1}{2}}$ at 6406 cm^{-1}or 1.5606 $\pm$0.0017 μm . From analysis of the NRL *SKYLAB* data, Feldman *et al.*[9] derived an intensity for the ^2D UV line of 1.9×10^{12} photons cm^{-2} s^{-1} sr^{-1}. Since it shares the same upper level as the 2.217 μm line, we infer an IR line intensity of 9×10^{11} photons cm^{-2} s^{-1} sr^{-1} $(= 3 \times 10^{-6} B_{\odot})$ from the Einstein A values of 1.84 and 0.868 s^{-1} respectively for the UV and the IR lines. It should be pointed out that the UV intensity was accurate only to a factor of two, and that the deduced path length was considered to be too large by a factor of 3 or 4.

Less promising is the 1.56 μm line mentioned above, where the corresponding UV line intensity was determined to be 10^{12} photons cm^{-2} s^{-1} sr^{-1}. However, since the IR/UV Einstein A coefficient ratio is less than 1%, the inferred IR line intensity is only 6×10^9 photons cm^{-2} s^{-1} sr^{-1}.

The most common situation where the IR and UV pair shares the same upper level is when the IR transition is forbidden and the UV allowed (electric dipole). Since the Einstein A ratio is typically 10^{-7} or 10^{-6}, the inferred IR line intensity is likely to be insignificantly small. For example in Fe XV, the EUV resonance line $^3P_1 -^1 S_0$ at 417.245 Å has an A value of 3.38 $\times 10^7$ s^{-1} (since the upper level is mixed with the $^1P_1$ level through the spin-orbit interaction). However the IR line $^3P_1 -^3 P_0$ at 1.739 μm has an A value of only 3.6 s^{-1}. The TN data gives for the EUV line an intensity of 339 ergs, cm^{-2} s^{-1} sr^{-1} = 7.1×10^{12} photons cm^{-2} s^{-1} sr^{-1}. We therefore infer an IR line intensity of 7.2 $\times 10^5$ photons cm^{-2} s^{-1} sr^{-1}, comparable to the Rydberg IR line intensity in Section 2.

There are circumstances where an IR line intensity can be inferred from another IR (or visible) line belonging to the same spectroscopic term if we assume statistical equilibrium within the multiplet. For example, the ground transition $^3P_1 -^3 P_2$ in Fe

XI at 7891.8 Å has been observed (Jefferies *et al.*, [13] to have an intensity of 4.94 ergs cm^{-2} s^{-1} sr^{-1}= 2.0×10^{12} photons cm^{-2} s^{-1} sr^{-1}. Its Einstein A value is 43 s^{-1} while that of the other line within the same multiplet, $^3P_0 - ^3P_1$ at 6.082 μm , is 0.223 s^{-1}. Assuming statistical equilibrium which reduces the intensity of the IR to the visible line by a factor of three, we obtain an inferred intensity of 3.5×10^9 photons cm^{-2} s^{-1} sr^{-1} for the 6 μm line.

Unfortunately the above assumption may not be reliable. Returning to Fe XV, we use the above procedure to re-examine the IR line intensity. The multiplet companion line, $^3P_2 - ^3P_1$ at 7058.6 Å has been observed (Jefferies *et al.* 1971) to have an interpolated intensity of 8.0×10^{11} photons cm^{-2} s^{-1} sr^{-1}. From the Einstein A ratio of 38:3.6 and the statistical weight ratio of 5:3, we infer an intensity of 6.2×10^{10} photons cm^{-2} s^{-1} $sr^{-1}$for the $^3P_1 - ^3P_0$ line. This value is five orders of magnitude larger than our previous value inferred directly from the EUV data. We therefore conclude that the assumption of statistical equilibrium within the multiplet sublevels of a spectroscopic term to be unreliable.

4. Conclusion

We have investigated the question of what IR coronal lines are intense enough to be observable using atomic theory and existing coronal spectra ranging from the visible to the extreme UV. Of particular interest are the analogues of the photospheric 12 μm lines in the coronal ions. Theory allows us to calculate all properties of these lines covering the IR to the EUV for coronal Fe ions. While there is some evidence in the EUV data for one such line in Fe XIV, the inferred IR line intensity is too low to be presently observable by 5 orders of magnitude. We present arguments to show that the above result applies to all coronal ions, and therefore it is not fruitful to search for the analogue of the photospheric 12 μm lines. As these are the strongest allowed transitions in the mid-IR, it appears unpromising to examine allowed transitions for magnetic-sensitive lines in the corona.

The known coronal lines are all forbidden and from ground spectroscopic terms. We have investigated forbidden lines from excited spectroscopic terms which are necessarily weaker. We have shown that their intensities can be inferred from those observed UV lines which share the upper level of the IR lines. In particular, the 2.218 μm line in Fe XII is found to have an intensity of 9×10^{11}photons cm^{-2} s^{-1} sr^{-1}, more than an order of magnitude brighter than the weakest observed coronal line. We have not uncovered other Fe ion lines of comparable intensity, but are hopeful of finding such lines in other elements by this procedure.

5. Acknowledgments

We thank the NASA-ASEE for the award of a Summer Faculty Fellowship to E.S.C. at the Goddard Space Flight Center, Greenbelt, MD, where most of this work was done. We are grateful to R.J. Thomas and A.K. Bhatia for enlightening discussions.

6. References

1. Bhatia, A.K. & Kastner, S.O.: 1993, *J. Quant. Spectrosc. Radiat.Transfer* **49**, 609.
2. Bhatia, A.K., Kastner, S.O., Keenan, F.P., Conlon, E.S. & Widing, K.G.: 1994, *Astrophys. J.***427**, 497.
3. Carlsson, M., Rutten, R.J. & Shschukina, N. G., 1992, *Astron. Astrophys.*, **253** 567.
4. Chang, E. S.: 1994, in Rabin,D.M., Jeffereies, J.T. & Lindsey, C. (eds.) *Infrared Solar Physics:IAU 154*, Kluwer Academic Publishers, Dordrecht, 297.
5. Chang, E.S., Avrett, E. H., Mauas, P.J., Noyes, R. W. & Loeser, R., *Astrophys. J. (Letters)***379** L79.
6. Deming, D., Hewagama, T., Jennings, D. E., and Wiedemann, G. 1991, in Solar Polarimetry, Proceedings of the Eleventh National Solar Observatory / Sacramento Peak Summer Workshop, Sunspot, New Mexico, edited by Laurence J. November, pp. 341-354.
7. Edlén, B.:1942, Z. Astrophys. **22**, 30.
8. Edlén, B.:1964, Handbuch der Physik, Bd. XXVII, 80.
9. Feldman, U.,Cohen, L. & Doschek, G.A.: 1983, *Astrophys. J.***273**, 822.
10. Fraga, S. Karwowski, J. & Saxena, K.M.S.: 1976, *Handbook of Atomic Data, Physical Sciences Data 5* pub.:Elsevier,p 321.
11. Green, L.C., Rush, P.P. & Chandler, C.D.: 1957, **3**, 37.
12. Hewagama, T., Deming, D., Jennings, D. E., Osherovich, V., Wiedemann, G., Zipoy, D., Mickey, D. L., and Garcia, H., 1993, Ap. J. (Suppl.) 86, 313:332.
13. Jefferies, J. T., Orrall, F.Q. & Jirker, J. B.: 1971, *Solar Phys.***16**, 103.
14. Johansson, S., Nave, G., Geller, M., Sauval, A.J., Grevesse, N., Schoenfeld, W.G., Chang, E.S. & Farmer, C.B., 1994 *Astrophys. J.***429**
15. Kastner, S.O.: 1993, *Solar Phys.* **143**, 197.
16. Kaufman, V. & Sugar, J. :1986, *J.Phys.Chem.Ref. Data*, **15**, 321.
17. Markiewicz E, McEachran R.P., & Stauffer A.D.: 1981, *J.Phys. B* 14, 949.
18. Olsen, K. H.,Anderson,C.R.,Stewart,J.N.: 1971, *Solar Phys.* **21**, 360.
19. Penn, M.: 1994, *15th NSO/Sac Peak Workshop* .
20. Solanki, S.: 1993, Solanki, S.K. 1993, Space Science Reviews **63**, 1.
21. Thomas, R.J., Neupert W.M.: 1994, *Astrophys. J. Suppl.***91**, 461.
22. Thomas R.J.: 1994, private communication.

SOME QUESTIONS OF SPECTROSCOPIC INTEREST IN INFRARED SOLAR PHYSICS

S.O. KASTNER
1-A Ridge Road, Greenbelt, MD 20770 U.S.A.

ABSTRACT

Three distinct topics will be discussed, which are, however, related to each other through some common aspects. These are (1) the photoexcitation-by-accidental-resonance of neutral oxygen Ly-β, a PAR process which will be referred to below as a pumping process; (2) a question of whether deuterium lines are present in an early chromospheric spectrum, and the general interpretation of that spectrum; and (c) the observability of an unusual forbidden line in highly ionized Fe XIV.

1. O I pumped by H Ly-β

Figure 1 gives a Grotrian diagram of the main transitions in oxygen, and Figure 2 details the particular levels and transitions involved in the pumping process.

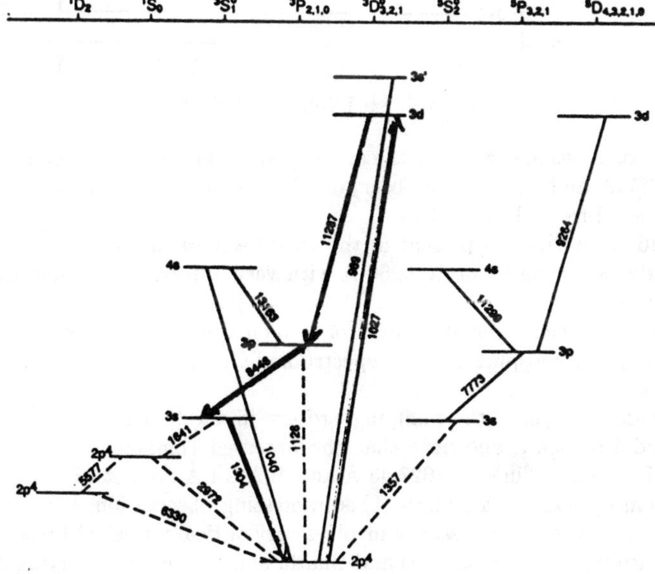

Figure 1: O I energy level diagram.

There have been three main studies of this process in solar or stellar chromospheres,[7,14,3] which have, however, dealt only with the tertiary cascade line at 1304 Å.

This is the line least sensitive to the PAR process, especially if density is appreciable, and the results have consequently not provided definite proof of the process.

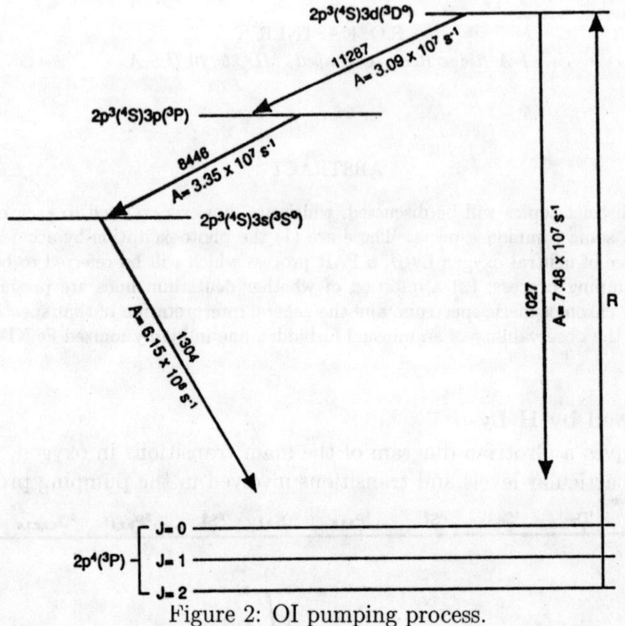

Figure 2: OI pumping process.

So the next question is whether the more sensitive primary and secondary cascade lines at 11287 Å and 8446 Å are observable in the Sun. The following facts about this are gathered from the literature.

The 8446 Å multiplet is present in the out-of-eclipse chromospheric spectrum,[13] which includes wavelengths up to 9266 Å with very approximate visual estimates of relative intensity.

The primary line emission at 11287 Å or 1.13 microns is not present in the Fourier-transform outer chromospheric eclipse spectrum of Olsen, et al.,[12] where it might have been expected.

Some evidence against the pumping process in the Sun has been suggested by Feldman and Doschek[5], who note that the observed component intensities of the resolved O I resonance lines at 1027.43 Å and 1028.14 Å are not similar as would be expected from optically thick lines. This interesting observation, however, assumes that the pumping process necessarily involves very optically thick O I resonance lines.

Before turning to work which Anand Bhatia and I have been carrying out on this topic, it is worth noting that Meier[10] carried out a detailed analysis of the tertiary cascade multiplet at 1304 Å observed in the terrestrial dayglow spectrum and concluded the pumping process was producing it. However, here again the near-infrared primary and secondary line emissions which would provide more definitive proof have not been treated or even observed, I believe.

Neutral atoms are difficult to treat theoretically because of their many-particle nature, the nuclear central force being weak in comparison to the interactions between electrons. However, we have constructed an oxygen model including 13 levels or terms which we think is the most complete to date, and have used it to produce expected line intensity ratios over a range of electron densities, temperatures and pumping rates. The next step was to compare with observations. Not surprisingly, there are very few quantitative observations of oxygen line ratios in astrophysical sources despite the great number of qualitative observations; as just mentioned, for the Sun there are only the approximate Pierce estimates. We, therefore, looked at classical novae in which the 8446 Å line becomes strong during the so-called Orion phase, and found six which could be compared against our results. Figure 3 shows how the observed ratio I(8446)/I(7774), a convenient ratio because reddening doesn't have to be taken into account and instrumental sensitivity is constant, varies with time in the novae and how it compares with the calculated grid of values. One sees that the novae values rise substantially higher than the zero-pumping contour. This is satisfying quantitative verification of a long-held belief that the hydrogen-oxygen PAR process operates in novae.

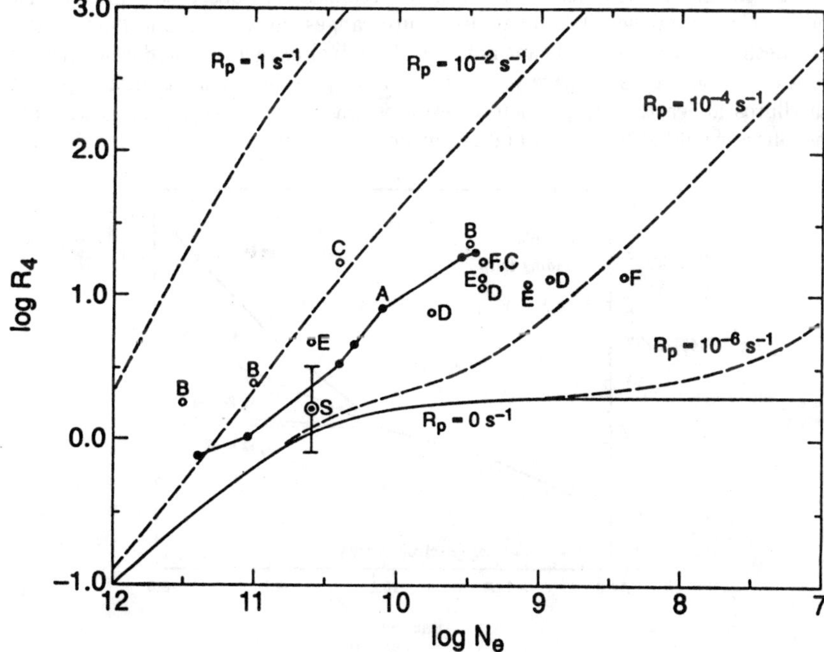

Figure 3: O I line ratio vs. electron density.

In the same figure, a single point is included to represent the Pierce observation for the Sun; its location is thus only approximate and can only suggest that the pumping process may exist in the solar chromosphere, without being definitive.

The figure illustrates further that time-dependent observations, such as are available in the case of the novae, are much more informative that static time-independent observations, for the purpose of investigating physical processes.

It is therefore of interest to speculate what might be gained by observing oxygen line ratios during a solar flare, most feasibly in a decaying phase in which both density and temperature are likely to decrease with time. Rough simulations are shown in Figures 4 and 5, of expected time-dependent behaviors of the 8446/7774 and 11287/7774 ratios. In these figures, the solid lines represent constant pumping rates. One sees that the pumping process is not detectable at the higher densities; on the other hand this means that the line ratios can therefore serve as density/temperature diagnostics. Since the ambient hydrogen Ly-β intensity is likely to decrease also with time, dashed lines are drawn in to represent such a decrease. The figure does not take into account excitation by the flare continuum radiation which will be present also, but such excitation may be comparable for the two lines in the case of 8446 Å 7774 Å. It is seen that the ratio 11287/7774 is somewhat more sensitive to the pumping process as expected. Whether such tracks are realized and observed will obviously depend on the actual density and temperature ranges involved in the flare, and of course whether there is a significant amount of neutral oxygen in the flare or near it. If they are, it would be a positive verification of the presence of the pumping process in solar flares at least. The problem of establishing its presence in the static solar chromosphere would still remain, and is touched on below.

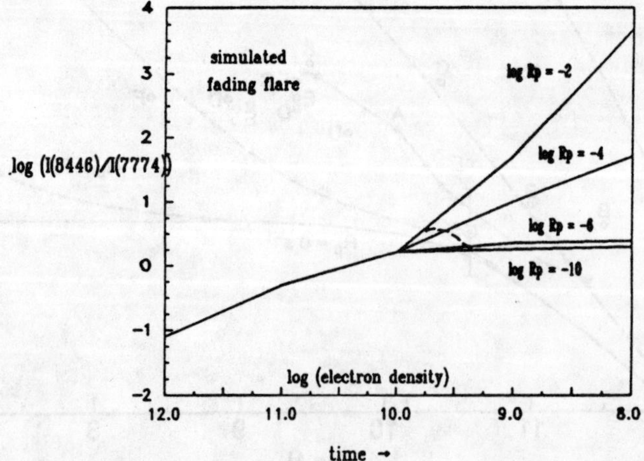

Figure 4: Evolution of OI line ratio during a solar flare.

It is stressed that information on this process gives, at the same time, knowledge of the local hydrogen radiation fields, difficult to obtain by other methods because of optical thickness in the hydrogen.

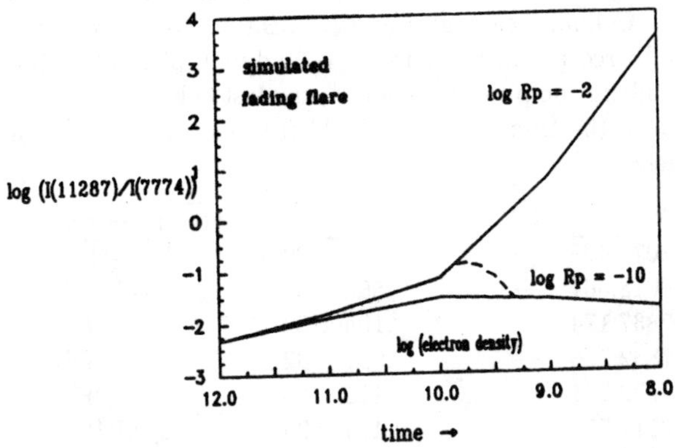

Figure 5: Evolution of OI line ratio during a solar flare.

2. The Pierce Chromospheric Line List; Deuterium or not Deuterium?

Pierce[13] produced a well-resolved chromospheric spectrum by setting a slit at the edge of the McMath Solar Telescope's large solar image, and carried out a careful wavelength calibration to produce the extensive line list published in the Astrophysical Journal Supplement. He also proposed assignments for many of the lines, based on the MIT Wavelength Tables. These assignments, many of which invoke low-abundance heavier elements such as samarium, prasodymium, etc. because of the laboratory emphasis of the MIT Tables, do not appear to have been critically examined, and I compared the wavelength list against the alternative list of Striganov and Sventitskii[15] which by contrast emphasizes lines of the lighter, astrophysically more abundant elements, and with more recent line lists of those elements. Partial results of this re-examination are shown in Tables 1 and 2 which compare the Pierce wavelengths with known wavelengths of carbon and magnesium. It is seen that the heavy element assignments can be discarded in favor of the lighter element classifications, in most cases listed in the tables. It is likely, therefore, that most of, and perhaps all of, the heavy element assignments associated with the Pierce list need to be discarded and replaced by more plausible classifications (this statement is supported by isolated studies such as that of Holweger and Werner[8] who showed that only two out of nine proposed tungsten line classifications in the photospheric spectrum are acceptable). This should be regarded as a positive step forward, rather than a problem, because the lighter atoms are far easier to deal with for diagnostic purposes than the heavier atoms. A fuller tabulation is in progress.

Table 1: Comparison of chromospheric wavelengths with laboratory wavelengths of neutral carbon[a]. [a]Column headings are: λ_{AKP}, chromospheric wavelengths listed by Pierce[13]; ID_{AKP}, Pierce's identifications; λ_{LAB} and multiplet, laboratory wavelengths and multiplet numbers compiled by Moore[11]; the last column gives the upper configuration and term of the transition. [b]Embedded in emission band. Table continued on next two pages.

λ_{AKP}	ID_{AKP}	λ_{LAB}	multiplet	upper term
8078.425	—	.480	30.01	2p5s($^3P^0$)
8018.567	—	.564	31	2p4d($^3D^0$)
7837.174	—	.105	32	2p4d($^3P^0$)
7832.706	—	.629	32	2p4d($^3P^0$)
7132.061	—	.112	26	2p4d($^3F^0$)
7111.553	—	.475	26	2p4d($^3F^0$)
6711.347	—	.291	21.01	2p5s($^3P^0$)
6688.827	—	.787	33	2p6s($^3P^0$)
6665.868	—	.884	34	2p5d($^3D^0$)
6098.949	—	.923	37	2p6d($^3D^0$)
6086.661	Co I	.686	38	2p7s($^1P^0$)
6070.780	—	.833	39	2p6d($^3P^0$)
5984.261	Co I	.261	26.09	2p6s($^1P^0$)
5976.67[b]	Fe I	.678	26.09	2p6s($^1P^0$)
5805.779	Fe I, La I	.801	40	2p8s($^3P^0$)
5805.206	Ni I	.192	18	2p4p(^3P)
5773.137	Fe II ?	.116	18	2p4p(^3P)
5668.924	Ce II	.951	22.06	2p5d($^1P^0$)
5603.692	—	.733	29.12	2p8s($^3P^0$)
5545.022	Gd II	.071	26.14	2p7s($^3P^0$)
5300.142	—	.118	22.08	2p7s($^3P^0$)

Table 1: Cont.

λ_{AKP}	ID_{AKP}	λ_{LAB}	multiplet	upper term
5280.201	A1 II	.238	22.09	2p7s($^1P^0$)
5268.972	—	.956	22.10	2p6d($^1P^0$)
5153.583	—	.571	26.22	2p8d($^3D^0$)
5017.066	—	.090	18.03	2p4f ' [2.5]
4949.545	—	.576	26.26	2p11d($^3D^0$)
4893.420	Y I ?	.429	3.01	2p4p(^3D)
4812.918	Ti I, Cu II	.916	5	2p4p(^3S)
4796.103	—	.082	18.04	2p5p(^3D)
4795.867	Co I	.878	18.04	2p5p(^3D)
4783.850	—	.795	18.04	2p5p(^3D)
4775.909	—	.907	6	2p4p(^3P)
4766.647	V I, Cr I	.676	6	2p4p(^3P)
4762.25^b	Mn I	.314	6	2p4p(^3P)
4478.813	Gd II	.825	18.06	2p5f[2.5]
4478.574	—	.588	18.06	2p5f[2.5]
4477.462	Y I	.472	18.06	2p5f[2.5]
4467.341	—	.309	18.07	2p5f'[3.5]
4466.48^b	Fe I	.476	18.07	2p5f'[3.5]
4463.864	—	.886	18.08	2p5f'[2.5]
4961.267	—	.300	18.08	2p6p(^3D)
4342.201	Gd II	.194	18.10	2p6p(^3P)
4341.623	—	.640	18.10	2p6p(^3P)
4223.390	—	.360	18.11	2p6f[2.5,3.5]
4222.606	—	.631	18.11	2p6f[2.5,3.5]
4222.433	Ce II	.466	18.11	2p6f[2.5,3.5]
4213.063	Ce II	.074	18.12	2p6f'[3.5]
4212.347	—	.342	18.12	2p6f'[3.5]
4211.818	—	.817	18.13	2p6f'[2.5,1.5]

Table 1: Cont.

λ_{AKP}	ID_{AKP}	λ_{LAB}	multiplet	upper term
4211.119	—	.115	18.13	2p6f'[2.5,1.5]
4209.691	Mo II, V II	.710	18.13	2p6f'[2.5,1.5]
4146.242	Cr,Ce II	.264	18.15	2p7p(^{3}P)
4073.490	Ce II	.464	7	2p5p(^{3}D)
4070.987	Ni II,Cr II	.970	7	2p5p(^{3}D)
4066.701	Sm II ?	.752	7	2p5p(^{3}D)
4025.205	—	.221	7.01	2p5p(^{3}P)
3991.320	—	.337	7.02	2p5p(^{1}D)
3769.724	CN(2,2)	.708	7.03	2p6p(^{1}P)
3762.265	—	.251	7.04	2p6p(^{3}D)
3756.547	Cr II, CN	.522	7.04	2p6p(^{3}D)
3613.231	Cr II	.235	7.06	2p7p(^{1}P)
3609.549	Pd I	.562	17.17	2p10p(^{1}S)
3609.261	—	.226	7.07	2p7p(^{3}D)
3607.913	Cr I	.940	7.06	2p7p(^{1}P)
3590.96[b]	Fe I	.972	7.08	2p7p(^{3}P)

In the course of making this general comparison, three lines were noted in the Pierce list which coincide with deuterium Paschen or Balmer line wavelengths. They are listed in Table 3. The most striking wavelength agreement is present for the observed chromospheric line at 9226.527 Å, for which there is also no apparent alternative assignment. In the case of the other two lines, agreement is less striking though not unreasonable within the accuracy of the Pierce wavelengths, and also there are alternative candidates though somewhat unlikely ones.

Because of its importance for theories of stellar nucleosynthesis, deuterium has been looked for in the Sun by some workers including Dr. Beckers, with only upper limits being set in agreement with the accepted view that deuterium is destroyed in the solar interior. However, there are indications that it may be produced locally in active regions or flares. It is therefore interesting that the Pierce list does emphasize lines from active regions.

It would also be interesting to know if the line at 9226.53 Å appears in flare spectra of the kind observed by Dr. Neidig and coworkers.

An argument against these lines being deuterium lines, on the other hand, is that other Paschen or Balmer lines of deuterium might also be expected in the Pierce list, but do not appear to be present.

Summing up, the Pierce list of chromospheric and active region line wavelengths is unique, but one needs photometrically calibrated intensities of the lines in order to make positive identifications, and to be able to make progress in physical analysis of the chromospheric layers. One example has just been given above in the matter of the hydrogen-oxygen PAR process. Modern detectors can do better in the infrared range where the photographic plate sensitivities fall off rapidly. It would seem useful for example, to have reliable intensities of the magnesium lines in Table 2, to help possibly in clarifying the infrared magnesium line emissions dealt with by Dr. Chang and coworkers. I believe it would therefore be valuable to repeat and extend the Pierce observations, including spatial resolution if possible, with equipment available now. The people here at this meeting can tell me whether such experiments are feasible with present or planned coronagraphs, and whether on the ground or in airborne experiments. It will probably take such equipment to establish the origin of the lines in Table 3, for example, by spatially resolving their source regions.

Table 2: Comparison of chromospheric wavelengths[a] with laboratory wavelengths[b] of neutral magnesium. [a] λ_{AKP}, chromospheric wavelengths listed by Pierce[13]; ID_{AKP}, Pierce's assignments. [b] λ_{LAB}, laboratory wavelengths compiled by Kaufman and Sugar[9]; the last column gives the upper configuration and level/term of the transition. [c] emission band. Table 2 continued on next page.

λ_{AKP}	ID_{AKP}	λ_{LAB}	upper level term
9255.764	CN(1,0)	9255.778	3s5f($^1F_3$)
8717.823	—	8717.825	3s7d($^3D_3$)
8712.689	—	8712.689	3s7d($^3D_2$)
8710.160	—	8710.175	3s7d($^3D_1$)
8707.171	—	8707.14	3s12d($^1D_2$)
8609.769	—	8609.71	3s13d($^3D_2$)
8473.719	—	8473.694	3s9s($^3S_1$)
8468.885	—	8468.845	3s9s($^3S_1$)
8310.3[c]	—	8310.264	3s8d($^3D_3$)
8305.612	—	8305.596	3s8d($^3D_2$)
8303.331	—	8303.313	3s8d($^3D_2$)
8159.133	—	8159.132	3s10s($^3S_1$)
8054.210	—	8054.232	3s9d($^3D_3$)
8049.833	—	8049.854	3s9d($^3D_2$)
7881.113	—	7881.135	3s10d($^3D_3$)
7759.316	—	7759.297	3s11d($^3D_3$)
7387.038	—	7387.004	3s9p($^1P_1$)
7193.189	Fe II	7193.172	3s9f($^1F_3$)
7060.394	—	7060.409	3s10f($^1F_3$)
6319.646	—	6319.716	3s6p($^3P_2$)
5167.3[c]	La II, Mg I	5167.322	3s6p($^3P_2$)

Table 2: Cont.

λ_{AKP}	ID_{AKP}	λ_{LAB}	upper level term
4621.389	—	4621.299	3s4s($^1S_0$)
4351.843	Pr II	4351.906	3s6d($^1D_2$)
4165.027	—	4165.101	3s8s($^1S_0$)
3858.927	Cr I, CN	3858.860	3s13d($^1D_2$)
3854.944	CN	3854.965	3s4p($^3P_1$)
3853.977	—	3853.960	3s4p($^3P_2$)
3848.8^c	Sm II, CN	3848.914	3s4p($^3P_1$)
3838.3^c	Mg I, Tm II	3838.29	3s3d(^3D)
3832.3^c	Mg I, CN, Pd I	3832.30	3s3d(^3D)
3829.3^c	Mg I, CN, Fe I	3829.355	3s3d($^3D_1$)
3821.915	Fe II	3822.00	3s16d($^1D_2$)
3813.991	Gd II, Zr II	3814.02	3s17d($^1D_2$)
3797.233	CN	3797.22	3s20d($^1D_2$)
3101.719	Ce II	3101.796	3s4d($^1D_2$)
3092.935	Nd II	3092.984	3s4d($^3D_2$)

Table 3: Chromospheric lines at deuterium wavelengths[a,b]. [a]wavelengths in Angstroms, [b]Balmer transitions are denoted by B_i, [c]Pierce, 1968.

deuterium transition	P_ζ 3d(^{2}D)-9f(^{2}F)	Bλ 2p(^{2}P)-5d(^{2}D)	B_η 2p(^{2}P)-9d(^{2}D)
laboratory wavelength	9226.505	4339.287	3834.342
Einstein solar shift	0.020	0.009	0.008
expected solar wavelength	9226.525	4339.296	3834.350
AKP^c wavelength	9226.527	4339.326	3834.324
other candidates: wavelength	—	4339.317	3834.364
element/ion	—	Ce	Mn

3. An Unusual Near-Infrared Forbidden Line in Fe XIV

Bhatia and I have been interested in the spectrum of highly ionized aluminum-like Fe XIV because it has been clear since the work of Blaha[2] that besides its well-known strong resonance lines, many of its weaker lines must also be present and observable in the solar EUV spectrum. For this purpose, we carried out a study of its spectrum which included the higher three-electron 3s3p3d configuration. In this configuration, there is a quartet F level with J-9/2 which can only decay by a magnetic dipole transition to the next quartet F level with J=7/2, with a very small transition rate of about 12 sec^{-1}. The J-9/2 level is therefore metastable and can achieve a relatively high population.

The much more well-known forbidden line of Fe XIV is the green coronal line at 5303 Å, between the two levels of the $3s^23p$ ground configuration. It is interesting therefore to consider in what ways these very different forbidden lines, which might be referred to as a "high" forbidden line and a "low" forbidden line, may behave under given conditions.

Figure 6 shows the predicted relative intensities of the two forbidden lines as a function of electron density, compared also with the weak but observable intercombination line at 484.60 Å. One sees that the variations with density are quite different as expected, with the infrared forbidden line reaching a pronounced maximum at an electron density of 10^{11} cm^{-1}. The maximum relative intensity reached in the figure is small, but is a lower limit because only collisional excitation has been considered. In a recombining flare plasma, for example, recombinations from more highly ionized stages are likely to add to the population of the metastable quarter $F_{9/2}$ level.

A next question then is what is the expected wavelength of this line. As seen in the figure, an approximate wavelength obtained from theoretical calculations is about 1.25 microns. Recently a more precise wavelength of 1.202 microns is available from the work of Churilov and Levashov.[3] The wavelengths of this transition in chromium, titanium and calcium are also available from this work and lie respectively at 1.911, 3.245 and 5.866 microns. It is possible that the relative intensities of this transition in these less-ionized species may be greater than in Fe XIV because the upper levels are likely to be still more metastable. With respect to observability in flares, the strength of the transition in Ni XVI might reach its maximum at a higher density more characteristic of flares; but its exact wavelength is not known yet, being something around 7600 Å or less. We will be examining this transition in more detail, in these other ions.

The interest in this type of "high" forbidden transition lies in two aspects. First, such a forbidden line forms an interesting counterexample to the conventional picture of forbidden line quenching, which postulates that forbidden lines are observable below a "critical density" where collisional de-excitation becomes greater than radiative de-excitation, and disappear when this density is reached.

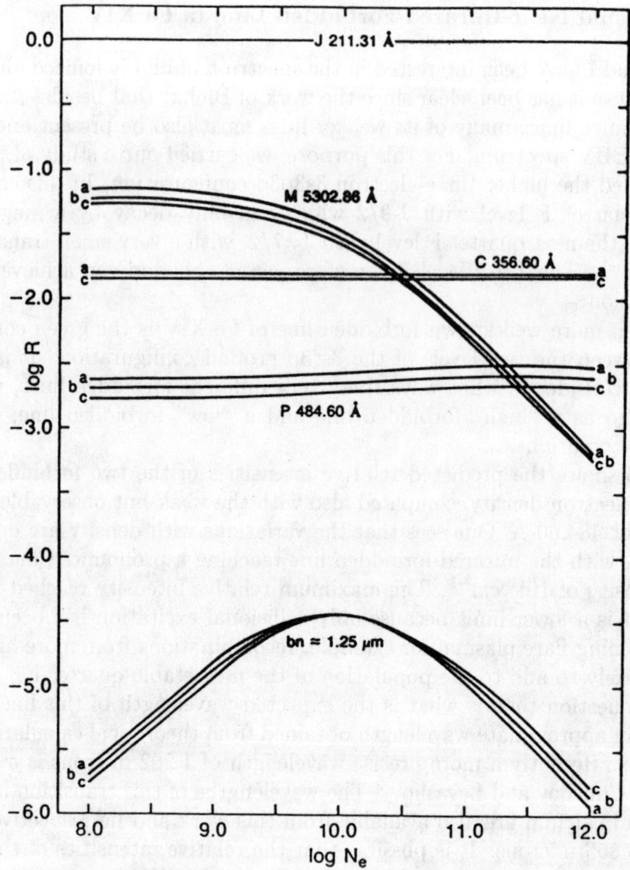

Figure 6: Relative intensities of Fe XIV emission lines vs. electron density.

Second, the time dependences of the "high" and "low" forbidden lines may be appreciably different in flares. Golub et al[6] have made a similar comment in connection with soft X-ray and EUV lines. This may be useful for diagnostic purposes. In very rapidly varying sources such as the impulsive phases of flares, in fact, it may be that steady-state populations are not attained and one has to take into account the wide range of excitation and de-excitation rates of the atomic system (this non-equilibrium situation differs from the steady-state or equilibrium assumption of Golub et al[5]). A simulation of the time-dependence of the ratio of the two forbidden lines, in flares of different durations ranging from 100 seconds to as short as one second, is shown in Figure 7. The flare conditions simulate a parabolic rise and fall, from an initial density of 10^9 cm^{-3} and temperature of 2 x 10^6 K to a maximum density of 10^{12} cm^{-3} and temperature of 2 x 10^7 K. One sees how the time variation of the ratio begins to depart from the steady-state (i.e. long-duration) locus as the flare duration shortens, lagging behind initially and remaining higher as the flare subsides. The non-equilibrium behavior arises essentially when the flare duration becomes short enough that it begins to approach the order of magnitude of the lifetime of the metastable levels. It may therefore be more pronounced and observable in the less-ionized atoms mentioned above where the metastable level lifetimes are probably longer than in Fe XIV.

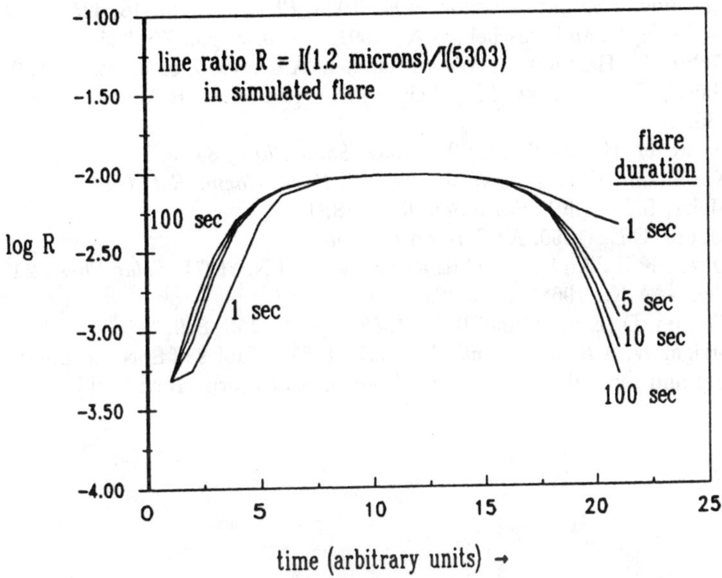

Figure 7: Ratio of Fe XIV forbidden lines during a solar flare.

For these reasons, I believe it might be worthwhile to search for this transition in Fe XIV and the other aluminum-like ions, in near-infrared solar flare spectra.

To conclude, I have attempted to illustrate the following general points: time-dependent solar phenomena can provide more information on physical processes than static sources; and early one-of-a-kind solar observations such as those of Pierce are worth closer examination, and especially are worth repeating and extending with modern instrumentation. Some particular spectroscopic lines of interest have been discussed, from these points of view, which may be of interest to observe.

4. References

1. Bhatia, A.K., Kastner, S.O., Keenan, F.P., Conlon, E.S. and Widing, K.G.: 1994, *Ap. J.*, **427**, 497.
2. Blaha, M.: 1972, *Astron. & Astrophys.*, **16** 437.
3. Carlsson, M. and Judge, P.G.: 1993, *Ap. J.*, **402**, 344.
4. Churilov, S.S. and Levashov, V.E.: 1993, *Phys. Scripta*, **48**, 425.
5. Feldman, U. and Doschek, G.A.: 1991, *Ap. J. Supp.*, **75**, 925.
6. Golub, L., Hartquist, T.W. and Quillen, A.C.: 1989, *Solar Phys.*, **122**, 245.
7. Haisch, B.M., Linsky, J.L., Weinstein, A. and Shine, R.A.: 1977, *Ap. J.*, **214**, 785.
8. Holweger, H. and Werner, K.: 1982, *Solar Phys.*, **81**, 3.
9. Kaufmann, V. and Sugar, J.: 1986, *J. Phys. Chem. Ref. Data*, **15**, 321.
10. Meier, R.R.: 1991, *Space Sci. Rev.*, **58**, 1.
11. Moore, C.E.: 1959, *NBS Technical Note 36.*
12. Olsen, K.H., Anderson, C.R. and Stewart, J.N.: 1971, *Solar Phys.*, **21**, 360.
13. Pierce, A.K.: 1968, *Ap. J. Supp.*, **17**, 1.
14. Skelton, D.L. and Shine, R.A.: 1982, *Ap. J.*, **259**, 869.
15. Striganov, A.R. and Sventitskii, N.S.: 1968, "Tables of Spectral Lines of Neutral and Ionized Atoms", (KFK/Plenum Data corp., New York).

New Observations of IR Coronal Emission Lines

M.J. Penn
National Solar Observatory/Sacramento Peak
National Optical Astronomy Observatories†
PO Box 62, Sunspot, NM 88349, USA

ABSTRACT

Predicted infrared forbidden coronal emission lines are reviewed. Eclipse observations of the infrared solar coronal spectrum are also reviewed, and new ground-based coronagraph observations obtained at NSO/Sac Peak are reported. The infrared [Si X] coronal emission line is measured, and new observations of the infrared [Fe XIII] emission lines are presented. Upper limits are set for coronal emission near 1266 nm and 1523 nm and for other lines in the 1160 to 1260 nm region. The measurement of the coronal electron density via the two [Fe XIII] lines is reviewed. The most up-to-date calculations are applied to NSO/Sac Peak data as well as to data from the 1991 total eclipse taken on Mauna Kea. Emission from the [Fe XIII] 1080 nm line and the [Si X] emission line is correlated with the coronal electron density as determined from the [Fe XIII] line ratio. However, the intensity does not appear to vary as N_e^2 as expected from collisional excitation. It is shown that line-of-sight integration in the optically thin corona removes the expected N_e^2 behavior while preserving the accuracy of the N_e measurement.

1. What can be expected?

A major source of noise in ground-based coronagraph studies of the Sun is due to the sky scattered background. Since the sky is blue at a good coronal sight (Rayleigh scattering dominates) the sky scattered background is greatly reduced at infrared wavelengths compared to visible wavelengths. For this and other reasons ground-based studies of the solar corona will have better signal to noise when infrared wavelengths are used.

Several emission lines are expected to exist in the infrared spectrum of the solar corona. Infrared wavelengths for transitions in the lower levels of ionized metals have been calculated by many authors[1,2,3,4,5] with the most comprehensive published list by Kaufman and Sugar[6]. Intensities for some IR lines have been estimated by a few authors[7,8,9] but intensity calculations are generally lacking.

The emission lines of the corona provide diagnostics of the physical conditions present there. Velocities are given by wavelength shifts of these lines. The coronal electron density can be derived from measurements of line ratios. Temperatures can be measured by studying the relative intensities of species with slightly different ionization energies. And finally, the magnetic field of the corona can be measured

† Operated by the Association of Universities for Research in Astronomy, Inc. (AURA) under cooperative agreement with the National Science Foundation

with the polarization displayed by emission lines. Given the observational advantages of using IR emission lines and the potential diagnostics, a detailed study of the IR solar coronal spectrum is warranted.

2. What has been seen?

While several IR coronal lines have been measured in night-time studies[10] few IR observations of the solar corona have been made since the idea was first suggested 50 years ago[11]. The [Fe XIII] line pair at 1075 and 1080 nm have been studied with silicon detectors by several groups[12], but in the near-IR wavelength region from 1000-2200 nm there have only been four spectra of the solar corona published[13,14,15,16]. All of these spectra have been taken during a total eclipse, three from airborne experiments. While a few emission lines have been observed, the line identifications are dubious[17] and none of the spectra are calibrated so the line intensities cannot be determined with confidence[18].

Recently several ground-based coronagraphic observations have been made from NSO/Sac Peak.[19,20] More precise [Fe XIII] wavelengths have been measured, placing the emission lines at 1074.617 ± 0.005 nm and 1079.783 ± 0.006 nm. The [Si X] transition discoved by Münch et al.[14] has been detected from the ground; the central wavelength is measured to be 1430.084 ± 0.006 and the [Si X] line is measured with an intensity similar to the [Fe XIII] 1080 nm line. Coronal emission in lines seen by Olsen et al.[16] at 1266 and 1523 nm are at least seven times fainter than the [Fe XIII] line at 1075 nm. Finally, a new coronal spectrum taken on 13 June 1994 (see Figure 1) shows no hot or cold lines from 1160 to 1260 nm in a region of the corona where [Fe XIII] 1075 nm emission was observed at 3×10^{-6} $B_{\odot}$ and He I 1083 nm emission was observed at 100×10^{-6} $B_{\odot}$. Note that the absorption lines in Figure 1 are known atmospheric or photospheric lines; the spectrum is normalized by the measured mean solar brightness, not the photospheric spectrum. The noise in the continuum, the linear decrease in the continuum, and the gap at 1230 nm are due to the marginal sky conditions during these observations. Further work is in progress.

3. What about the physics?

Coronal electron density measurements will be discussed in this section, and Kuhn[21] will address magnetic field measurements.

Several papers[22,23,24] have addressed the determination of coronal electron density from the ratio of the two [Fe XIII] emission lines. The work of Flowers and Pineau des Forêts[24] uses statistical equilibrium calculations, including electron and proton collisions with several energy levels of the ion, to compute the behavior of the line ratio with local electron density and photospheric radiation intensity. This work, although more than 20 years old, represents the most recent calculations published, and the results are used to interpret 1991 eclipse data from Hawaii[12] and NSO/Sac Peak data.

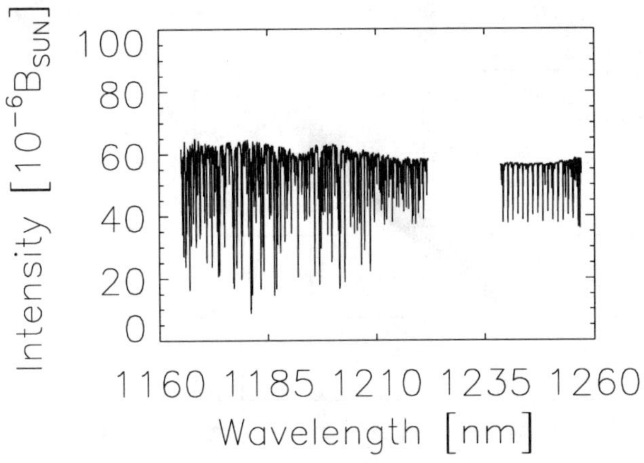

Figure 1: The observed coronal spectrum on 13 June 1994 between 1160 and 1260 nm. No lines brighter than the observed [Fe XIII] or He I emissions appear in this region of the coronal spectrum.

Electron densities computed from the line ratio at various coronal positions span the range of 2×10^7 to 3×10^8 cm^{-3}. While these values are in agreement with previous measurements[25], it would be useful to test the validity of these measurements. Are these densities correct?

A simple test is to compare the emission line intensities with the computed electron densities. Because collisions are the dominant excitation mechanisms for coronal emission lines, the line intensities are expected to vary as N_e^2. However, as seen in Figure 2, the [Fe XIII] 1080 nm emission line appears to vary as $N_e^{\sim0.5}$ in the data from the 1991 eclipse. In this same data, the [Fe XIII] 1075 nm emission line shows an even shallower slope. And finally coronagraph observations of the [Si X] emission[20] (which has an ionization potential similar to Fe XIII) show the line intensity varies as $N_e^{\sim0.4}$. This simple test has an unexpected result in both data sets.

With simulated data the computations of Flowers and Pineau des Forêts[24] give reasonable results; the factor of four between the measured and expected slopes is not caused by incorrect density calculations. The line intensities must be the source of the discrepancy. The most obvious source of confusion with the line intensities is due to the line-of-sight (LOS) integration through the optically thin corona. Two models of the solar corona were made to investigate how LOS integration influences both the line intensity and the line ratio.

First a trivial case is examined; a two component coronal model is constructed. Various LOSs in the corona are filled with random amounts of vacuum and corona at

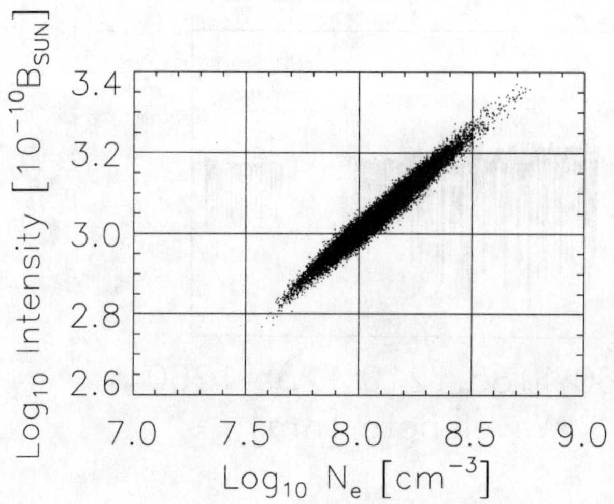

Figure 2: Observed behavior of [Fe XIII] 1080 nm line emission versus the coronal electron density computed from the line ratio. Note that the slope is about 0.5

one density, N_e. It is easily seen that LOS integration cannot affect the N_e computed with the line ratio method; only one electron density is measured.

A second case is examined; in this case the cross-section of a large coronal loop is constructed with a Gaussian electron density distribution peaked in the center of the loop. The LOS integrated line intensities are computed, and the line ratio is used to compute the "observed" electron density. This case showed interesting results. First, the "observed" N_e was within a factor of two of the LOS mean of the "actual" N_e and the "observed" density showed the same spatial behavior as the LOS mean of the "actual" density. The line ratio is not very sensitive to LOS integration effects, and gives a density close to the LOS mean density of the structure.

However the line intensity was strongly affected by the LOS integration. In the model, the local line intensity was set to vary approximately as the local electron density squared, consistent with collisional excitation. (This is not exactly correct for both emission lines, as one line varies as $N_e^{\sim 1.5}$ and the other varies as $N_e^{\sim 2.0}$. However these differences are small compared to the effects discussed.) After LOS integration is performed, the "observed" line intensity is compared to the "observed" N_e. Both emission lines showed "observed" variations as $N_e^{0.5}$ with this Gaussian model. LOS integration can produce observations which show an unexpected behavior of observed line intensity with observed electron density.

4. What's next?

The search for near-IR emission lines in the solar corona will continue with efforts from the coronagraph at NSO/Sac Peak as well as a ground-based experiment during the 3 November 1994 total eclipse in Chile. Predictions of a relatively bright [Fe XII] emission at 2218 nm[8] and questions about emission from cold gas[9] in the O I line at 1129 nm will be tested in the next few months.

The interpretation of [Fe XIII] electron density measurements will continue with more modeling of LOS integration effects. In particular, the different behavior of the two [Fe XIII] lines will be examined for diagnostic potential of the density variations along the coronal LOS. Comparisons of the line emission with the continuum emission observed during the 1991 eclipse will be made hoping to make similar diagnostics, although the interpretation of this comparison is difficult since different coronal temperatures are observed with these two techniques.

5. References

1. F. Rohrlich and C. Pecker, *ApJ* **138** (1963) 1246.
2. B. Edlén, *Sol. Phys.* **24** (1972) 356.
3. M.W. Smith and W.L. Wiese, *J. Phys. Chem. Ref. Data* **2** (1973) 85.
4. M. Kafatos and J.P. Lynch, *ApJ Supp.* **42** (1980) 611.
5. S.O. Kastner and A.K. Bhatia, *ApJ* **309** (1986) 883.
6. V. Kaufman and J. Sugar, *J. Phys. Chem. Ref. Data* **15** (1986) 321.
7. G. Münch, *ApJ* **145** (1966) 237.
8. E. Chang, *these proceedings*
9. S.O. Kastner, *these proceedings*
10. M.A. Greenhouse, U. Feldman, H.A. Smith, M. Klapisch, A.K. Bhatia and A. Bar-Shalom, *ApJ Supp.* **88** (1993) 23.
11. P. Swings, *PASP* **56** (1944) 80.
12. M.J. Penn, J. Arnaud, D.L. Mickey and B.J. LaBonte, *ApJ* **436** (1994) 368.
13. J.H. Taylor and R.M. MacQueen, *Appl. Opt.* **3** (1964) 1506.
14. G. Munch, G. Neugebauer, and D. McCammon, *ApJ* **149** (1967) 681.
15. J. Mangus and R. Stockhausen, *NASA report X-614-66-29* (1966).
16. K.H. Olsen, C.R. Anderson, and J.N. Stewart, *Sol. Phys.* **21** (1971) 360.
17. S.O. Kastner, *Sol. Phys.* **143** (1993) 197.
18. M.J. Penn and J.R. Kuhn, *Sol. Phys.* **151** (1994) 51.
19. M.J. Penn, J.R. Kuhn, J. Arnaud, D.L. Mickey and B.J. LaBonte, *Proceedings of the Second SOHO Workshop, eds. Fleck, Noci, and Poletto* (1994) 185.
20. M.J. Penn and J.R. Kuhn, *ApJ* **434** (1994) 807.
21. J.R. Kuhn, *these proceedings*
22. R.A. Chevalier and D.L. Lambert, *Sol. Phys.* **10** (1969) 115.
23. J.B. Zirker, *Sol. Phys.* **11** (1970) 497.
24. D.R. Flower and G. Pineau des Forêts, *AA* **24** (1973) 181.
25. P.L. Byard and K.E. Kissel, *Sol. Phys.* **21** (1971) 351.

LOW BACKGROUND SCIENTIFIC ARRAYS FOR THE 1-5 MICRON REGION

A.M. Fowler
National Optical Astronomy Observatories *
Tucson, AZ 85726

ABSTRACT

Although the development of two dimensional arrays for infrared astronomy began in the late 1970's with the General Electric 32 x 32 InSb Charge Injection device, the first really useful device generally available to the astronomy community was the Santa Barbara Research Center (SBRC) 58 x 62 InSb array. It was soon followed by the Rockwell NIC-MOS family of MCT (HgCdTe) arrays. Both companies utilized "source follower per detector" (SFD) architecture and indium bump technology which has become the standard for low background IR focal planes. Low background focal planes use the SFD design because injection type circuits have high noise and dark current and CCD's do not perform well at low temperatures. Over the years, three detector technologies have produced devices of interest to scientific users. These are Platinum Silicide (PtSi), Indium Antimonide (InSb), and Mercury Cadmium Telluride (MCT). With the introduction of the 1K x 1K family of high performance arrays in InSb and MCT, interest in PtSi devices has waned due to its low quantum efficiency. The paper discusses the performance, advantages, and disadvantages of these materials but concentrates on the current and next generation (256 x 265 and 1024 x 1024) SBRC InSb focal planes.

1. Introduction

The intended focus of this talk was state of the art low background, high performance, infrared focal planes for the 1-5 micron region. But after listening to many of the previous speakers, it became apparent that the InSb arrays I am discussing can be applied to solar problems from 0.6 microns out to 5.5 microns. They will not compete with CCD's in their primary range of 0.4 to 0.9 microns but are certainly competitive at the important 10830 Å line. In an era of limited funds, the fact that an InSb focal plane can cover the entire range is certainly of importance. The new generation of 1024 x 1024 focal planes has the higher spatial resolution which limited their usefulness when compared to CCD's. The evolution of infrared array technology has progressed at a very rapid rate. In the early 80s there were a few arrays in the 32 x 32 format, but the introduction of the 58 x 62 InSb focal plane by SBRC and a similar device, the 64 x 64 MCT focal plane by Rockwell in the mid-80's represented the start of the current revolution in infrared astronomy. At the present time, both technologies are providing 256 x 256 devices to users, and the next generation of 1024 x 1024 devices is not far behind. Since the next speaker will concentrate on the

*Operated by the Association of Universities for Research in Astronomy, Inc.(AURA) under cooperative agreement with the National Science Foundation.

MCT devices available from Rockwell, I will direct my talk to the InSb technology available from SBRc which is widely used at NOAO. Figure 1 shows the growth of pixels available to astronomers over that past decade and a half and projects focal planes of 4K x 4K before the end of the century. In fact, discussions are now taking place about butting the current generation of 1K x 1K device to produce that focal plane.

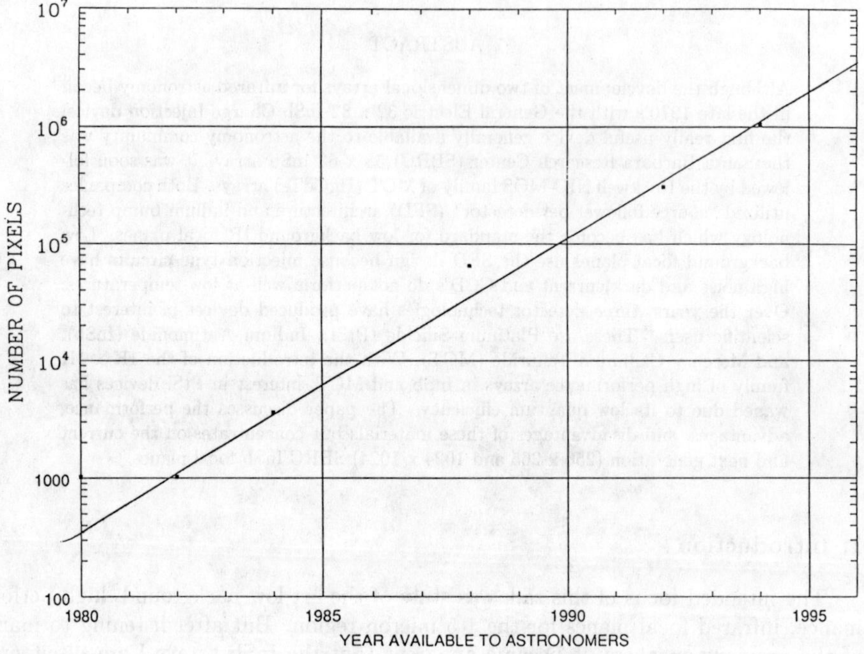

Figure 1: Growth rate of Infrared Arrays from 1980 to 1996 showing the logarithmic growth in the number of pixels versus time.

Large focal planes could more easily be made using a monolithic technology like CCD's, but the narrow bandgap of useful infrared detector materials has made this a nearly impossible task. Therefore, the development of the indium bump hybrid focal plane in the mid-70's has set the direction for all current high performance arrays. There is still work being done on monolithic arrays, but they cannot compete in performance and yield with the hybrid devices. Figure 2 is a cross section view of a typical hybrid focal plane. A significant problem with this approach is that the incoming photons must pass through the detector substrate material. In the case of InSb focal planes, this requires thinning the detector substrate material to 10 microns or less for acceptable quantum efficiencies, and for MCT it has meant growing the detector material on a transparent substrate material such as sapphire. These added steps have resulted in higher costs and lower overall yields. However, use of the hybrid approach has resulted in effective fill factors approaching 100% as none of the photon collecting area must be devoted to readout electronics.

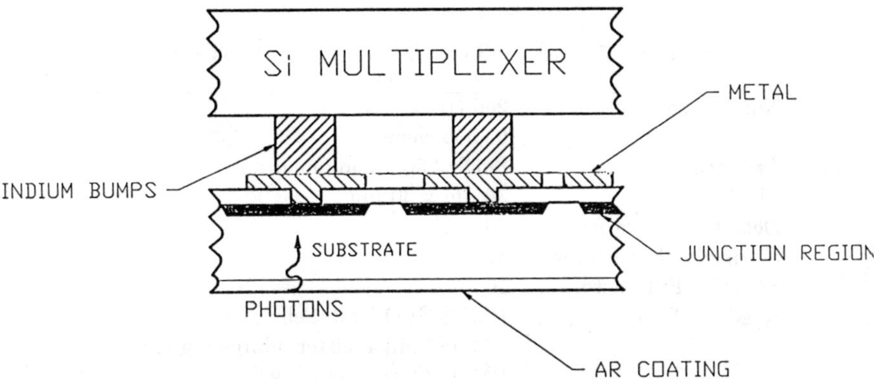

Figure 2: Typical IR Hybrid Array Cross Section. The substrate may consist of more than one layer depending on the technology.

As I previously mentioned, there are three technologies which can produce the large high performance focal planes demanded by astronomers today. The first to be available in large formats was PtSi. It had the advantage of being larger and cheaper for several years, as it was made using well-developed silicon processing technology. It is very stable, has few bad pixels, and is presently in use in the Kitt Peak infrared four color imaging camera SQID. The one, very critical and fatal flaw, is its very low quantum efficiency. Although it can be made in monolithic form, the added problem of less than 100% fill factor makes this architecture unsuitable for astronomical applications. Therefore, it is no longer considered for many high performance applications. The next, and still competitive material, is MCT. It has the advantage of operating at 77K with high quantum efficiency and low dark current. In addition, because its cutoff wavelength is 2.5 microns, instrument design is easier while still being able to do many of the interesting solar problems. It still exhibits a residual image problem which make spectroscopy difficult, and the QE is very temperature and wavelength dependent especially at the shorter wavelengths, such as 10830Å. The last and most universal material is InSb. It is sensitive over the wavelength range 0.6 to 5.0 microns, and with special AR coatings, and has very high QE over this range. It has low dark current (< 0.2 e/s) and no residual image effects. Disadvantages are its operating temperature, 35K, and its broad wavelength sensitivity, which dictates careful instrument design if the low dark current performance is to be achieved.

2. The SBRC 256 x 256 InSb Focal Plane Array

The current, readily available, InSb array is the 256 x 256 Sensor Chip Assembly (SCA). Its characteristics, which are typical of those measured by NOAO, are shown in Table 1.

Table 1: 256 x 256 InSb SCA CHARACTERISTICS

	Specification
Number of Pixels	256 (H) x 256 (V)
	65,536 elements
Architecture	Single 256 x 256 Quadrant
Pixel Size	30 μm x 30 μm
Detector Active Area	28 μm x 28 μm
Geometric Fill Factor	87%
Effective Fill Factor	100%
Readout Type	PMOS-SFD Unit Cell
	PMOS Shift-register addressing logic
	PMOS Output Drivers
Number of Outputs	4
Frame Rate	50 ms.
Reset Options	Destructive or non-destructive by pixel
IR Detector	Thinned InSb
Full Well Capacity	$\sim$ 250K electrons at 1V applied bias
Wavelength Range	0.6 - 5.5 microns
	With special AR coatings
Operating Temperature	35K
Noise	$<$ 30 e rms using Fowler 1[1]
Dark Current	$<\sim$ 0.2 e/sec
Quantum Efficiency	$>$ 80% 0.9 to 5 microns
Defective Pixels	$<$ 0.5%
	No Bad Rows or Columns

The addressing logic is row and column, two-phase, non-overlapping clock, shift registers, and the four outputs are organized as interleaved columns. It has both destructive and non-destructive resetting capability so all readout techniques of Ref. 1 can be utilized.

Diamond turning is now used to thin the InSb, increasing yield and improving uniformity. We have measured 5% uniformity (stdev/mean) with 99.3% of the pixel response falling within a $\pm$ 20% range. This is a substantial improvement over the older lapping technology.

Initially, the 256 x 256 SCA's exhibited a ring or picture frame of high dark current, as shown in Figure 3. Further testing showed this to be due to contamination diffusing in from the edges of the hybrid when it is not stored in a vacuum. Using an old trick to stabilize performance in previous InSb arrays, we found that vacuum baking the SCA at 80deg C for upwards of 2 weeks effectively eliminated the effect. Figure 4 demonstrates the improvement obtained by using this baking technique.

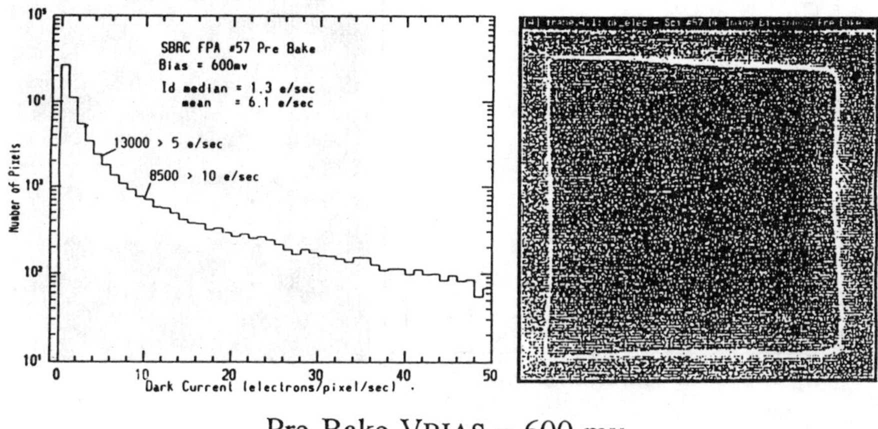

Pre-Bake VBIAS = 600 mv

Figure 3: Pre-Bake Dark Current Illustrating the "Picture Frame Effect" and the higher overall dark current due to contamination effects.

In the lab, we obtained readnoise measurements of 26e rms, but in our instruments at the observatory, values of 30-35e rms are more typical. We have applied Fowler sampling[1] to this device and achieved noise measurements of 10-15e rms. Our testing shows that the noise starts to increase after 16 pairs due to the dark current or LED effects generated by the reading process, so 8 pairs is considered optimum for Fowler sampling in our instruments. Under those conditions an improvement of approximately 2.5 is achieved. Detector node capacitance directly affects both readnoise and full well capacity, which is another critical factor. In our evaluation of the 256 x 256 SCA, we achieved the full well capacity performance shown in Table 2.

Obtaining these results required careful attention to the clocking required for this device. In particular, we found it necessary to clock Vgguc from its normal operating value of -1.5v to between -0.8 and -1.0v. This in effect turned off the current source driving the column bus, resulting in the bus maintaining the level set by the last row rather than collapsing towards 0v. It is this collapse towards 0v which discharges the detector node and reduces the full well capacity. Table 3 is a list of the voltages used at NOAO to run these SCA's.

The switching of V_{ddout} was done to reduce the LED effect from the output driver, a very common problem, from contributing to excess dark current during integration.

More detailed information on the evaluation and optimization of this device can be found in a prior paper[2] on this array.

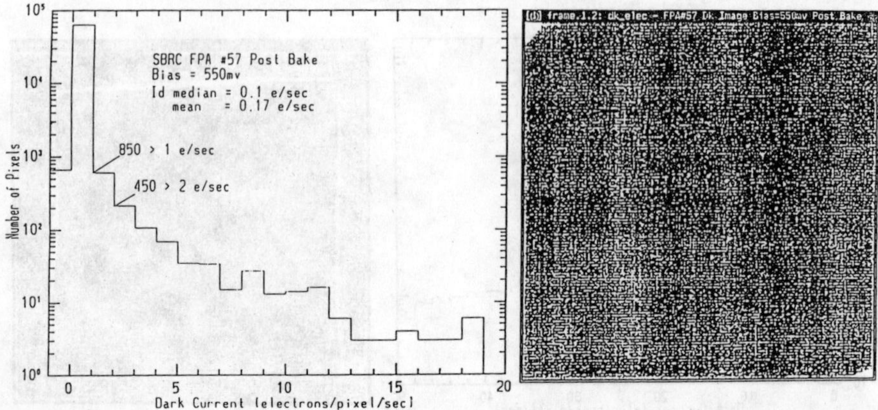

Post-Bake VBIAS = 550 mv

Figure 4: Post-Bake Dark Current Illustrating the Effect of Vacuum Baking in removing the "Picture Frame Effect" and lowering of the overall dark current by an order of magnitude.

Table 2: Full Well Capacity

Applied V_{bias}	Full Well Capacity
600 mv	$\sim$ 100,000 electrons
800 mv	$\sim$ 200,000 electrons
1000 mv	$\sim$ 260,000 electrons

Table 3: Operating Parameters

ϕ_1 fast	=	-2.8/-7.25	ϕ_1 slow	=	-3.0/-7.05
ϕ_2 fast	=	-2.8/-7.25	ϕ_2 slow	=	-3.0/-7.05
ϕ_{sync} fast	=	3.0/-6.2	ϕ_{sync}	=	-3.0/-6.0
ϕ_{rst}	=	-2.3/-6.0	V3	=	-2.71
V_{dduc}	=	-3.7V	V_{ssuc}	=	0.0 (Gnd)
V_{ddout}	=	-0.6 unless readig; then -1.0			
V_{gguc}	=	-1.0 unless reading; then -1.5			

3. The Next Generation InSb focal Plane Array

3.1. ALADDIN: The 1024 x 1024 InSb SCA Development Program

The ALADDIN program is a collaboration between the US Naval Observatory and NOAO to develop a 1024 x 1024 InSb SCA at SBRC. The program started in September 1993 and we received the first SCA in September 1994. This is the largest (physical size) hybrid focal plane array attempted to date. With 27 micron pixels it is over an inch square. Because of its size, a special leadless chip carrier (LCC) was designed to package the hybrid. Fortunately a standard burn-in test socket from Yamaichi can be used to mount the array. The anticipated characteristics for this device are given in Table 4.

A couple of notes concerning the specifications are in order. The pixel size was selected to be as large as possible. It was limited by detector material wafer size and the optimum arrangement of readout die on 4-inch wafers. As the pixel size is decreased the MTF is seriously degraded by the lateral diffusion of carriers between the surface, where they are created, and the junction where they are collected. Also, as the pixel size decreases, diffraction limited performance in the instrument design becomes more difficult. Unlike CCD's where the detector is put at the focus of the telescope, IR instruments require re-imaging optics and baffling to reduce background flux.

When the ALADDIN program was first initiated, we thought that producing readouts would be the most challenging task. In our first lot we yielded over 8% perfect (no bad rows or columns) die and over 20% allowing for some bad rows or columns. Another issue was the difference in thermal contraction between the silicon readout and the InSb detector material ($\sim$ 0.06%; 273K to 35K). But a nominal disadvantage of InSb hybrids (having to thin the detector material to less than 10 microns) is an advantage in this case. This thermal contraction stress is taken up between each of the indium bumps instead of accumulating over the whole device because the thin InSb material accommodates the thicker silicon readout. To date we have thermally cycled our first ALADDIN SCA over 25 times with no degradation. Some unexpected problems in hybridization have slowed the program, but they are not basic yield limiting issues. By early 1995, we expect to have determined the yield and thereby the cost of the ALADDIN devices.

Table 4: ALADDIN 1024 x 1024 InSb Focal Plane Characteristics
SCA Specifications

Number of Pixels	1024 (H) x 1024 (V) 1,048,576 elements
Architecture	4-Independent 512 x 512 Quadrants
Pixel Size	27 μm square
Effective Fill Factor	100%
Readout Type	PMOS SFD Unit Cell CMOS Shift-registers Control Logic PMOS or NMOS Output Drivers
Number of Outputs	32 (8 per quadrant)
Frame Rate	50 milliseconds
Reset Options	Destructive and non-destructive by rows
IR Detector	Thinned InSb
Full Well	250 x 10^3 electrons at 1.0v bias
Wavelength Range	0.6 - 5.5 microns with special AR coatings
Operating Temperature	35K
Dark Current	$<$ 0.1 e/sec
Noise	$<$ 25 e rms with Fowler Sampling
Quantum Efficiency	$>$ 80% 0.9 to 5 microns
Defective Pixels	$<$ 0.5% No Bad Rows or Columns

Because of the independent quadrant architecture of the design, all 124 pins on the LCC are used. It is not necessary to bring all 124 pins out of the instrument as over half can be combined to reduce that number to around 60. Table 5 is a list of pins by function and how they can be combined.

We are testing the first SCA with results apparently consistent with our design goals. The readouts have been tested more extensively and they are operating as expected with the readnoise potentially[3] being somewhat higher than expected. We expect to receive one more SCA before the end of 1994 and several more in early 1995. With these devices, we will get a better idea of the performance and the expected yield, making possible the estimate of production run size and cost to yield a specific number of science grade devices. This information will be used to determine the cost of future devices.

Table 5: ALADDIN 1024 x 1024 Electrical Interface Requirements

Signal Type	# of Pins on LCC	Min # of Pins	Voltage Range
Bias	2+4x4=18	5	0 to -4.0
Clocked Biases	4x4=16	4	0 to -6.5
D.C. Levels	3x4=12	3	0 to -6.5
Clock	8x4=32	8	0 to -6
Current Sources	1x4=4	1	-1 to -10 (μA)
Grounds	2+4=6	1	0
Outputs	8x4=32	32	0.6 to -3
Diagnostic	4x1=4	0	0 to -3
Totals	124	54	

4. Conclusions

From the remarks at this conference concerning the needs of solar astronomy, it is clear that InSb IR arrays can be very useful. They can compete with CCD's at visible wavelengths and are superior beyond 0.9 microns. They have the advantage over MCT of responding out to 5 microns but they must be cooled to 35K for proper operation. The arrays I have described are not suitable for solar imaging as the full well capacity is quite limited compared to injection type devices such as the Amber array (2 - 3 *times* 10^5 compared several $\times$ 10^{7th}). But for high resolution spectroscopy their performance is superior. The 256 square array is presently available off the shelf from SBRC but it will be some time before the 1024 square ALADDIN arrays are generally available.

5. References

1. Garnett, J.D. & Forrest, W.J.. 1993 "Multiply sampled read-limited and background-limited noise performance", *Proceeding of SPIE, Infrared Detectors and Instrumentation*, **1946**, 395-404
2. Fowler, A.M., Gatley, I., Vrba, F.J., Ables, H.D., Hoffman, A.W., & Woolaway, J.: "Performance of the Current 256 x 265 SBRC InSb Array and Status of the Next Generation 1024 x 1024 InSb Array", *Proceeding from the Conference on Infrared Astronomy with Arrays: The Next Generation, V190*, Kluwer Academic Publishers, Ian S. McLean, ed.
3. Fowler, A.M., Bass, D.L, Heynssens, J.B., Gatley, I., Vrba, F.J., Ables, H.D., Hoffman, A.W., Smith, M. & Woolaway, J.T., 1994: "Next generation in very large InSb arrays: ALADDIN, the 1024 x 1024 InSb focal plane array readout evaluation results", *Proceeding of the SPIE Infrared Spaceborne Remote Sensing II*, **2268**

Near-Infrared HgCdTe 1024 x 1024 FPA for Astronomy

K. Vural, L.J. Kozlowski, S.C. Cabelli, C.Y. Chen,
G.L. Bostrup, W.L. McLevige, R.B. Bailey, and D.E. Cooper
Rockwell Sci. Center, Thousand Oaks, CA

K. Hodapp and D. Hall
University of Hawaii, Institute for Astronomy, Honolulu, Hawaii 96822

ABSTRACT

Rockwell Science Center and the University of Hawaii have developed a short wavelength infrared (SWIR) HgCdTe 1024x1024 focal plane array (FPA) for ground based astronomy. The device is named the HgCdTe Astronomical Wide Area Infrared Imager (HAWAII). The first hybrids have been fabricated, characterized, and delivered to the University of Hawaii and used in the imaging of the Shoemaker-Levy 9 Comet collision with Jupiter. Performance highlights include FPA mean dark current < 0.1 e-/sec at 77K, pixel yield of 99%, mean quantum efficiency $>$ 60%, and mean read noise $<$9 e-. The device has a unit cell size of 18.5μmx18.5μm with a $>$95% fill factor. The band of response is 0.9-2.5μm. The multiplexer input is a source follower per detector (SFD) and there are four outputs. The multiplexers were processed at Rockwell's commercial fax/modem chip line resulting in a relatively high yield of devices with no column/row outages. The detectors were processed on 3" diameter PACE-I material which is HgCdTe grown by liquid phase epitaxy on sapphire substrate.

The Next Generation of Near-Infrared Solar Magnetographs

Douglas Rabin
National Solar Observatory, National Optical Astronomy Observatories

ABSTRACT

Several groups are now measuring photospheric magnetic fields using the two Zeeman-sensitive Fe I lines ($g = 3$ and 1.53) near 1.56 μm The scientific emphasis is shifting toward full Stokes polarimetry of sunspots and plages, and highly sensitive Stokes-V searches for intrinsically weak fields ($B \leq 500$ G) in network and internetwork areas. It is likely that the next generation of near-IR magnetographs will also concentrate on the 1.56 μm lines. Experience with the NSO Near Infrared Magnetograph has shown that, although several instrumental effects are smaller in the infrared than the visible, the same procedures will be necessary to assure high-quality Stokes measurements—e.g., correction for the telescope Mueller matrix, image stabilization, and rapid (≥ 30 Hz) polarization modulation. Thus, the quickest path to a sophisticated IR magnetograph may be to adapt an infrared camera and polarization optics to an existing spectrograph-based instrument (such as the HAO/NSO Advanced Stokes Polarimeter or the Zürich Imaging Polarimeter). Next-generation cameras will incorporate 1K$\times$1K arrays, so beamsplitting will be a viable alternative or addition to rapid modulation. A key technical objective for near-infrared magnetometry is the development of compact, filter-based instruments. I briefly describe the specifications of NIM-2, a prototype imaging polarimeter based on a rapidly-tunable Fabry-Perot etalon. The National Optical Astronomy Observatories are operated for the National Science Foundation by the Association of Universities for Research in Astronomy. Near-infrared magnetometry at NSO is supported by the the NASA Space Physics Division through the SR&T program in solar physics.

Infrared Coronal Magnetic Field Measurements

J.R. Kuhn
National Solar Observatory/Sacramento Peak
National Optical Astronomy Observatories†
PO Box 62, Sunspot NM 88349 USA

ABSTRACT

A spectro-polarimetric technique for observing coronal magnetic fields using IR arrays is described. The method is useful for minimizing systematic errors due to seeing and a non-uniform array flat-field, while observing faint diffuse coronal emission. Empirical results are presented from observations of the infrared Fe XIII line at a wavelength of 1074.7nm

1. Measuring Coronal Magnetic Fields

While there is no doubt that magnetic fields control the structure and dynamics of the corona, there is also no routine technique for directly measuring this field. Clearly these are difficult observations, since the mean magnetic field strength is only of order 10G at 0.1 $R_\odot$ above the limb.[1] Potential field extrapolations of realistic photospheric magnetic sources[2] yield a similar magnitude, although above active regions the field should be higher.

Bastian[3] describes how microwave (gyrosynchrotron) measurements have been used to infer the magnetic field above active regions. Measurements of the Faraday rotation of the carrier signal from the Helios spacecraft have also been used to infer the magnitude of the coronal field at several solar radii from the limb.[4] While these are important and interesting results, it is difficult to assess how the radio measurements will constrain coronal models since these methods yield only occasional localized B field estimates.

Optical measurements of coronal fields have not been proven. There are several Zeeman splitting measurements of prominence magnetic fields[5,6] but the coronal problem is much harder. Arnaud and Newkirk[7] successfully used the resonance polarization phenomena to detect the coronal magnetic field orientation using the FeXIII line at 1074.7nm although the field could not be determined. This coronal emission line is often as bright as the coronal red line[8] and with recent IR array detector advances offers real potential for measuring coronal magnetic fields from Zeeman observations.

2. Observing Technique

Our first attempt at measuring the circular polarization of the FeXIII 1074.7 line was made using the Evans coronagraph and the 128×128 pixel HgCdTe IR array

† Operated by the Association of Universities for Research in Astronomy, Inc. (AURA) under cooperative agreement with the National Science Foundation

described in McPherson et al[9]. This is far from a photon counting system but the detector has very deep wells (3×10^7e) which simplified the data analysis (long integration times were possible). The technique described here will benefit greatly from the use of the NICMOS3 camera system we've recently completed.

The IR array allows simultaneous measurements in the spatial and spectral dimensions. With a Lande factor of 1.5 and a magnetic field of 10G the Zeeman splitting of this infrared line is only about 0.1pm. With the Littrow spectrograph at the Evans telescope and our detector such a field will shift the polarized line profile by only about 0.02 pixels. Clearly careful control of systematic errors is critical. Also, since long integration times are required (compared to atmospheric seeing timescales), we must simultaneously image spectra in right- and left-circularly polarized light. We can control the remaining systematic, spatial calibration and flat field variations, by optically "switching" the right and left polarized beams. This must be done without mechanically shifting the optical paths. In effect we perform a double-differential measurement.

Our optical system used a liquid crystal variable retarder (switching between $\pm\lambda/2$) with its optical axis rotated 45° from one of the linear polarization axis of the Wollaston prism that followed it. This optical arrangement yielded two displaced spectra in right and left circular polarized light over two halves of the array. The polarized beams could be switched by electrically reversing the state of the variable retarder. Figure 1 shows an example of two data frames. The bright emission line in each of the 4 spectra is the FeXIII 1074.7nm line. Successive bright regions in this figure show RCP, LCP, LCP and RCP. The spatial extent (horizontal) of each of these spectra is about 1' across 40 pixels and the dispersion (vertical) is about 0.02nm/pixel. Each of the two data frames recorded here were obtained with a one minute integration.

3. Analysis

The first step in the reduction was to extract the right and left emission line profile segments from each pair of frames. This yields 4 spectra (two in RCP and two in LCP) that are registered to within a pixel. After dark correcting we then spatially averaged each spectrum. For the $\lambda/2$ retarder state the left and right spectra $d_{L,R}(\lambda)$ are given by

$$d_{L,R}(\lambda) \;=\; g_{L,R}(\lambda)I(\lambda \pm \delta - v)$$

where δ represents the magnetically induced lineshift, and $g_{L,R}$ is the effective detector/telescope gain variation in the left and right spectra. After reversing the retarder state the new spectra satisfy

$$d'_{R,L}(\lambda) \;=\; g_{L,R}(\lambda)I(\lambda \pm \delta - v).$$

Note that the ratio of the ratios of the d functions can be combined to yield a new

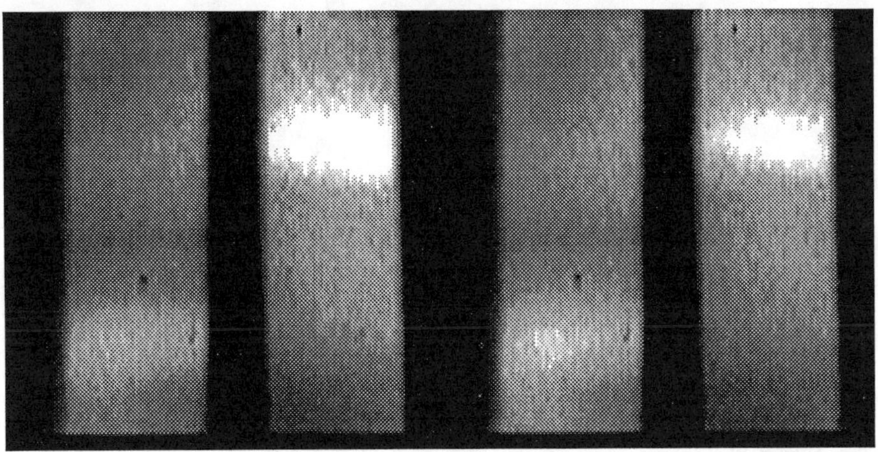

Figure 1: Infrared FeXIII Polarized Coronal Emission Line Spectra

function of wavelength, K,

$$K(\lambda) = \frac{d_L/d_R}{d'_L/d'_R} = -4\delta\frac{d\ln I}{d\lambda} + C.$$

This is an interesting function in that the derivative of the mean line profile should be proportional to K. It is also robust in that we can use the full line profile to derive the Zeeman shift. A least squares fit of K to the profile derivative is proportional to the splitting parameter, δ.

Figure 2 shows the result of this analysis. The dotted line is the mean FeXIII line profile, the dashed line is the numerical derivative of the profile and the solid line shows the function $K(\lambda)$. Fitting the derivative function to K yields a coefficient $4\delta < 0.072$ pix. This corresponds to an upper limit on the field measurement of 40G.

4. Conclusions

This limit of 40G is close to being interesting for coronal measurements near active regions. It is likely that an improvement by nearly an order of magnitude can be realised by using a quieter detector (NICMOSIII) although the systematics may require additional work. Nevertheless we are optimistic that IR Zeeman coronal B-field measurements will be possible in the near future that are accurate to a few gauss.

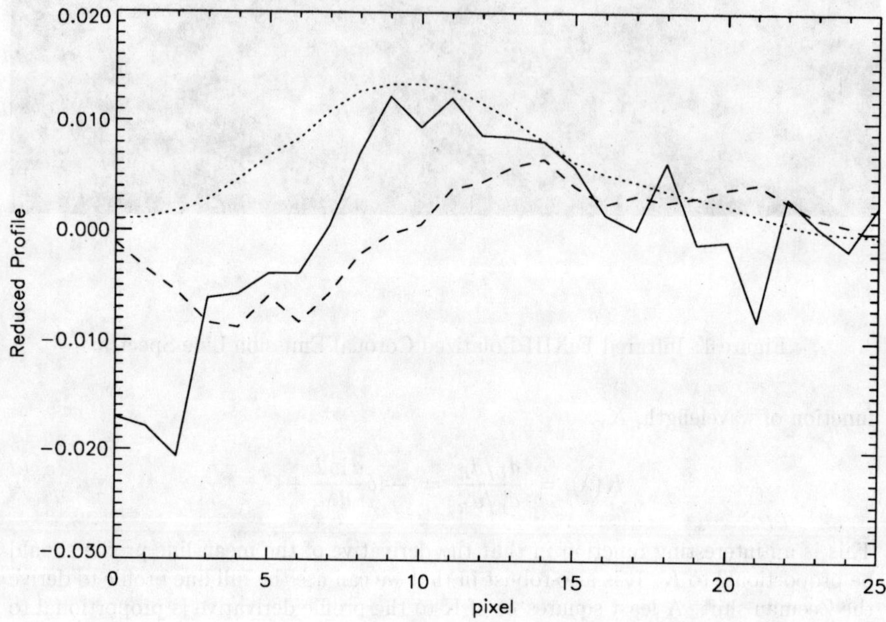

Figure 2: Line profile (dotted), Derivative (dashed), and K(λ) (solid) functions

Of course these prospects should improve if strong coronal emission lines further into the IR are discovered.

5. Acknowledgements

The help of the observing and support staff at the Evans Telescope is gratefully acknowledged.

1. Dulk, G.A., McLean, D.J. *Sol. Phys.* **57** (1978) 279.
2. Hoeksema, J.T. and Scherrer, P.H. *Sol. Phys.* **105** (1986) 205.
3. Bastian, T. (1995) *these proceedings*
4. Patzold, M., Bird, M.K., Volland, H., Levy, G.S., Seidel, B.L., Stelzried, C.T. *Sol. Phys.* **109** 91.
5. Harvey, J *Ph.D. Thesis* (Univ. of Colorado, 1969)
6. Nikol'skii, G.M., Kim, I.S., Koutchmy, S., Stepanov, A.I., Stellmacher, G. *Soviet Astr.* **29** (1985) 669.
7. Arnaud, J. and Newkirk, G.,Jr. *Astron. Astrophys.* **178** (1987) 263.
8. Penn, M.J. and Kuhn, J. R. *ApJ* **in press** (1995)
9. Mcpherson, M., Lin, H., Kuhn, J.R. *Sol. Phys* **139** (1991) 255.

CFHT'S IMAGING FOURIER TRANSFORM SPECTROMETER

D. A. Simons, Jean-Pierre Maillard, C. C. Clark, S. Smith, J. Kerr, and S. Massey
Gemini Project

ABSTRACT

We have recently developed a Fourier imaging spectrometer that is now in use at the Canada-France-Hawaii Telescope. This instrument, the only one of its kind in the world, is capable of recording both images and spectra of sources in the 1-2.5 micron wavelength range across its 24" field of view. The spatial resolution of images is seeing limited with 0.33 arcsec/pixel sampling. It works by coupling the facility Fourier Transform Spectrometer (Maillard and Michel 1982, Instrumentation for Astronomy with Large Optical Telescopes, IAU Coll. No. 67, ed. C. M. Humphries, pub. D. Reidel, 92, p. 213-222.) and "Redeye" infrared camera (Simons et al., 1993, SPIE, Infrared Detectors and Instrumentation, Vol. 1946, pp. 502-512) to work as a single instrument. Specifically, the CFHT f/35 focal plane is reimaged onto an infrared array through the FTS optics by a special optical interface, creating a pair of complementary images that modulate in intensity as the interferometer is stepped through a scan. The camera records an image at each interferometer step and uses a Hg:Cd:Te NIMCOS3 infrared array with 256x256 pixels on 40 micron centers. In its current configuration the overhead between steps is only 2 sec. From the recorded data, interferograms and spectra can be extracted through straightforward aperture photometry of complementary regions in the field of view. Alternatively, it is possible to invert an entire raw data cube, making a four-dimensional processed cube (x, y, wavenumber, intensity) from which monochromatic images can be extracted and manipulated. Some of the unique advantages of this instrument over other imaging spectrometers in use at observatories include: (1) Easily tuned spectral resolution across the 1 - 2.5 micron range up to R approx. 104, (2) Both emission and absorption line observations are practical, (3) Wavelength calibration is intrinsic to the data. Successful observations have been made to date of various astronomical targets, including planets and emission nebulae.

THE NSO/NASA HE I 1083.0 nm VIDEO FILTERGRAPH/MAGNETOGRAPH

HARRISON P. JONES
NASA/Goddard Space Flight Center
Laboratory for Astronomy and Solar Physics
Southwest Solar Station, c/o NSO, PO Box 26732, Tucson, AZ 85726 USA

and

JOHN W. HARVEY
National Solar Observatory†, *PO Box 26732, Tucson, AZ 85726 USA*

ABSTRACT

The 1083.0 nm Video Filtergraph/Magnetograph (VFM) is a new instrument for high spatial and temporal resolution imagery of chromospheric and, indirectly, lower transition-region and coronal material above solar active regions. The instrument was developed jointly by NASA/Goddard Space Flight Center (GSFC) and the National Solar Observatory (NSO) for operation at the NSO/Kitt Peak Vacuum Telescope (KPVT). The principles of operation, based on a three-mode, dual channel, five-element birefringent filter are described. The instrument is undergoing testing and integration at the telescope.

1. Introduction

Diachronic observations of the 1083 nm line of He I show a richly structured and time-variable phenomenology which is thought to map solar structure at the base of the transition region modulated by coronal radiation incident from above (cf. Andretta[1]; Fontenla *et al.*[2]; Avrett *et al.*[3]). Thus, network, filaments, plage, and coronal holes can all be visible in one image. When viewed by other means, most of these active-sun structures exhibit transient and spatial behavior beyond the resolution capabilities of the spatially scanned, spectrograph-based instruments (previously the Diode Array Magnetograph; currently, the spectromagnetograph (SPM)) at the KPVT with which most of the 1083 nm data have been obtained. We have therefore developed a new filter-based instrument, the VFM, to improve the KPVT's ability to observe rapid, small-scale solar activity in the 1083 nm line. Although not yet operational, the instrument has been built and filter performance has been verified. All optical and electronic components are on hand, and final installation at the telescope, testing, and software development are in progress.

†The National Solar Observatory is a Division of the National Optical Astronomy Observatories which is operated by the Association of Universities for Research in Astronomy, Inc., under cooperative agreement with the National Science Foundation

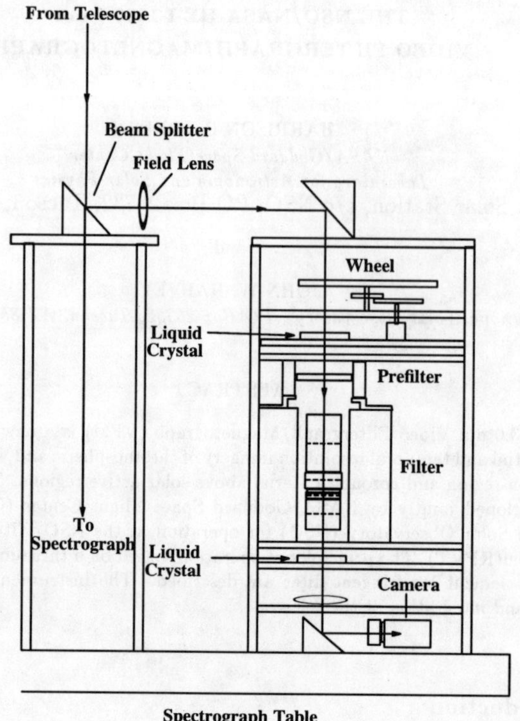

Figure 1: Layout of the VFM

2. Instrument Description

Figure 1 shows the schematic layout of the VFM which is mounted on the observing table of the 10-m Littrow spectrograph at the KPVT but which makes no other use of the spectrograph. A beam-splitter, optimized for 1083 nm, allows light to reach the SPM for simultaneous observations, although no spatial scanning will be permitted during data acquisition by the VFM. After passing through suitable optics which produce and fold a telecentric beam, the light passes through modulating elements, a blocking filter, a narrowband birefringent filter, and a camera lens onto the CCD detector. Frames are digitized at video rates and processed on-line with equipment similar to the SPM but, as discussed below, a by a much simpler processing algorithm.

The filter itself, shown schematically in more detail in Figure 2, consists of five calcite elements separated by polarizers and quarter-wave plates in a conventional

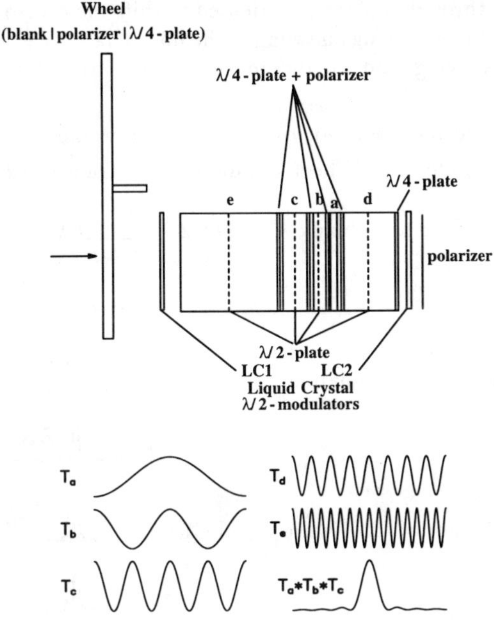

Figure 2: Lyot Filter Schematic

Lyot configuration. The four thickest elements are split and separated by half-wave plates for wider acceptance angles. The ideal transmission functions for each element are channel spectra as shown in Figure 2, and the transmission of the three central thinnest elements is fixed as the product of the respective individual functions without phase shift. One of two ferroelectric liquid crystal half-wave retarders is used to modulate the transmission function of the entire assembly at video rates in one of three modes determined by an aperture wheel by shifting the phase of one of the two thickest elements. The final data products are the accumulated differences and sums of the two modulated states.

The six possible transmission functions and their use are shown in Table 1. In the equivalent width mode, the wheel is positioned with an open aperture so that the thickest element has no effect. Element d is modulated by the second liquid crystal between a centrally tuned state which transmits the 1083 nm line and a state shifted by a half wave which transmits two continuum bands on either side of the line. The difference of these two images divided by either the continuum or, as in the current SPM algorithm, the sum is closely analogous to the equivalent width and will produce a signal much like the familiar diachronic data but with much faster cadence and much higher spatial integrity. In the Doppler and magnetic modes, element d is fixed and centrally tuned while the transmission is modulated by shifting the phase of the thickest element, e, by the first liquid crystal. For velocity, the wheel is positioned so that light passes through a polarizer oriented to shift the transmission of element e by a quarter wave, thus sampling one wing of the line; a half-wave shift of transmission samples the opposite wing, and the difference is a measure of line-of-sight velocity.

Table 1: Operation Modes for the 1083 nm Filtergraph/Magnetograph

Mode	Wheel	Modulator	Transmission
Eq. Width	blank	LC2	Line Cont
Doppler	polaroid	LC1	Red Blue
Magnetic	1/4-wave	LC1	$R^+ + B^-$ $B^+ + R^-$

The magnetic mode makes use of a scheme proposed by Ramsey[4]. The wheel places a quarter-wave plate in the beam which translates the two states of circular polarization into two orthogonal linear states. Thus in one phase of the liquid crystal left-circularly polarized light sees a transmission function centered on the blue wing of the line and right-circularly polarized light sees a transmission function centered on the red wing. Rotating the polarization directions by π with the liquid crystal interchanges the transmission-circular polarization combinations, and the difference of the two signals serves as a conventional double side-band detection of longitudinal magnetic field. Many details of the formation of the 10830 line remain to be settled, but virtually all models place the dominant region of formation of the line in an optically and geometrically thin region near the top of the chromosphere and base of the transition region with no contribution from the photosphere or lower chromosphere. Thus the instrument will only sense the high chromospheric field–a feature of considerable physical interest.

The data acquisition and filter control systems are sketched in Figure 3. The camera (Cohu model 6712) is a frame-transfer CCD which produces analog video in CCIR format and provides the timing input for the data and control systems. The data is digitized in the image processing board (Datacube Max-20) where even and odd frames are separately accumulated for an operator-specified number of frames. The two accumulation buffers are appropriately flat-fielded, differenced, summed and ratioed to provide the final data project which can be stored on tape or disk. The image processor board translates the camera's video synchronization signals into a

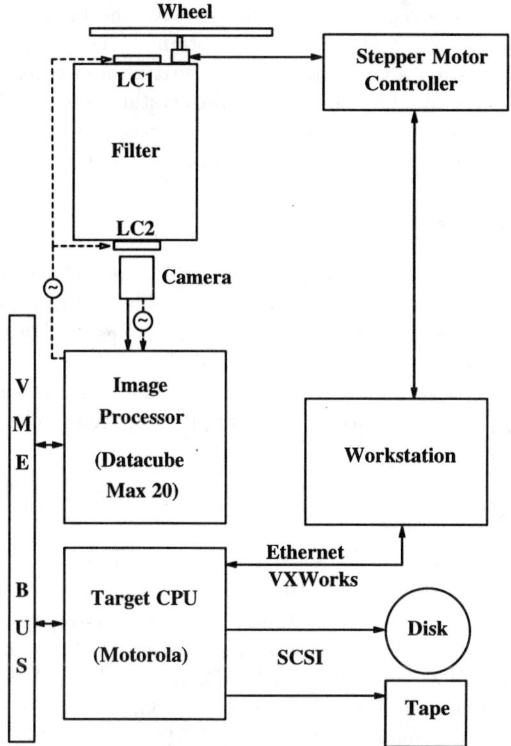

Figure 3: Control System Schematic

modulation timing clock for the liquid crystals at the video frame rate. A board-level CPU (Motorola) which serves as a VXWorks (Wind River Systems) target controls the image processor board. A single host workstation provides the operator interfaces both to the VFM and the SPM. The target computer has ethernet, SCSI, and serial communications links with its host and peripherals (disk and heliscan tape drive). A separate controller sets the position of the filterwheel using serial commands from the host workstation.

3. Status

All of the mechanical components of the VFM have been fabricated and assembled, and the Lyot filter has been assembled and tuned to 1083 nm at the operating temperature of its enclosing oven. The filter and camera have been installed in the housing which is in place on the spectrograph table at the KPVT. The beam splitter can be inserted and acceptable solar images with the broad central transmission of combined elements a, b, and c have been obtained. The data system is on line, and the basic accumulation algorithm has been written and tested. In the near future, we will install the exit polarizer, the liquid crystals, and the filter wheel elements. We will also align the entire optical train more carefully, install control, test and standardize flat-fielding procedures, and improve the control software.

4. Acknowledgements

We thank T. Gull for making available GSFC resources in the final development stages of this project as well as M. Amato, C. Shields, and L. White for their assistance in producing the final design and technical drawings. Software was written by J. Schwitters, and the instrument was fabricated and assembled by NSO and NOAO Engineering and Technical Services personnel. Partial funding for this instrument was provided by NASA Space Physics Supporting Research and Technology Task 170-38-52-14.

5. References

1. V. Andretta, PhD Thesis, University of Naples (1994).
2. J. M. Fontenla, E. H. Avrett, and R. Loeser, *Astrophys. J.* **406** (1993) 319.
3. E. H. Avrett, J. M. Fontenla, and R. Loeser, in Proceedings, IAU Symposium 154, *Infrared Solar Physics*, eds. D. Rabin, J. T. Jefferies, and C. Lindsey (Kluwer Academic Publishers, 1994), p. 35.
4. H. E. Ramsey, *Solar Phys.* **21** (1971) 54.

INFRARED CAMERA INSTRUMENTATION

ROBERT F. KURTZ
Infrared Laboratories, Inc.
1808 E. 17th St., Tucson, AZ 85719

ABSTRACT

Designs for focal plane array cameras in the band of 0.8 - 28 micrometers that have been built by Infrared Laboratories are presented. A variety of cameras have been constructed utilizing HgCdTe, InSb, Si:Ga, and Si:As BIB focal plane arrays. These instruments incorporate cryogenic, optical, electronic, and array subsystems integrated to function as a system. A number of these designs are presented in this paper as a means of showing the technologies

1. NICMOS 3 HgCdTe System

Figure 1 is a side section of a pumped LN2 dewar housing a NICMOS 3 HgCdTe array camera. The LN2 container cools a shield surrounding the optics module. There is an additional LN2 container that cools the optic mounting plate and the shield around it as well as all the optical components. The use of dual LN2 containers enables the user to pump down the inner container to approximately 55K if desired. The use of copper wool inside the container ensures a stable temperature when the liquid freezes as it is pumped down to low vapor pressures. Without copper wool affixed to the bottom plate, the solid nitrogen tends to lose contact with the cold plate which may result in the cold plate warming up.

The system contains cooled refractive optics. Optical materials were chosen to ensure chromatic aberrations were minimized. A motor driven v-groove for a coronagraph mode is located on the front near the camera aperture. There are also two cooled filter wheels and a Lyot stop wheel. The array is mounted on a z-stage to allow for focusing of the focal plane. The entire optics chain is enclosed in order to minimize stray radiation. The inside of the shield on the optics plate has been sandblasted and anodized to also help minimize stray radiation. The optics lens mounting tubes are anodized black on the inside to minimize reflections. All the optical components are mounted using a spring arrangement to allow for differential thermal expansion during cooling and warm-up. Both of the filter wheels, Lyot stop wheel and aperture wheel are driven by stepper motors which are controlled through an RS232 computer port.

Figure 2 is a bottom view showing the rectangular shape of the case extension which has been mounted on our standard 10-inch dewar. The motor drives and electrical connectors have been mounted on the top of the rectangular case extension. The only feed-thru on the sides of the case is the control for focusing of the array. This was done to allow the maximum area for mounting the camera to the telescope.

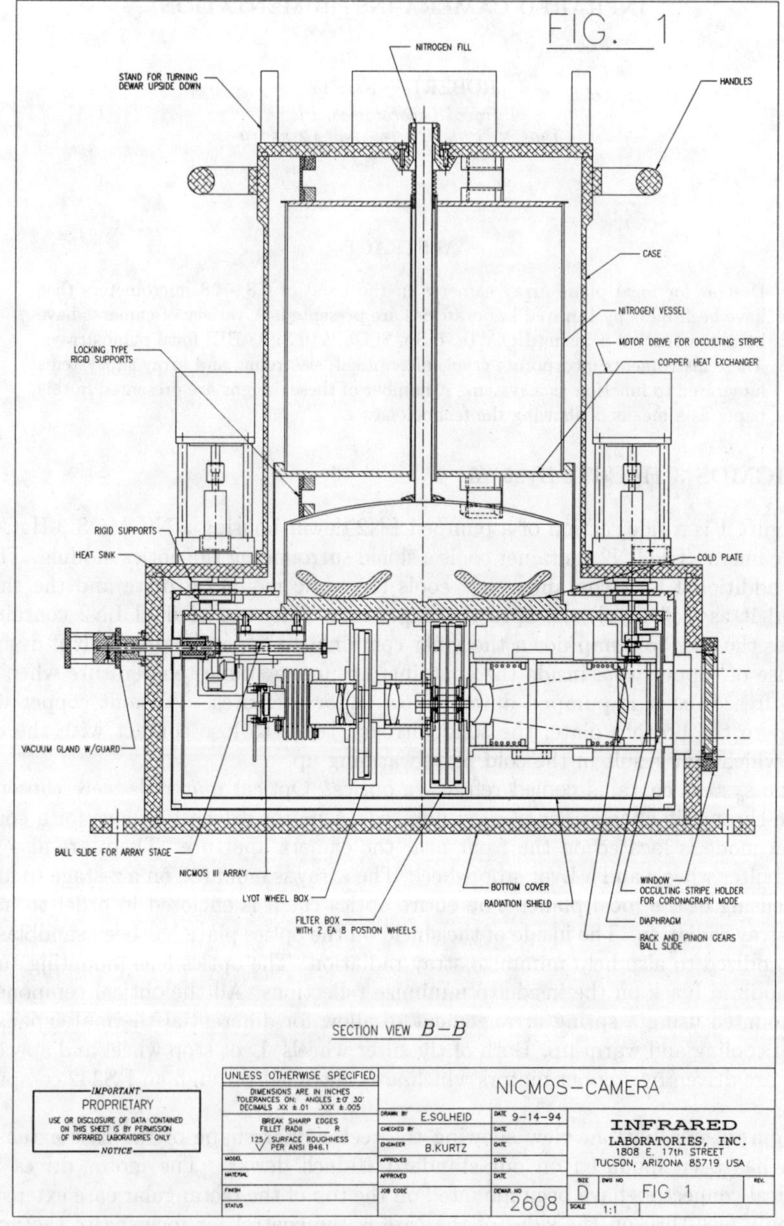

Figure 1: Block diagram of the MCE-3 system

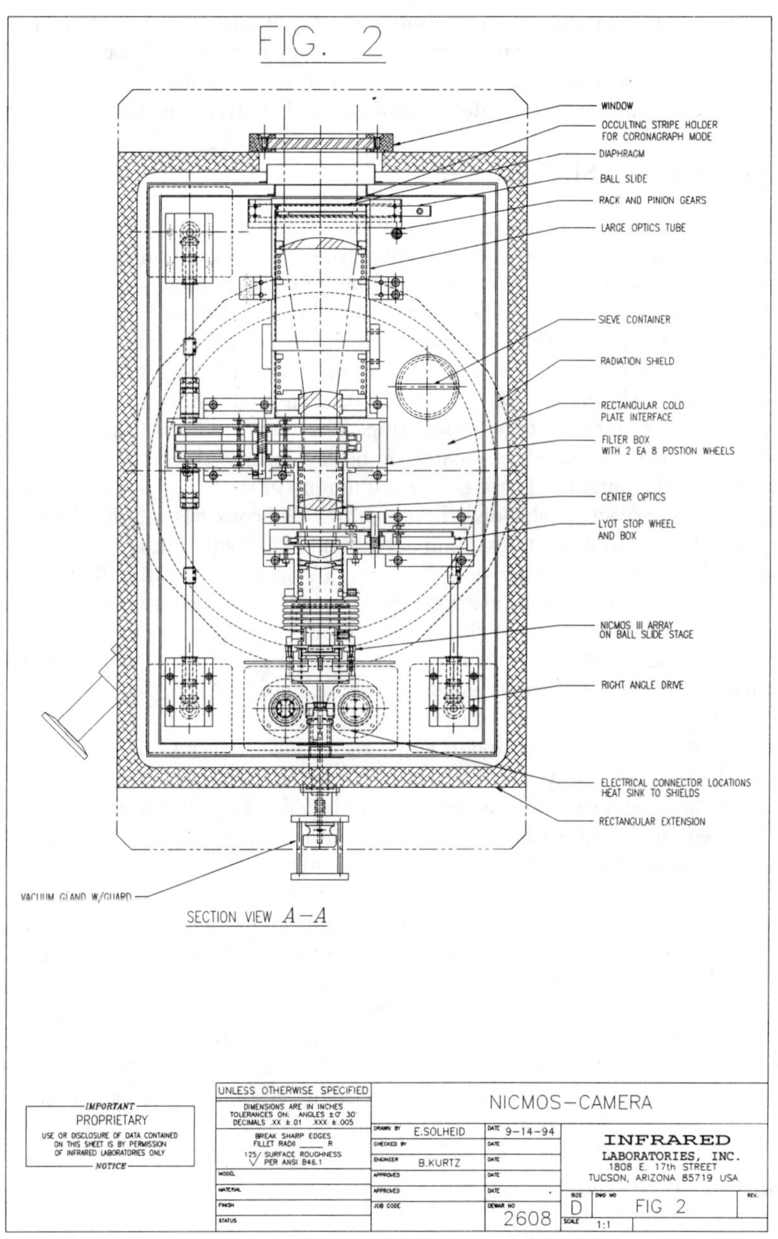

Figure 2: Section View A - A

Since this is a nitrogen dewar, a molecular sieve container is mounted inside the cold shields to insure a high vacuum while the system is cold. All basic materials are aluminum for light weight and strength. Most of the components holding the optics are made of aluminum to keep differential thermal contractions to a minimum.

2. Dual InSb and Si:Ga System

Figure 3 shows the bottom view of a system utilizing InSb and Si:Ga arrays. Light from the telescope enters the system through the window and then to a dichroic which splits the two wavelengths. The energy for the InSb channel then goes to a filter wheel next to an optics wheel box. The optics wheel has four different optics modules that are rotated into place for the different magnifications of the system. The InSb array is mounted on a z-stage with a manual control to allow for focusing of the system. The Si:Ga channel is similar in that it also has its own filter wheel, four optics modules, and array mount on the z-stage for focusing. The electrical leads from both arrays are brought out through heat sinks located on both shields of the system to the hermetic connector on the outside. These cables are unique in that they use a combination of coax as well as cryogenic shielded ribbon cable. The coax has a Teflon outer coating which is cut away in a small area and stycast into the heat sink to prevent any stray radiation from entering the system along the Teflon outer coating. If this procedure is not performed, the Teflon coating acts as a light pipe and allows stray radiation to enter the low background environment. Again, as in a prior camera, the shield is sandblasted and anodized to have a rough absorbing surface. The optical changing path leading to the two arrays is completely enclosed. All optics changing wheels, filter wheels and z-stages are motor driven on this system.

This system is unique in that the cooling is done with a Gifford-McMahan two-stage Helium cooler. This can be seen in Figure 4. The first stage is connected to the outer shield and operates at approximately 65K. The second stage of the cooler is connected to the optics mounting plate which is cooled to approximately 10.5K. The CTI cold head is mounted utilizing shock absorbing mounts at the case with a flexible welded bellows for the vacuum seal. Bellows type cooper mounts are used at the cold stages to minimize vibrations by a factor of approximately 300. This system allows vibrations to be tuned by the selection of rubber shock absorbing mounts used. Cooling of the system usually takes between 12 and 20 hours depending on the heat load and mass on the two stages. In order to reduce the cool-down time to about six hours, a LN2 continuous flow heat exchanger can be added to the first stage shield. The CTI Gifford-McMahan has been shown to operate for 5000 hours without maintenance. The use of the vibration de-coupling system minimizes the disturbance of the optics in this system to less than 3 microns.

3. Si:As BIB System

Figure 5 is a side view of a Si:As BIB camera. This was constructed in a standard

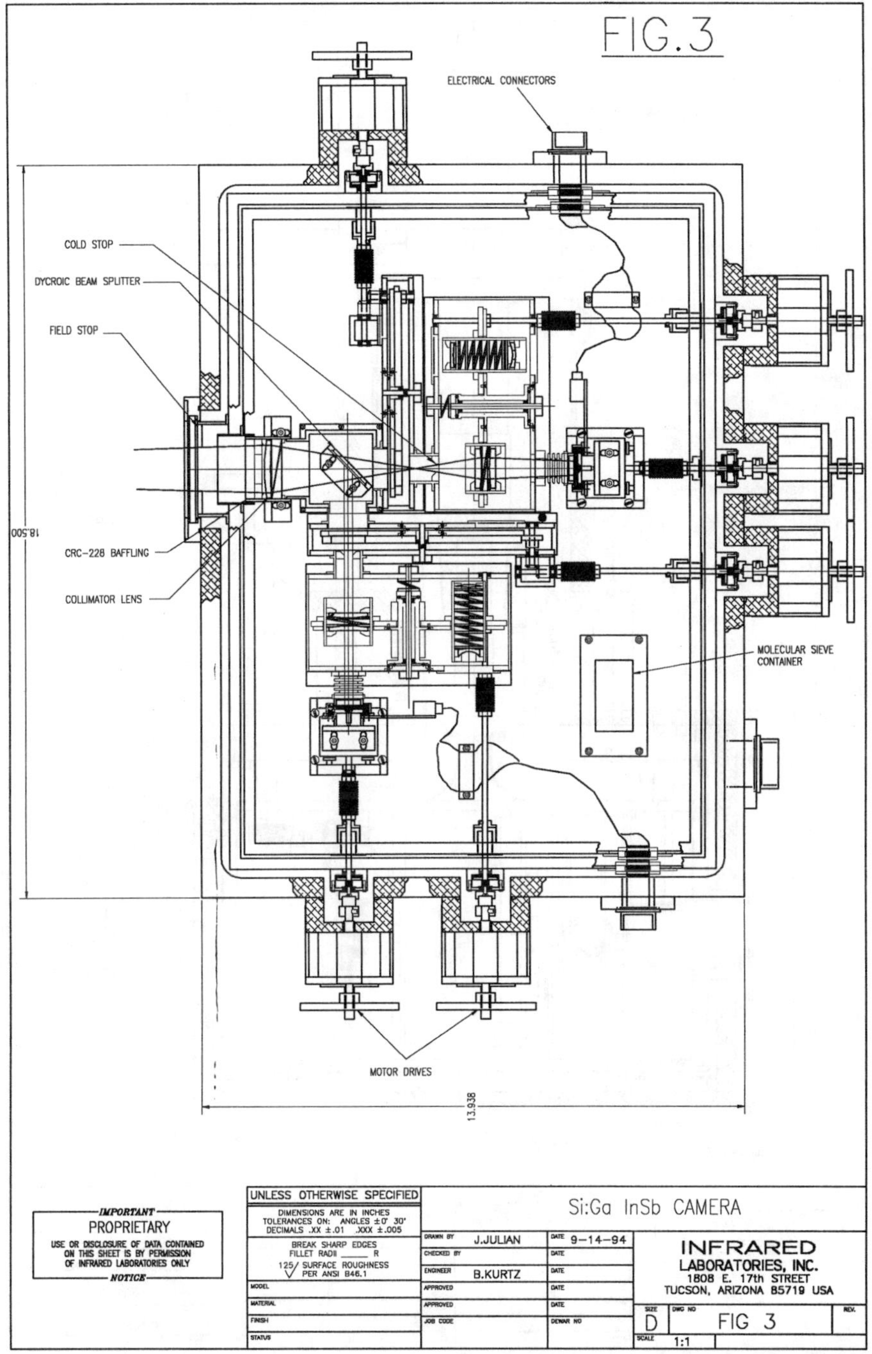
FIG.3
ELECTRICAL CONNECTORS
COLD STOP
DYCROIC BEAM SPLITTER
FIELD STOP
18.500
CRC-228 BAFFLING
COLLIMATOR LENS
MOLECULAR SIEVE CONTAINER
MOTOR DRIVES
13.938
IMPORTANT
PROPRIETARY
USE OR DISCLOSURE OF DATA CONTAINED ON THIS SHEET IS BY PERMISSION OF INFRARED LABORATORIES ONLY
NOTICE
UNLESS OTHERWISE SPECIFIED
DIMENSIONS ARE IN INCHES
TOLERANCES ON: ANGLES ±0' 30'
DECIMALS .XX ±.01 .XXX ±.005
BREAK SHARP EDGES
FILLET RADII ______ R
125 SURFACE ROUGHNESS PER ANSI B46.1
MODEL
MATERIAL
FINISH
STATUS
Si:Ga InSb CAMERA
DRAWN BY J.JULIAN
DATE 9-14-94
CHECKED BY
ENGINEER B.KURTZ
APPROVED
JOB CODE
DEWAR NO
INFRARED LABORATORIES, INC.
1808 E. 17th STREET
TUCSON, ARIZONA 85719 USA
SIZE D
DWG NO FIG 3
REV.
SCALE 1:1

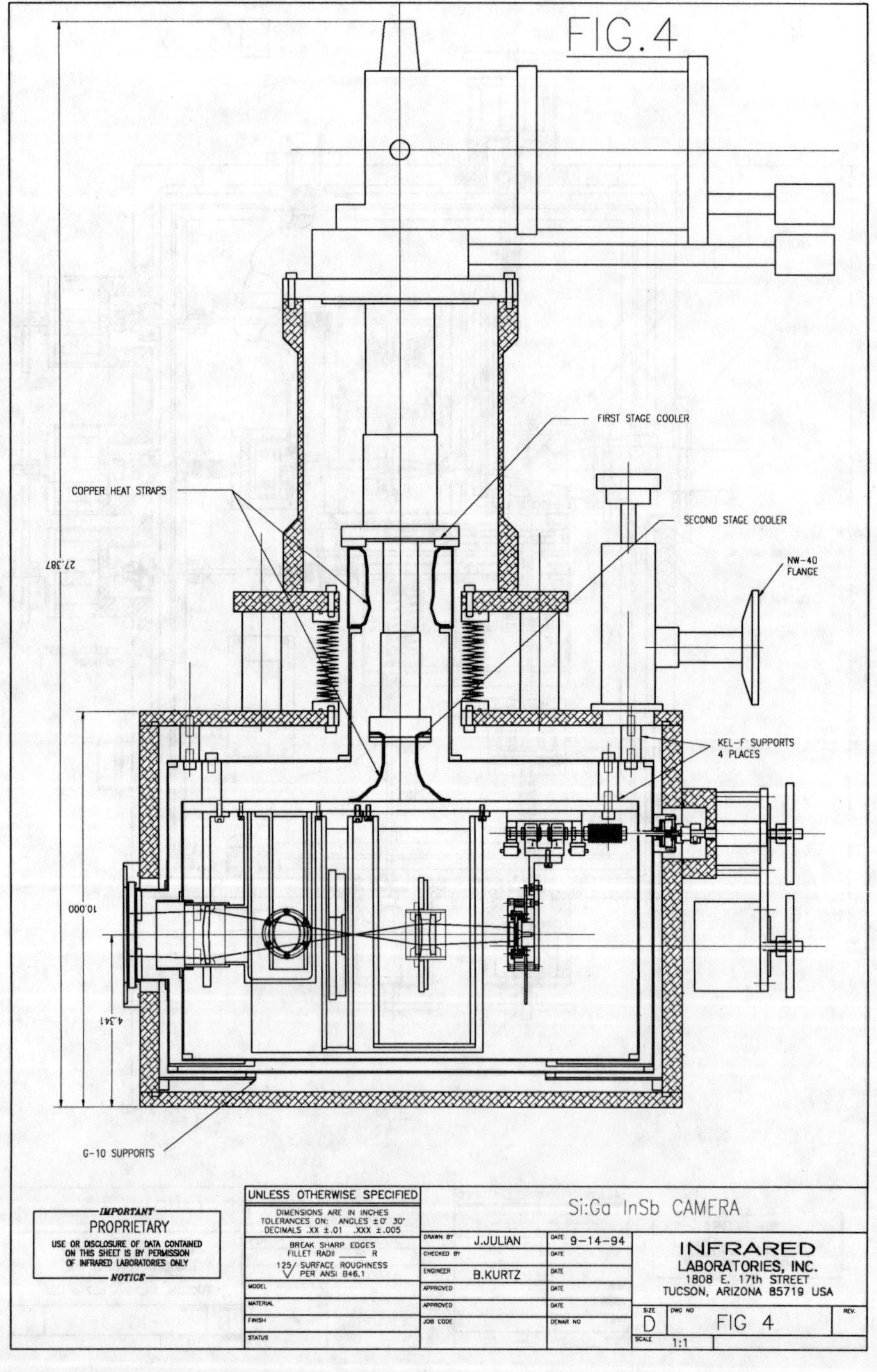
FIG. 4
FIRST STAGE COOLER
COPPER HEAT STRAPS
SECOND STAGE COOLER
NW-40
FLANGE
27.387
KEL-F SUPPORTS
4 PLACES
10.000
4.341
G-10 SUPPORTS
UNLESS OTHERWISE SPECIFIED
DIMENSIONS ARE IN INCHES
TOLERANCES ON: ANGLES ±0° 30'
DECIMALS .XX ±.01 .XXX ±.005
BREAK SHARP EDGES
FILLET RADII ____ R
125/ SURFACE ROUGHNESS
PER ANSI B46.1
Si:Ga InSb CAMERA
DRAWN BY J.JULIAN DATE 9-14-94
CHECKED BY
ENGINEER B.KURTZ
IMPORTANT
PROPRIETARY
USE OR DISCLOSURE OF DATA CONTAINED ON THIS SHEET IS BY PERMISSION OF INFRARED LABORATORIES ONLY
NOTICE
INFRARED
LABORATORIES, INC.
1808 E. 17th STREET
TUCSON, ARIZONA 85719 USA
SIZE D
FIG 4
SCALE 1:1

HDL-10 Infrared Laboratories dewar. This system has a LN2 container that cools a shield around the LHe container which has the optics and array mounted on it. An additional LHe cooled radiation shield has been added around the optics and the array to minimize the amount of stray radiation. This dewar, as well as the others, have fiberglass straps between the top plate of the dewar and the LN2 container. Straps are also used between the nitrogen and helium containers. This makes a very rigid support for the cryogen containers and also preserves the alignment of the system during transportation and tipping of the telescope. With the rigid support system, movement of the cold plate is limited to a few microns while tilting the dewar 45 degrees in any direction from vertical.

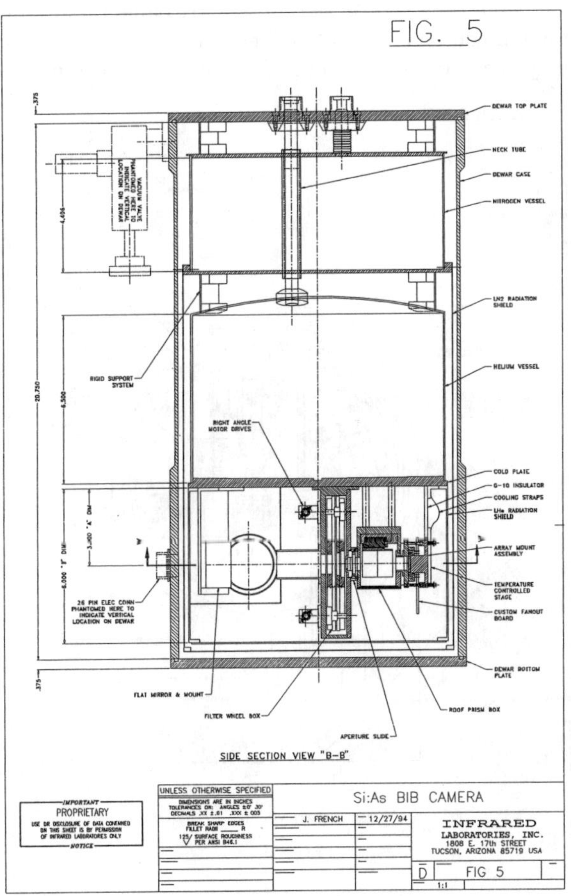

Figure 6 is the bottom view of the BIB camera. The telescope beam is focused on all aperture located inside the dewar. This is an all reflective design because of the long wavelength thermal response of this camera. The optical mirrors are diamond turned aluminum. The material goes through a stabilization process before turning to eliminate figure changes after cooling due to stresses in the metal. The surfaces are also polished after turning to eliminate the grating scattering effect of the diamond turning tool marks. There is a powered optic and turning mirror which focuses the beam onto a Lyot stop through two filter wheels. In order to obtain more than one magnification, a set of right angle mirrors can be inserted into the beam to increase the path length. The magnification switching is done with a manual control. The two filter wheels are stepper motor controlled. Because this is a long wavelength camera, and a large amount of thermal radiation enters the camera, we have utilized an extremely long baffle tube at 77K as well as a 4K baffle tube to decrease the solid angle that the Helium container sees of ambient radiation. This camera design offers a fairly compact system that fits into one of our standard helium dewars.

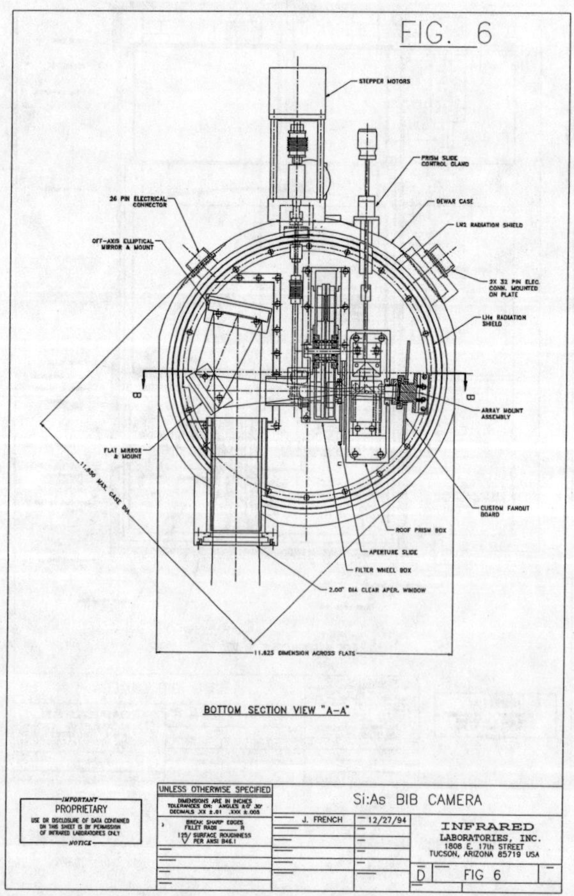

4. Camera Control Electronics

Figure 7 is a block diagram of a third generation modular electronic system for the control of FPA camera systems. This system is currently under development. This design is currently under development. This design allows the system to be upgraded from four inputs to up to as many as 32 inputs. This feature allows the user to implement a new or different array without purchasing a complete new electronics system. The fiber optic interface gives complete AC and DC ground isolation between the telescope electronics and the control room instrumentation. It also allows the camera to be operated remotely from the control location. The data link will sustain data rates as high as 80 megabytes per second.

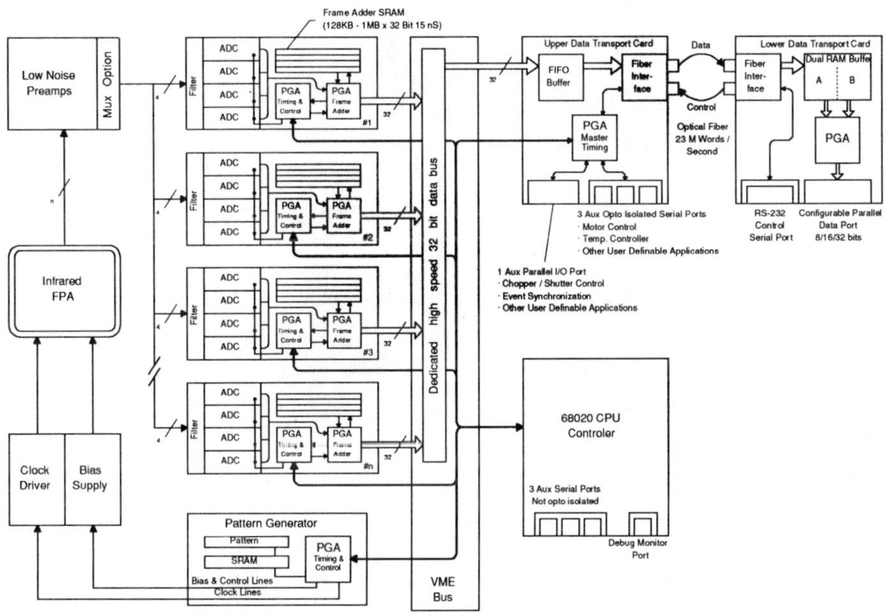

TUNABLE FIELD-WIDENED ETALONS FOR THE VISIBLE AND INFRARED

W. J. Rosenberg, J. K. Kumer, I. Zayer
Lockheed Palo Alto Research Laboratories

ABSTRACT

We are investigating wide field tunable etalons for the visible and infrared and have demonstrated the effectiveness of the field widening technique. We have employed a hybrid spacer consisting of a high index solid wafer and a thin air (or vacuum) layer. This design allows for the tuning of the etalons by moving one mirror in the air/vacuum while retaining most of the field advantage of a solid etalon with a high index spacer. We have fabricated fixed-frequency, field widened etalons in both spectral regions. The visible etalon, designed to operate at the HeNe wavelength of 632.8 nm, has a field-of-view (FOV) characteristic of a solid etalon with a spacer index of 1.6. The infrared etalon, using germanium for operation at 5.0 microns, has a FOV characteristic of a spacer index of 3.5. Both etalons have internal reflections consistent with the uncoated surface of the field widening wafer. It should be possible to virtually eliminate the reflections by means of anti-reflection coatings. There is some scattering and a slight spectral nonuniformity in the visible etalon which is attributable to the cement layer. The infrared etalon, whose surfaces are joined by optical contacting Hafnium-Oxide layers, appears more uniform. We are pursuing the fabrication of tunable versions of these etalons. We will investigate the use of Zinc-Selenide or Cleartran-Zinc-Sulphide for etalons This work has been supported by Lockheed Independent Research funding.

RADIO DIAGNOSTICS OF CONDITIONS IN THE SOLAR ATMOSPHERE

T. S. Bastian
National Radio Astronomy Observatory
Socorro, NM 87801, USA

ABSTRACT

Radio emission from the Sun provides a powerful tool for probing a variety of physical parameters in the solar atmosphere, including the electron number density, the effective temperature, and the magnetic field strength and orientation. The sources of continuum opacity from submillimeter to decimeter wavelengths are reviewed, and a variety of observational data are briefly discussed. It is pointed out that semi-empirical models of the chromosphere disagree to varying degrees with available radio data. Similar inconsistencies arise when comparing radio and UV emission from the prominence/corona transition region, or radio and SXR emission from solar active regions. A proper accounting of the highly inhomogeneous nature of the solar atmosphere may alleviate these problems.

1. Introduction

Radio wavelengths span a large spectral domain – using a relaxed definition, one might include all wavelengths longward of a few hundred microns. For the purposes of this review, however, I shall confine myself to wavelengths relevant to the low-solar atmosphere – that is, wavelengths of a few hundred microns to a meter, or so, which probe the temperature minimum region up to the low corona. Furthermore, I shall confine myself to radio diagnostics of the non-flaring Sun; the discussion will include both the quiet Sun and active regions.

The discussion is restricted to continuum radio emission for the simple reason that for wavelengths longward of a few $\times 100\mu$m, no spectral lines have been detected! In the late-1960's Dupree[20] suggested that high-n states of certain ions would be over-populated as a result of dielectronic recombination and that radio recombination lines (RRLs) might be detectable at millimeter wavelengths. Berger and Simon[11] and Shimabukuru and Wilson[65] conducted searches for RRLs near a wavelength of 3 mm and detected nothing. Berger and Simon revised Dupree's estimates of the expected line-to-continuum ratios downward by factors of 100–1000. Greve[26,27] then showed that when the effects of pressure and/or Zeeman broadening are included in the relevant calculations, the line-to-continuum ratios again decrease by significant factors. Interest in the subject was briefly renewed in the late-1980s when interferometric observations became possible at millimeter wavelengths. A survey of RRLs between 85–115 GHz carried out with the BIMA array by A. Grossman and S. White again yielded no detections (private communication).

Finally, discussion of polarimetry will be confined to Stokes parameters I and V. While some solar radio sources are expected to be intrinsically linearly polarized, the linearly polarized component is subsequently washed out by strong differential Faraday rotation as the radiation propagates through the relatively dense, magnetized corona. Under certain special circumstances, it may be possible to detect linearly polarized radiation from the Sun over a narrow bandwidth[3], but this awaits confirmation. In most circumstances, the observed radiation is unpolarized or circularly polarized, and any information concerning the magnetic field is embodied in the Stokes I and V parameters.

A review of this kind cannot be complete in any sense of the word. I have attempted to convey a sense of what kinds of observations are possible in the radio regime, what the relevant emission mechanisms are. In §2 I discuss continuum emission in the submillimeter and millimeter wavelengths regimes. In §3 I discuss centimeter and decimeter wavelengths. I conclude in §4.

2. Submillimeter and Millimeter Wavelengths

Observations at millimeter and submillimeter wavelengths offer several distinct advantages over those in visible or ultraviolet wavelengths. First, the sources of continuum opacity at these wavelengths – collisions between electrons and positive ions and with neutral hydrogen atoms – are well-understood. Because hydrogen is the dominant ion, we refer to H and H^- free-free opacity. Approximate expressions for the absorption coefficients are

$$\kappa_{ff}(\mathrm{H}^-) = 1.37 \times 10^{-29} \frac{n_e n_{HI}}{\nu} \left[1 + \frac{3.18 \times 10^{11}}{\nu} \left(1 - \frac{594.15}{T_e}\right)\right] \quad (1)$$

and

$$\kappa_{ff}(\mathrm{H}) = 1.77 \times 10^{-14} \frac{n_e n_p}{\nu^2 T_e^{3/2}} g(\nu, T_e) \quad (2)$$

where ν is the cyclic frequency of the radiation, n_e, n_p, and n_{HI} are the number densities of electrons, protons, and neutral hydrogen atoms, respectively, T_e is the effective temperature of the particles, and $g(\nu, T_e) \approx (\sqrt{(3)}/\pi) \log[5 \times 10^7 (T_e/\nu)]$ is the Gaunt factor for $T_e \lesssim 3.6 \times 10^5$ K. The relative contributions of H and H^- free-free opacity to the total is, of course, wavelength dependent. On the basis of the semi-empirical chromospheric models of Vernazza, Avrett, and Loeser[70] (hereafter, VAL), for example, H and H^- free-free opacity contribute roughly equally at an optical depth $\tau \approx 1$ when the wavelength $\lambda = 1$ mm. However, H free-free opacity contributes the bulk of the opacity when $\lambda = 3$ mm, and it completely dominates for wavelengths longward of a few millimeters. It is clear that higher frequencies probe denser regions of the solar atmosphere. In the case of H free-free absorption it is also clear that cool material contributes more strongly than hot material.

A second advantage of observations in the millimeter and submillimeter wavelength regime is that the continuum forms under conditions of local thermodynamic equilibrium (LTE), so the source function is Plankian. Third, since $h\nu/k_BT \ll 1$, the Rayleigh-Jeans approximation holds and the observed flux is linearly proportional to the brightness temperature of the emitting material.

The primary disadvantage to working in this wavelength regime is that existing single dish instruments are diffraction limited to an angular resolution $\theta \approx 25''\lambda/D$ where λ is the wavelength in mm and D, is the aperture in units of 10 m. Interferometric techniques are still relatively immature at millimeter and submillimeter wavelengths. To date, such observations have been largely confined to active phenomena, employing sparse arrays of antennas[76].

2.1. Center-to-Limb Brightness and Limb Extension Measurements

Two classes of experiments have generally been performed with available single dishes. First, the brightness distribution has been measured in one or two dimensions in order to establish the nature of the limb brightening[8,28,32,33,35,36,46,56,57,63,64,69]. Early efforts tended to find little or no evidence for limb brightening except, perhaps, in a narrow annulus at the extreme limb. These results were in conflict with prevailing chromospheric models, leading various authors[5,17,67] to suggest that the chromosphere was "rough" due to the presence of inhomogeneities.

The second type of experiment commonly performed at these wavelengths is to measure the position of the solar limb, most often during a partial or total eclipse of the Sun when the moving knife-edge of the moon can be used to gain an increase in the effective angular resolution in one dimension[9,29,40,48,66,74]. These experiments have shown that the solar limb is far more extended than would be expected on the basis of chromospheric models computed under the assumption of hydrostatic equilibrium. Again, the presence of inhomogeneities (e.g., spicules) has been proposed by a number of authors as a means of accounting for the observed limb extensions.

In recent years, mapping observations have tended to reveal the presence of a general limb brightening at millimeter wavelengths[31,47,49], although not always to an extent that the observations are fully compatible with chromospheric models. The most recent work in this area is that of Lindsey and Roellig[51], Bastian, Ewell, and Zirin[6], and Lindsey *et al.*[54]. In contrast to most previous work, these authors have taken pains to deconvolve the telescope beam – typically characterized by a diffraction limited core and broad wings – from the one- and two-dimensional maps. Near the limb the wings of the beam fall on the sky, leading to a diminished brightness. Beam deconvolution is therefore critical to center-to-limb brightness measurements.

Bastian *et al.*[6] used the Leighton telescope at the Caltech Submillimeter Observatory (CSO) to map the Sun in two-dimensions at 850μm prior to the eclipse of 11 July 1991. Using a fourth-order polynomial fit to the data, they find a quiet Sun center-to-limb brightening of 16%. They also computed the dispersion in brightness from the center to the limb of the solar disk. They conclude that while the degree of limb brightening and the dispersion in brightness is broadly consistent with VAL, the center-to-limb brightness profile differ from VAL in detail. On the basis of a two-dimensional map of a subregion of the solar disk, made at 850μm with the James Clerk Maxwell Telescope (JCMT), Lindsey and Jeffries[52] reached a similar conclusion regarding the dispersion in brightness. In contrast to Lindsey *et al.*[49], Bastian *et al.*find no evidence for an azimuthal dependence for the center-to-limb variation in brightness.

Lindsey *et al.*[54] have analyzed 350, 850, and 1200μm maps of the partial solar disk acquired by the JCMT. They find significant limb-brightening in all three bands. The degree of limb brightening is again comparable with that expected from VAL, but the brightness inside the limb is lower than expected from VAL at 350μm, and is higher than expected at 850 and 1200μm. While the 850μm limb brightening is higher (24%) than that reported by Bastian *et al.*, part of the difference may be accounted for by the choice of fitting function. Using the functional form fitted by Lindsey *et al.*[54], the data of Bastian *et al.*[6] yield a limb brightening of 20%.

In contrast to recent limb brightening measurements, the most recent limb extension experiments[10,21,60] have confirmed previous work, showing that the solar limb is significantly extended at millimeter and submillimeter wavelengths. Braun and Lindsey[12] and Roellig *et al.*[60] have suggested that spicules can account for the observed limb extension. They introduced a spicular component in an *ad hoc* fashion and were able to reproduce the observed variation of limb extension as a function of wavelength between 200μm and 3 mm. However, no models have yet been proposed which introduce inhomogeneities in a self-consistent way[6].

In summary, observations at millimeter and submillimeter wavelengths offer a convenient, linear, thermometer with which to probe the structure of the solar chromosphere. Limb extension measurements, and one- and two-dimensional mapping experiments place strong constraints on chromospheric models. While recent mapping experiments appear to be in rough agreement with the prevailing models in terms of the magnitude of the center-to-limb brightening observed, and in the dispersion of brightness, the two disagree in detail. Furthermore, no models have been able to account for both the limb extension measurements and the observed limb-brightening profiles in a self-consistent way. As will be discussed in §3, microwave observations stand in more stark disagreement with the models.

2.2. Prominence Observations

Several other phenomena have been investigated in the millimeter and submillimeter wavelength regimes. Harrison *et al.*[30] mapped a solar prominence at a wavelength of 1300μm with the JCMT near the time of the eclipse of July 1991. Bastian *et al.*[7] have mapped the same prominence in the 850 and 1250μmbands using the CSO. Bastian *et al.*[7] demonstrate that the prominence emission is optically thin $\tau_{850} = 0.10 - 0.13$ and they infer mean column emission measure of $\langle n_e^2 L \rangle = 1.15 - 2.02 \times 10^{29}$ cm^{-5}. If $L \approx 45,000$ km (as inferred from Hα filtergrams from previous days) and a filling factor $f = 0.03 - 0.3$ is assumed, then $n_e \sim 1 - 3 \times 10^{10}$ cm^{-3}. Under somewhat different assumptions, Harrison *et al.*[30] obtain $n_e \sim 7 \times 10^9$ cm^{-3}.

Bastian *et al.*[7] have also analyzed the nature of the 850μm brightness near Hα filaments and, more generally, near magnetic neutral lines. They find that the brightness near filaments and neutral lines is comparable to or below the brightness of the quiet Sun near the center of the disk. Furthermore, regions associated with filaments and neutral lines do not appear to participate in the general limb brightening observed for the quiet Sun. The brightness decrement in these regions cannot be accounted for in terms of filament cavities[37] nor by the presence of cool absorbing material[37,73]. Bastian *et al.*[7] suggest that the diminished brightness associated with filaments and neutral lines is the result of filament channels, whose distinguishing characteristics are an absence of spicules and network structure. Schmahl, Bobrowsky, and Kundu[61] and Gary, Zirin, and Wang[24] have reached a similar conclusion based on observations at longer wavelengths.

Millimeter- and centimeter-wavelength observations have been used to constrain models of filaments. These will be briefly discussed in §3.

2.3. Sunspot Observations

Lindsey and Kopp[53] have studied the brightness in sunspot umbrae and penumbrae at 350, 850, and 1250μm. Previous work has indicated umbral brightness temperatures which are comparable with that of the quiet Sun[50,52]. In contrast, the superior angular resolution of the JCMT and careful deconvolution of the telescope beam allowed Lindsey and Kopp to show that sunspot umbrae are 1000-2000 K lower in brightness than the quiet Sun. On the other hand, sunspot penumbrae range in brightness from values comparable to the quiet Sun up to 1000 K greater than the quiet Sun, comparable with that in plages.

3. Centimeter and Decimeter Wavelengths

At centimeter and decimeter wavelengths, the magnetic field plays an important role in the emission and absorption of radiation because the frequency of the radiation becomes more nearly comparable to the electron gyrofrequency or low harmonics thereof. That is, ν/ν_B is no longer necessarily negligible, as it is at millimeter and submillimeter wavelengths. H free-free opacity continues to be an important source of opacity at centimeter and decimeter wavelengths, but in the presence of a magnetic field the absorption coefficient (Eqn. 2) must be divided by $\mu_{o,x}(1 \pm (\nu_B/\nu)|\cos\theta|)^2$, where $\mu_{o,x}$ is the refractive index of the ordinary (o-) and extraordinary (x-) magnetoionic modes, $\nu_B = eB/2\pi m_e c = 2.8 \times 10^6 B$ is the electron gyrofrequency, and θ is the angle between the wave vector and the local magnetic field; the $-$ sign refers to the x-mode, and the $+$ sign to the o-mode. Sources which are optically thick to free-free absorption are unpolarized. For optically thin sources, the degree of circular polarization is $\rho_C \approx 2\cos\theta(\nu_B/\nu)$.

In regions of the solar atmosphere where the magnetic field strength is sufficiently high gyroresonance opacity becomes important. Radio waves with frequencies which are low integer multiples of ν_B (i.e., $\nu = s\nu_B$; $s = 1, 2, 3, 4, ...$) may be resonantly absorbed and emitted by electrons. Since, under quiet Sun conditions, the electron distribution function is a Maxwellian, one generally refers to thermal gyroresonance absorption. The thermal gyroresonance absorption coefficient is complicated. However, in the case of quasi-longitudinal propagation ($\theta \lesssim 60°$), with $\mu_{o,x} \approx 1$ and $s \geq 2$, the absorption coefficient simplifies[18]

$$\kappa_{gr} = \left(\frac{\pi}{2}\right)^{5/2} \frac{\nu_p^2}{\nu c} \frac{2}{}\left(\frac{s^2\beta_\circ^2 \sin^2\theta}{2}\right)^{s-1} (1 - \sigma|\cos\theta|)^2 \tag{3}$$

Here $\nu_p^2 = n_e e^2/\pi m_e$ is the square of the electron plasma frequency, $\beta_\circ = v_{th}/c = (kT_e/mc^2)^{1/2}$, θ is the angle between the wave vector and the magnetic field; $\sigma = -1$ for the x-mode and $\sigma = +1$ for the o-mode. The assumptions embodied in the above approximations are not extreme and the expression gives one the ability to place meaningful constraints on local conditions in the source. A detailed treatment of a given source requires care, of course.

The polarization properties of gyroresonance sources depend on s and θ. The critical angle $\alpha_{o,x}$ at which $\tau_{gr} = 1$ in a given harmonic and mode is given by[78]

$$\alpha_x \approx \frac{\sqrt{2}}{s\beta_\circ}\left(2\frac{s!}{s^2}\frac{\nu c}{L_B\nu_p^2}\right)^{\frac{1}{2s-2}}; \qquad \alpha_o \approx \frac{\sqrt{2}}{s\beta_\circ}\left(8\beta_\circ\frac{s!}{s^2}\frac{\nu c}{L_B\nu_p^2}\right)^{\frac{1}{2s}} \tag{4}$$

where the previous assumptions remain in place and $\nu = s\nu_B$. For $\alpha \lesssim \alpha_{o,x}$ at a given harmonic, the optical depth drops precipitously. For conditions typical of the corona in solar active regions one finds that the medium is optically thick to gyroresonance absorption in the x-mode when $s = 3$ and in the o-mode when $s = 2$. Furthermore, it is found that $\alpha_x < \alpha_o$ for the relevant values of s. Hence, under many circumstances, a gyroresonance source is expected to be polarized in the sense of the x-mode in the quasilongitudinal approximation, with a high degree of circular polarization possible.

In contrast to observations in the millimeter and submillimeter regimes, continuum observations at centimeter and decimeter wavelengths yield constraints not only on the effective temperature and number density, but also the magnetic field. However, in order to fully exploit radio observations to constrain the magnetic field polarimetric observations are needed with adequate spatial *and* spectral resolution. These conditions have been difficult to fulfill in practice.

3.1. Quiet Regions

Microwave and decimetric observations of the quiet Sun date back to the 1950's, although it was not until the mid-1970s, with the advent of large interferometric arrays such as the Westerbork Synthesis Radio Telescope (WSRT) and the Very Large Array (VLA), that relatively high-angular resolution observations became possible. Kundu *et al.*[38] were the first to exploit the WSRT to observe the supergranular network at centimeter wavelengths. Gary and Zirin[23] used the VLA to image a quiet regions at both 6 and 20 cm. They found a good correspondence between the supergranular network and the 6 cm map; while a general correspondence was also found at 20 cm, the 20 cm sources were more diffuse and the contrast between the network and cell interiors was significantly reduced in comparison to 6 cm. A comparison of the brightness increment observed between the network and cell interiors showed general agreement with model (VAL) calculations[16] for many of the network sources[23].

A very interesting aspect of the observations of Gary and Zirin[23] is the comparison between the average 6 cm map (10.5 hr) with maps made over shorter times (2.5 hr). The former displays the supergranular network while maps made over short timescales do not. Instead, maps made on a short timescale show the presence of discrete microwave sources. These sources are all located within the average supergranular network. The implication is that the supergranular network is traced out by transient sources at microwavengths.

More recently Gary *et al.*[24] observed the quiet Sun at 3.6 cm. A good correspondence was again found with the supergranular network and the brightness increment between cell interiors and the and the network was again consistent with VAL for many elements. However, the absolute brightness of the 3.6 cm measurements are not in agreement with VAL, the models producing substantially more brightness than observed.

Well-calibrated brightness measurements of the quiet Sun at radio wavelengths have been problematic over the years. A wide variety of instruments and techniques have been employed, leading to a large scatter in the reported values. The situation was remedied by Zirin, Baumert, and Hurford[79], who used the Owens Valley frequency-agile interferometer to establish the spectrum of the average quiet Sun between 1.7–30 cm, calibrated against the moon. These authors again find a significant discrepancy between the observed brightness temperature spectrum and that resulting from the VAL models which imply significantly higher brightnesses than are observed.

In summary, microwave observations of the quiet Sun have drawn attention to the fact that chromospheric models, largely based on UV and XUV observations over the relevant heights, are at odds with the radio observations. Either our understanding of the relevant physics is incomplete, or assumptions about the structure of the solar chromosphere are wrong, or both.

3.2. Prominences and Filaments

Prominences and filaments have also been observed at centimeter and decimeter wavelengths since the 1950's. Filaments are associated with depressions in the radio brightness. While early observations suffered from poor angular resolution, Raoult, Lantos, and Fürst[59] were able to demonstrate that for wavelengths $\lambda \lesssim 6$ cm "radio filaments" do not differ substantially in size from their Hα counterparts. More recent observations have exploited the WSRT and the VLA[22,24,34,39,58]. These studies have demonstrated that while there is a good association between Hα filaments and depressions in the radio brightness at centimeter wavelengths, the correlation is one-way in the sense that while all Hα filaments have associated depressions in radio brightness, not all radio depressions have an Hα counterpart.

Observations of prominences and filaments in the millimeter and centimeter wavelength bands have most commonly been interpreted in terms of prominence/corona transition region models[14,15,22,39,58] wherein the cool, dense prominence is assumed to be optically thick. It is further assumed to be surrounded by a thin sheath through which the temperature increases to coronal values. While these models have been successful in accounting for the radio spectrum of Hα filaments between 3 mm and 20 cm, they have been inconsistent with UV observations to the extent that the models imply far less UV radiation than is observed[58]. In this respect, prominence models suffer a problem similar to that outlined for the quiet Sun.

3.3. Active Regions

When observed at centimeter wavelengths, a typical active region consists of bright compact sources embedded in diffuse emission. The compact components are generally associated with sunspots and possess brightness temperatures $T_B \sim$ a few $\times\, 10^6$ K. They are due to gyroresonance emission at low harmonics of the electron gyrofrequency[1,2,41,68]. The diffuse emission at these wavelengths has a lower brightness ($T_B \lesssim 10^6$ K), is associated with chromospheric plages, and results from thermal H free-free emission. The gyroresonance components are optically thick at coronal heights at these wavelengths, whereas the diffuse free-free component is not. At longer wavelengths, the diffuse component predominates and compact sources are not seen, in general. This is because the free-free opacity is sufficient to render the corona optically thick[19,42] at wavelengths $\gtrsim 10 - 20$ cm at heights of $\sim$ a few $\times 10^4$ km above the photosphere[4] although it has sometimes been suggested[13,77] that gyroresonance may also contribute to the total emission at these wavelengths.

Joint radio/soft X-ray observations have been a particularly important tool for the study of solar active regions. A key issue involves the the interpretation of the observed radio brightness temperatures in the diffuse (free-free) component and the electron temperatures inferred via X-ray observations. Comparisons made between VLA observations at 6 and/or 20 cm and SMM/XRP observations reveal brightness temperatures that are significantly lower than the effective temperature of the X-ray emitting material[13,43,55,62,75]. In order to reconcile the lack of correlation between the 6 and 20 cm radio sources and/or the discrepancies between observed and expected radio brightness temperatures, it has been suggested that cool plasma ($T_B \lesssim 5\times 10^5 K$) in the form of layers[13,55,62] or sheaths around hot loops[43] may be present in the solar corona which absorbs the underlying radio emission and reduces the observed temperatures.

Recently, Vourlidas[71] and Vourlidas and Bastian[72] have used multiband VLA observations to show that radio observations of solar active regions may be reconciled with soft X-ray observations by assuming the coronal medium above active regions is thermally and spatially inhomogeneous. They have developed a simple slab model wherein many layers are assigned temperatures between $10^5 - 10^7$ K in a random fashion. The density of these layer is constrained by requiring the model differential emission measure $\phi(T)$ to conform to $\phi(T) = \alpha T^\beta$ up to some temperature T_{max}, beyond which it turns over sharply. The slab model yields brightness temperature spectra which are consistent with that observed in those areas of solar active regions emitting thermal free-free emission.

Gary and Hurford[25] have obtained broadband, spectroscopic, microwave observations of solar active regions with the OVRO solar array. Broadband imaging spectroscopy is a particularly powerful tool for extracting the physical parameters of interest from the observed brightness. Briefly, the spectrum obtained in a given resolution element is composed of contributions from free-free and gyroresonance absorption, with one typically dominating over the other. In the optically thick part of the spectrum, $T_B \approx T_e$, irrespective of the emission mechanism. If the spectrum is due to free-free absorption, $\tau \approx 0.21\nu^{-2}T_e^{-3/2} \int n_e^2 dl$ so that $\int n_e^2 dl \approx 4.8\nu_{ff}^2 T_e^{3/2}$ cm^{-5}, where ν_{ff} is the spectral turnover frequency (where $\tau \sim 1$). If the spectrum is dominated by gyroresonance absorption, one can deduce the maximum coronal magnetic field strength along the line-of-sight from the break in the brightness temperature spectrum to low values ($B_{max} = \nu_{gr}/2.8 \times 10^6 s$). An upper limit to the column emission measure is derived from gyroresonance spectra by calculating the maximum value for which $\nu_{ff} < \nu gr$. Performing this analysis for all resolution elements spanning an active region, Gary and Hurford have constructed two-dimensional maps of T_e, column emission measure, and B in the low-corona.

Recent advances in the nature of the gyroresonance component have been made by Lee, Hurford, and Gary[44], and Lee, Gary, and Hurford[45] who obtained broadband, spectroscopic microwave imaging of an isolated sunspot as it rotated from the Sun's center to the limb, again with the OVRO solar array. These authors were able to determine the magnetic field strength and radial variation at coronal heights, showing that the field was more radially confined than would be expected on the basis of an extrapolation of a dipole model. The height of the resonant layer was also determined, as were the magnetic scale height, and the electron number density above the layer.

An integrated model of active regions which accounts for the free-free radio and SXR emission, and the thermal gyroresonance component, remains a goal for future work.

4. Concluding Remarks

This brief review has attempted to outline some of the diagnostic uses of radio waves for probing the nature of the solar atmosphere. Continuum opacity mechanisms have been discussed and examples of the manner in which radio observations are exploited in practice have been presented.

At millimeter and submillimeter wavelengths, the continuum arises under conditions of LTE. Such observations may therefore be used to probe the thermal structure of the solar atmosphere from the vicinity of the temperature minimum up into the low-to-middle chromosphere. At centimeter and decimeter wavelengths, H free-free opacity continues to play an important role. In addition, gyroresonance opacity plays an important role wherever the magnetic field strength is sufficiently high, offering a means of constraining the magnetic field strength and orientation in the upper layers of the solar atmosphere.

While radio observations provide the means of placing useful constraints on a variety of physical parameters in the outer atmosphere of the Sun for a number of phenomena (quiet Sun, filaments, active regions), radio observations also present a problem. In particular, the observed radio emission falls short of that expected from semi-empirical models of the solar chromosphere and transition region and from models of prominences. The shortfall apparently increases with radio wavelength. Obviously, this difficulty must be confronted in future modeling efforts. A proper accounting of inhomogeneities in the solar atmosphere is likely to play a key role in reconciling observations at radio, UV, EUV, and X-ray wavelengths.

I thank D.E. Gary for critically reading the review. The National Radio Astronomy Observatory is operated by Associated Universities, Inc., under cooperative agreement with the National Science Foundation.

1. Alissandrakis, C.E., Kundu, M.R., and Lantos, P., *Astron. Astrophys.* **82** (1980) 30.
2. Alissandrakis, C.E., and Kundu, M.R., *Astrophys. J.* **253** (1982) L49.
3. Alissandrakis, C.E., and Chiuderi-Drago, F. 1994, *Astrophys. J. Lett.* **428** (1994) L73.
4. Aschwanden, M.J., and Bastian, T.S., *Astrophys. J.* **426** (1994) 434.
5. Athay, R.G. 1959, in *Paris Symposium on Radio Astronomy* (Stanford Univ. Press, Stanford, 1959), 98.
6. Bastian, T.S., Ewell, M.W., and Zirin, H., *Astrophys. J.*, **415** (1993) 364.
7. Bastian, T.S., Ewell, M.W., and Zirin, H., *Astrophys. J.*, **418** (1993) 510.
8. Beckman, J.E., and Clark, C.D., *Solar Phys.* **29** (1973) 25.
9. Beckman, Lesurf, J.C.G., and Ross, J., *Nature* 254 (1975) 38.
10. Belkora, L., Hurford, G.J., Gary, D.E., and Woody, D.P. *Astrophys. J.*, **400** (1992) 692.
11. Berger, P.S., and Simon, M., *Astrophys. J.* **171** (1972) 191.
12. Braun, D., and Lindsey, C.A., *Astrophys. J.* **320** (1987) 898.
13. Brosius, J.W., Willson, R.F., Holman, G.D., and Schmelz, J.T., *Astrophys. J.* **386** (1992) 347.
14. Butz, M., Fürst, E., Hirth, W., and Kundu, M.R., *Solar Phys.* **45** (1975) 125.
15. Chiuderi, C., and Chiuderi-Drago, F., *Solar Phys.* **132** (1991) 81.

16. Chiuderi-Drago, F. Kundu, M. and Schmahl, E.J., Sol. Phys. **85** (1983) 237.
17. Coates, R.J., *Astrophys. J.* **128** (1958) 83.
18. Dulk, G.A., *Ann. Rev. Astron. Astrophys.* **23** (1985) 169.
19. Dulk, G.A., and Gary, D.E., *Astron. Astrophys.* **124** (1983) 103.
20. Dupree, A.K., *Astrophys. J.* Lett., **152** (1968) L125.
21. Ewell, M.W., Jr., Zirin, H., Jensen, J.B., and Bastian, T.S., *Astrophys. J.* **403** (1993) 426.
22. Gary, D.E., in *Coronal and Prominence Plasmas* (NASA CP 2442, 1986) 121.
23. Gary, D. E., and Zirin, H., *Astrophys. J.* **329** (1988) 991.
24. Gary, D. E., Zirin, H., and Wang, H., *Astrophys. J.* **355** (1991) 321.
25. Gary, D. E., and Hurford, G. J., *Astrophys. J.* **420** (1994) 903.
26. Greve, A., *Solar Phys.* **44** (1975) 371.
27. Greve, A., *Solar Phys.* **52** (1977) 423.
28. Hachenberg, O. Steffen, P., and Harth, W., *Solar Phys.* **60** (1978) 105.
29. Hagen, J.P., Swanson, P.N., Haas, R.W., Wefer, F.L., and Vogt, R.W., *Solar Phys.* **21** (1971) 286.
30. Harrison, R.A., *et al.*, *Astron. Astrophys.*, **274** (1993), L9.
31. Horne, K., Hurford, G.J., Zirin, H., and De Graauw, T., *Astrophys. J.* **244** (1981) 340.
32. Kawabata, K., Fujishita, M., Kato, T., Ogawa, H., and Omodaka, T., *Solar Phys.* **65** (1980) 221.
33. Kundu, M.R., *Solar Phys.* **21** (1971) 130.
34. Kundu, M.R., in *Coronal and Prominence Plasmas* (NASA CP 2442, 1986), 109.
35. Kundu, M.R., and Liu, S., *Solar Phys.* **49** (1976) 267.
36. Kundu, M.R., Liu, S., and McCullough, T.P. Sol. Phys. **51** (1977) 321.
37. Kundu, M.R., Fürst, E. Hirth, W., and Butz, M., *Astron. Astrophys.* **62** (1978) 431.
38. Kundu, M.R., Rao, A.P., Erskine, F.T., and Bregman, J.D., *Astrophys. J.* **234** (1979) 1122.
39. Kundu, M.R., Melozzi, M., and Shevgaonkar, R.K., *Astron. Astrophys.* **167** (1986) 166.
40. Labrum, N.R., Archer, J.W., and Smith, C.J., Sol. Phys. **59** (1978) 331.
41. Lang, K.R., and Willson, R.F., *Astrophys. J.* **255** (1982) L111.
42. Lang, K.R., Willson, R.F., and Rayole, J., *Astrophys. J.* **258** (1982) 384.
43. Lang, K.R., Willson, R.F., Smith, K.L., and Strong, K.T., *Astrophys. J.* **322** (1987) 1035
44. Lee, J.W., Hurford, G.J., and Gary, D.E., *Solar Phys.* **144** (1993) 45.
45. Lee, J.W., Gary, D.E., and Hurford, G.J., *Solar Phys.* **144** (1993) 349.
46. Lindsey, C.A., and Hudson, H.S., *Astrophys. J.* **203** (1976) 753.
47. Lindsey, C.A., Hildebrand, R.H., Keene, J., and Whitcomb, S.E., *Astrophys. J.* **248** (1981) 830.
48. Lindsey, C.A., et al., *Astrophys. J.* **264** (1983) L25.
49. Lindsey, C.A., De Graauw, T., De Vries, C., and Lindholm, S., *Astrophys. J.*

277 (1984) 24.
50. Lindsey, C.A., *et al.*, *Astrophys. J.* **353** (1990) L53.
51. Lindsey, C.A., and Roellig, T.L., *Astrophys. J.* **375** (1991) 414.
52. Lindsey, C.A., and Jeffries, J.T., *Astrophys. J.* **383** (1991) 443.
53. Lindsey, C.A., and Kopp, G., *Astrophys. J.* (1995) submitted.
54. Lindsey, C.A., Kopp, G., Clark, A., and Watt, G., *Astrophys. J.* (1995) submitted.
55. Nitta, N., *et al.*, **374** (1991) 374.
56. Noyes, R.W., Beckers, J.M., Low, F.J., *Solar Phys.* **3** (1968) 26.
57. Righini, G., and Simon, M., *Astrophys. J.* **203** (1976) L95.
58. Rao, P.A., and Kundu, M., *Solar Phys.* **55** (1977) 161.
59. Raoult, A., Lantos, P., and Fürst, E., *Solar Phys.* **61** (1979) 335.
60. Roellig, T.L., et al., *Astrophys. J.* **381** (1991) 288.
61. Schmahl, E.J., Bobrowsky, M., and Kundu, M., *Sol. Phys.*, **71** (1981) 311.
62. Schmelz, J.T., Holman, G.D., Brosius, J.W., and Gonzalez, R.D., *Astrophys. J.* (1992) .
63. Shimabukuru, F.I., *Solar Phys.* **12** (1970) 438
64. Shimabukuru, F.I., *Solar Phys.* **18** (1971) 247
65. Shimabukuru, F.I., and Wilson, W.J., *Astrophys. J.* **183** (1973) 1025.
66. Shimabukuru, F.I., Wilson, W.J., Mori, T.T., and Smith, P.L., *Solar Phys.* **40** (1975) 359.
67. Simon, M., and Zirin, H., *Solar Phys.* **9** (1969) 317.
68. Strong, K.T., Alissandrakis, C.E., and Kundu, M.R., *Astrophys. J.* **277** (1984) 865.
69. Suzuki, I., Kawabata, K., Ogawa, H., *Solar Phys.* **46** (1976) 205.
70. Vernazza, J.E., Avrett, E.H., and Loeser, R., *Astrophys. J.* **45** (1981) 635 (VAL).
71. Vourlidas, A., Masters Thesis (NMIMT, 1983).
72. Vourlidas, A., and Bastian, T.S., *Astrophys. J.* (1995) submitted.
73. Vršnak, B., *et al.*, *Solar Phys.* **137** (1992) 67.
74. Wannier, P.G., Hurford, G.J., and Seielstad, G.A., *Astrophys. J.* **264** (1983) 660.
75. Webb, D.F., Holman, G.D., Davis, J.M., Kundu, M.R., and Shevgaonkar, R.K., *Astrophys. J.* **315** (1987) 716.
76. White, S.M., and Kundu, M., *Solar Phys.* **141** (1992) 347.
77. White, S.M., Kundu, M.R., Gopalswamy, N., *Astrophys. J.* **78** (1992) 599.
78. Zheleznyakov, V.V., *Radio Emission of the Sun and Planets* (Pergamon, Oxford, 1971)
79. Zirin, H., Baumert, B.M., and Hurford, G. J., *Astrophys. J.* **370** (1991) 779.

SOLAR BURST EMISSIONS IN SUB-MM/FAR-IR CONTINUUM

PIERRE KAUFMANN
Center for Radio-Astronomy and Space Applications (CRAAE) and
Sao Paulo State University of Campinas (UNICAMP/NUCATE)
Sao Paulo, SP, Brazil

ABSTRACT

Few measurements on solar sub-mm and infrared continuum emissions were made in the past. Attempts to measure bursts emissions were inconclusive. The missing observational information in that part of the electromagnetic spectrum may bring essential contributions for the understanding of physical processes in quiet, active and flaring phenomena. Certain bursts exhibit spectra with fluxes increasing shorter mm-waves, suggesting a maximum in the sub-mm/IR range. At the other end, the spectral features of white light flare emission are not known in the IR. Subsecond burst time structures might become pronounced in the sub-mm-IR range. The fast time scales bring severe constraints for interpretation. One possibility assumes the production of multiple discrete compact synchrotron sources made up by ultra-relativistic electrons, with emission peaking somewhere in the sub-mm/IR part of the spectrum. Inverse Compton effect actuate in those sources, turning down the electrons' energies and producing concurrent X-ray spikes. This short lived "pre-impulsive" burst phase may be better investigated in sub-mm/IR observations. A new original project for a solar sub-mm-wave telescope was proposed to study these problems at two frequencies, with millisecond time resolution, high sensitivity and using multiple beams for dynamic imaging of bursts. It is being evaluated by Brazilian research agencies, and receives collaboration from Switzerland's University of Bern solar group and Argentina groups IAFE and CASLEO, for installation in El Leoncito site, Argentinean Andes.

The electromagnetic spectrum of solar emission in continuum remains almost unexplored in the range covering sub-mm to far infrared wavelengths (about 3000-20 microns). Several far IR observations were reported on emissions of the quiet Sun and quiescent active regions, such as on the brightness temperature of the center of the Sun;[9,10] on the 300 second oscillations;[12] on limb brightening, sunspots, prominences;[17,18] and emission from active centers.[8]

Attempts to measure flares in the sub-mm/IR continuum were fewer and inconclusive. It is possible that the 1200 microns variable emission of an active region, reported by Clark and Park,[3] was the first indication of important continuum emission, interpreted as being non-thermal.[1]

There are various indications that solar burst emission in the sub-mm/IR range can be significant and, in some cases, more intense than at microwaves, as shown in

Figure 1. The event observed by Kaufmann et al.[13] with high sensitivity and time resolution, exhibited flux increase at mm-waves only. A similar burst spectrum was suggested by Shimabukuro.[21] Composite spectra, with a suggested mm to sub-mm-waves component were found by several authors. Flat spectra in the microwave range are not uncommon. A statistical distribution of 19-35 GHz spectral indices observed by Bern University suggests that 75% of events have a flat or intensity increasing with frequency spectra.[5]

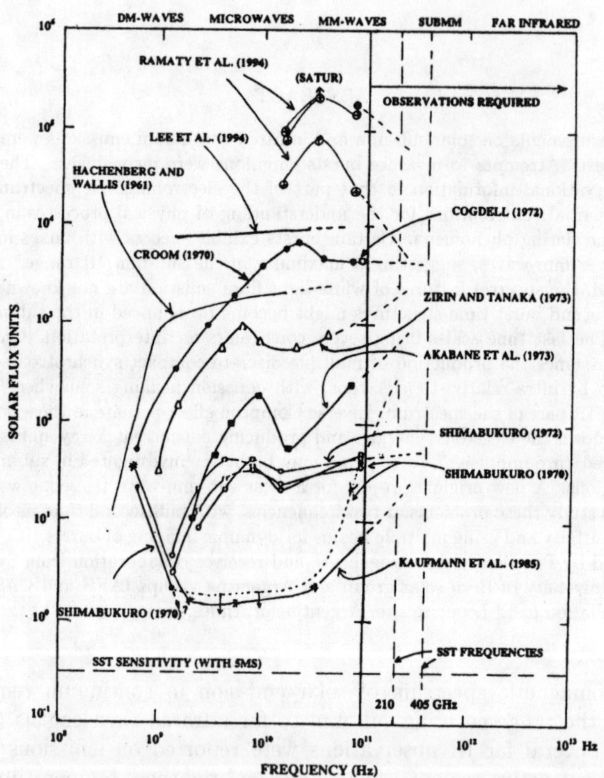

Figure 1: Examples of solar burst spectra with suggested important contribution in the sub-mm range. The frequencies to be used in the new solar sub-mm-wave telescope, and its sensitivity are shown.

Figure 2 shows the complete electromagnetic spectrum for a flare emission in continuum.[15] Various thermal and non-thermal models may fit the observed emission spectral features at frequencies lower and higher than the sub-mm/IR range. Their investigations depend critically on the still unknown continuum measurements in that range.

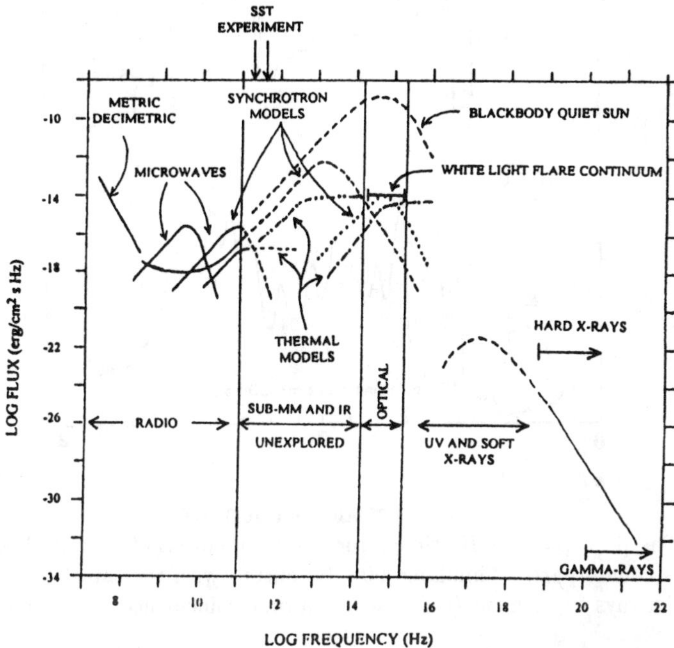

Figure 2: The complete electromagnetic spectrum of a solar flare continuum emission.

The most critical boundary condition is set by the time scales observed in continuum emission. The event observed at mm-waves only, shown in Figure 1,[13] displayed fast multiple spikes lasting less than 100ms and well correlated to hard X-ray fast time structures (Figure 3).

The burst time scales are too small to be explained by synchrotron (mm-sub-mm) or bremsstrahlung (X-rays). One possible interpretation was suggested assuming that bursts are initially made up by compact synchrotron sources with spectra peaking somewhere in the IR range, with lifetimes shortened by efficient inverse-Compton (IC) action, producing fast X-rays pulses.[14] Higher sensitivity solar burst measurements by BATSE experiment on GRO satellite are indeed showing that sub-second time structures are very common at hard X-rays.[19]

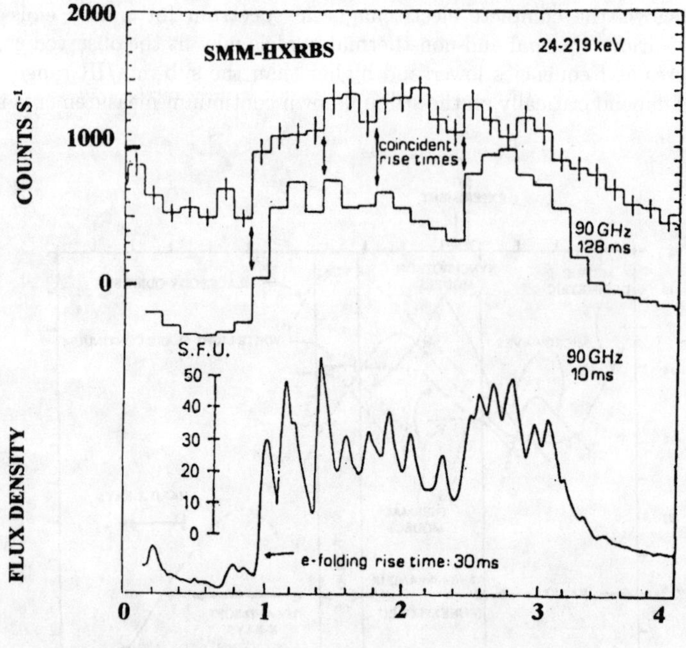

Figure 3: The principal 90 GHz time structure of the event of May 21 1984 and its hard X-rays counterpart.[13] Simulated 90 GHz output with the same time resolution of hard the X-rays experiment (top) indicate a time coincidence better than 100ms.[6]

This model can be reconciled with current interpretations on solar flare emission assuming the existence of a "pre-impulsive" burst phase (Figure 4a), with the primordial compact synchrotron source lasting less than 100ms, during which the IC quenching reduces the electrons' energies into mildly relativistic values, forming beams and producing the better known manufacturing such as the gyrosynchrotron emission and X-rays by bremsstrahlung. (Figure 4b).

Spectra taken on time intervals large compared to the lifetimes of the "pre-impulsive" phase burst sources, may exhibit both components, in the sub-mm-IR and in the microwave range. Such composite spectra are suggested for various examples of bursts measured in the past, at the microwave-mm waves range, shown in Figure 1.

A new project to investigate solar continuum emission at two sub-mm- wavelengths (210 GHz and 405 GHz) is being evaluated by Brazilian agencies. It will receive collaboration from Switzerland's University of Bern solar group, and from Argentina IAFE and CASLEO groups, and will be installed at El Leoncito Observatory in the Argentinean Andes.

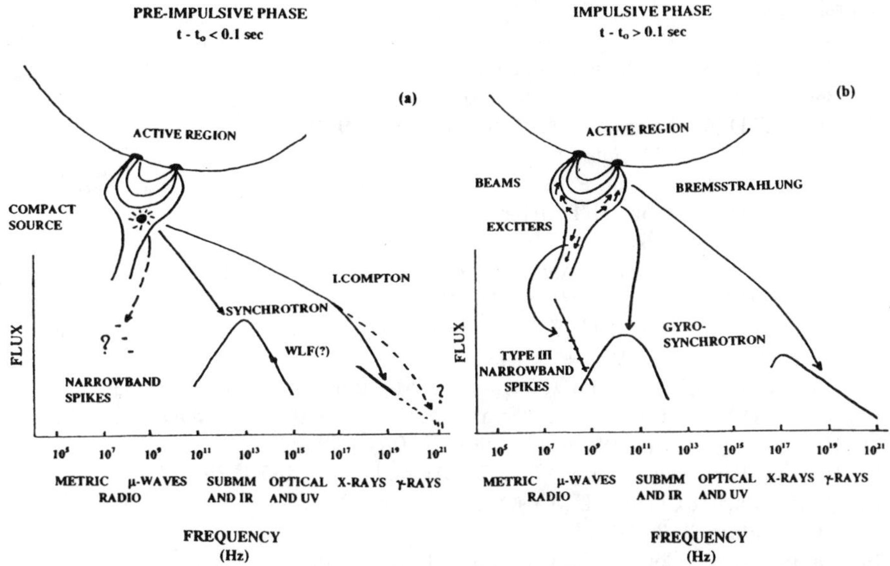

Figure 4: Model suggested sequence of phases for a simple solar burst. In the pre-impulsive phase (a) a compact synchrotron source produces a continuum spectrum centered in the sub-mm/IR. Inverse Compton effect in that source reduces the electrons' energies in less than about 100ms, producing the corresponding X-ray spike. Some gamma lines, narrowband decimetric/metric spikes [2] and white light emissions may also occur in this phase. The electrons' energies are reduced to mildly relativistic levels and will immediately produce the impulsive phase (b), which description is well known.

Acknowledgments

The author acknowledges AFOSR for the support received to attend this Workshop. CRAAE is jointly operated by Brazilian institutions USP, INPE, Mackenzie and UNICAMP.

References

1. Beckman, J.E.: 1968,*Nature*, **220**, 52.
2. Benz, A.O.: 1994, *Space Sci. Rev.*, **68**, 135.
3. Clark, C.D., and Park, W.M.: 1968, *Nature*, **219**, 922.
4. Cogdell, J.R.: 1972, *Solar Phys.*, **22**, 147.
5. Correia, E., Kaufmann, P., and Magun, A.: 1994, in "Infrared Solar Physics" (ed. by D.M. Rabin, J.T. Jefferies and C. Lindsay), Kluwer Acad. Publ., Dordrecht, NL, 125.
6. Costa, J.E.R., and Kaufmann, P.: 1986, *Solar Phys.*, **104**, 253.
7. Croom, D.L.: 1990, *Solar Phys.*, **15**, 414.
8. Degiacomi, K., Kneubëhl, F.K., Huguenin, D., and Müeller, E.A.: 1984, Int. J. Infrared and MM-Waves, 5, 643.
9. Eddy, J.A., Lna, P.J., and MacQueen, R.M.: 1969, *Solar Phys.*, **10**, 330.
10. Gezari, D.Y., Joyce, R.R., and Simon, M.: 1973, *Astron. Astrophys.*, **26**, 409.
11. Hachenberg, O., and Wallis, G.: 1961, *Zs.f.Ap.*, **52**, 42.
12. Hudson, H.S., and Lindsay, C.: 1974, *Astrophys. J.*, **187**, L35.
13. Kaufmann, P., Correia, E., Costa, J.E.R., Zodi Vaz, A.M., and Dennis, B.R.: 1985, *Nature*, **313**, 380.
14. Kaufmann, P., Correia, E., Costa, J.E.R., and Zodi Vaz, A.M.: 1986, *Astron. Astrophys.*, **157**, 11.
15. Kaufmann, P.: 1988, *Adv. Space Res.*, **8(11)**, 39.
16. Lee, J.W., Gary, D.E., and Zirin, H.: 1994, *Solar Phys.*, in press.
17. Lindsay, C., Hildebrand, R.H., Keene, J., and Whitecomb, S.E.: 1981, *Astrophys. J.*, **248**, 830.
18. Lindsay, C.: 1994, in "Infrared Solar Physics" (ed. by D.M. Rabin, J.T. Jefferies and C. Lindsay), Kluwer Acad. Publ., Dordrecht, NL, 85.
19. Machado, M.E., Ong, K.K., Emslie, A.G., Fishman, G.J., Meegan, C., Wilson, R., and Paciesas, W.S.: 1993, *Adv. Space Res.*, **13**, 9/175.
20. Ramaty, R., Schwartz, R.A., Enome, S., and Nakajima, H.: 1994, *Astrophys. J.*, in press.
21. Shimabukuro, F.I.: 1970, *Solar Phys.*, **15**, 424.
22. Shimabukuro, F.I., 1972, *Solar Phys.*, **23**, 169.
23. Zirin, H., and Tanaka, K.: 1973, *Solar Phys.*, **32**, 173.

ECLIPSE MEASUREMENTS OF THE DISTRIBUTION OF CO EMISSION ABOVE THE SOLAR LIMB

T.A. CLARK*
Department of Physics and Astronomy
University of Calgary, 2500, University Drive, N.W., Calgary, AB, T2N 1N4 CANADA

C. LINDSEY
Solar Physics Research Corporation
National Solar Observatory, PO Box 26732, Tucson, AZ 85726-6732, USA

D.M. RABIN and W.C. LIVINGSTON
National Solar Observatory, PO Box 26732, Tucson, AZ 85726-6732, USA

ABSTRACT

We observed the limb occultation of near IR solar CO vibration-rotation lines at 4.66 μm (2145 cm^{-1}) during the partial eclipse of 10 May 1994 at the 1.5-m McMath-Pierce Telescope on Kitt Peak. Limb profiles of continuum and CO line intensity have been obtained with an effective spatial resolution of $0.1''$ over a height range of $11''$ around the limb. CO lines change from absorption to emission at the solar limb and this emission extends beyond the limb by heights which depend linearly upon line strength. The effective limb for the strong ^{12}CO 3-2 R14 line occurs at only $0.56 \pm 0.04''$ above the IR continuum limb and no line emission is seen beyond about $1.1''$. This small limb extension of CO emission places important constraints on models which incorporate a cool component into a structured solar atmosphere.

1. Introduction

The enigma of cool carbon monoxide existing within a hotter chromosphere has puzzled solar observers for many years. The initial observation by Noyes and Hall (1972a,b)[1,2] of very low central intensity in near-IR vibration-rotation lines of CO has since been reinforced and extended by many subsequent experiments (e.g. Ayres, 1978,[3] Ayres and Testerman, 1981[4]). Since the vibration-rotation transitions are in LTE,(Ayres and Wiedemann, 1989[5]) these observations indicate very low CO gas temperatures at line source height. To account for these observations, solar atmospheric models have been developed in which cool CO gas is assumed to survive within supergranular cells. These cells are then surrounded by a hot and magnetically active chromospheric network which undergoes the 'classical' temperature inversion in the overlying chromosphere (Ayres 1990)[6].

*Visiting Astronomer, National Solar Observatory, National Optical Astronomy Observatories, which are operated for the National Science Foundation by the Association of Universities for Research in Astronomy.

One atmospheric characteristic which has eluded observational measurement until recently is the height distribution of this CO within the solar atmosphere. Within the last two years, the main spectrograph of the McMath-Pierce Telescope has been upgraded to measure the solar IR spectrum with high spectral and spatial resolution, initially with a single IR detector (Solanki et al., 1994[7]) but more recently with an IR array (Uitenbroek et al., 1994[8]). However, the spatial resolution of limb measurements at these wavelengths is limited to 1–2″ by seeing fluctuations in the Earth's atmosphere and by diffraction by the 1.5-m telescope aperture.

The present paper describes observations of the 10 May 1994 partial solar eclipse with the IR array on this spectrograph. The near–annular geometry of this eclipse provided the opportunity to obtain limb scans of CO line and continuum intensities with a resolution far better than that achievable by direct observation. In this case, the attainable resolution was governed by the acute inclination (6–12°) of the lunar limb to the solar limb at the chosen point of contact. A selection of seven contact points were monitored in sequence as the eclipse progressed and these measurements have provided excellent profiles of the height distribution of CO emission above the solar limb and its dependence upon line strength.

2. Observational Technique

For this experiment, the Amber 256^2 InSb array and the 13.7-m vertical spectrograph of the McMath-Pierce Telescope were configured to measure the solar IR spectrum over a wavelength range of 2.4 cm^{-1}. Before each contact, the spectrograph was rotated so that the entrance slit was perpendicular to the solar limb and centered at the point to be occulted. Each observing sequence was limited to 120 frames by the memory capacity of the array control computer. This limitation, when combined with the chosen readout rate of 3.76 frame/sec and lunar limb motions of 0.14 to 0.34″/sec, restricted observations to a height range of between 4″ and 11″ in lunar limb motion. The time required for rotation of the spectrograph, transfer of data to permanent storage and realignment of telescope to the new contact point dictated a cadence of one contact every 3 minutes as the Moon's silhouette traversed the solar disk. Of these contacts, the first four were measured in the spectral range 2142.3–2144.6 cm^{-1}, covering several ^{12}CO and ^{13}CO lines of varying strengths, while the remaining three contacts were measured over the range 2147.7–2150.1 cm^{-1}, approximately centered upon the Pfund β line. The array sampled the spectra at intervals of 0.0093 cm^{-1}. The inclination of the lunar limb to that of the Sun was 12° for the first and last contacts and 6° for the optimuum 3rd and 4th contacts. When combined with the 0.53″ slit-width broadened by telescope diffraction, the effective spatial resolution per frame at the solar limb was 0.1″ for the optimum contacts and 0.2″ for the worst contacts. Since data recording capacity was limited, it was important that the appropriate start time for each eclipse contact be anticipated accurately. This was done by viewing the progress of the occultation on the slit-jaw video monitor.

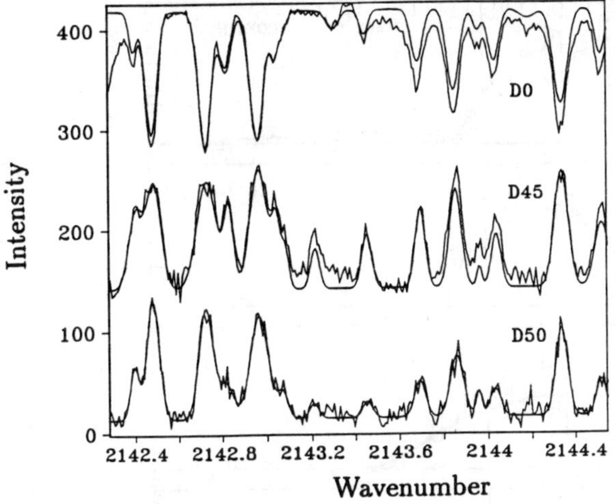

Figure 1: A series of three spectra of CO lines are shown as the eclipse proceeds, from an absorption spectrum inside the IR limb, D0, to emission spectra beyond the limb, D45 and D50.

3. Observations and Data Analysis.

While all seven planned contacts were observed successfully, the present analysis has concentrated upon the 3rd and 4th contacts, events C and D, for which the eclipse geometry was optimum and the passband included only CO lines. In each data sequence, the average of a series of dark frames, taken after the occultation was complete, was subtracted from each of the occultation frames to remove the effects of dark current from the images. The signal was then integrated along the slit direction at each wavelength to give the integrated limb spectrum for each frame. The sequence of these spectra with time represents the occultation curve with a resolution of about 0.1″, the limit set by geometry of the contact and the sampling time. The equivalent solar limb profile for each spectral element was obtained by simple differentiation of these curves.

4. Results

Fig. 1 shows spectra from the optimum contact, event D, with best-fit synthetic spectra superimposed upon them. The upper spectrum, D0, is the average spectrum over a 1″ region centered 2.5″ inside the IR continuum limb and shows an absorption spectrum with ^{12}CO and ^{13}CO lines in about the same intensity ratios as in the on-

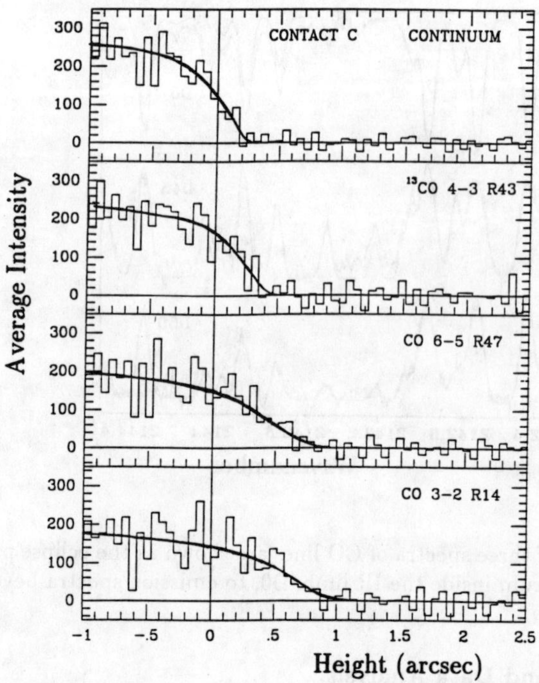

Figure 2: Limb profiles of the IR continuum and central intensities of several CO lines are shown, overlaid by eye estimates of the smooth limb curves. The position of the IR continuum limb is defined as the point of inflexion of the continuum limb profile.

disk ATMOS spectrum (Farmer and Norton, 1989[9]). The middle spectrum, D45, is an average over 0.4″ in height centered 0.2″ above the IR continuum limb. Relative to the strong ^{12}CO lines in this spectrum, the weak ^{13}CO lines at 2142.39, 2143.03, 2143.21, 2143.45 and 2144.035 cm^{-1} have become much more intense than in the on-limb spectrum, as expected from saturation of the strong lines. However, as can be seen in D50, the spectrum averaged over 0.4″ in height centered at 0.6″ above the continuum limb, the strength of these lines fades before the strong ^{12}CO lines. At 1″ beyond the limb, only the remnants of strong ^{12}CO lines remain.

Fig. 2 shows limb profiles for the third measured contact, event C. The upper graph shows the IR continuum, averaged over 2 regions where no lines are apparent on the ATMOS spectrum. The other 3 graphs show the profiles in the cores of 3 CO lines. These are averages over 3 columns in the detector array, a range of 0.028 cm^{-1} in wavenumber. Equivalent profiles for the 2 optimum contacts, C and D, were found to be virtually identical for these lines. To obtain representative values

Table 1: Limb Extension for Several CO Lines

Line	Limb Extension
^{12}CO 3-2 R14	0.56 ±0.040″
^{12}CO 6-5 R47	0.33 ±0.040″
^{13}CO 4-3 R43	0.13 ±0.035″

for the limb extension for each line, the line core profile was normalized to that of the continuum on the limb and offset in position to match the continuum profile. This offsetting procedure yielded the values of mean limb extension for the CO lines shown in Table 1. These follow an approximately linear relation with ATMOS line center intensity. When compared to previous non-eclipse measurements of line limb profiles, (Solanki et al., 1994[7], Uitenbroek et al., 19948) the eclipse data show smaller extensions beyond the limb and, more particularly, a much sharper drop-off with height above the continuum limb.

There is evidence in these data of a broadening of the CO lines as the extreme limb is approached, perhaps as a continuation of the trend reported by Ayres (1978)[3]. The broadening continues in the emission lines just above the IR continuum limb, but appears to be followed by a narrowing again in the tail of the limb distribution. Further analysis of this effect is being carried out.

5. Conclusions

The sharp truncation of CO emission profiles indicated by the occultation measurements imposes strong constraints on models that propose to account for the highly structured nature of the solar atmosphere. The rapid disappearance of CO at heights above 700 km suggests that there simply exists no chromospheric material at higher levels sufficiently cool to allow CO.

Ayres has used numerical simulation to explore the consequences of thermal bifurcation models on the behavior of CO lines near the extreme limb. In his latest models, Ayres (1995—these proceedings) has considered the eclipse results and arrives at limb profiles that more closely match those presented here.

A strong correlation is known to exist between the non-radiative energy flux that supports elevated chromospheric temperatures and the presence of strong magnetic fields. This correlation is sufficient to suggest that the role of magnetic fields is integral to non-radiative heating and that elevated chromospheric temperatures are *confined* to magnetic regions (e.g., Solanki and Steiner, 1990)[10].

The model of Steiner (1994)[11] represents the magnetic geometry of the solar atmosphere by a distribution of flux tubes concentrated at the boundaries of supergranules. These flux tubes are narrow in the photosphere but rapidly spread outward with increasing height to occupy the entire volume of the overlying chromosphere above about 700 km, the height of the magnetic canopy overlying the supergranule interior in the model. A feature of the Steiner model is a sharp temperature discontinuity at the boundary separating the cool supergranule photosphere, the top of which conduces an abundance of CO, and the overlying heated magnetic canopy. The idea of such a boundary appears to be supported by the rapid disappearance of CO emission at the relatively low levels indicated by our measurements. The total absence of CO emission detected above 1,200 km strongly suggests a middle chromosphere that is devoid of cool, non-magnetic material.

6. Acknowledgements

It is a pleasure to thank David Jaksha, Claude Plymate and Paul Hartmann for their invaluable assistance during this experiment and Kim Gillies for programming assistance prior to the eclipse. T.A. Clark acknowledges support for this research from the NSERC of Canada. C. Lindsey holds a visiting appointment at the National Solar Observatory and has benefited greatly from cooperation between NSO and the Solar Physics Research Corporation. This work was done under the support of NSF Grants ATM-9311451 and ATM-9122073.

7. References

1. Noyes, R.W. and Hall, D.N.B. *Ap. J.* **176** (1972) L89.
2. Noyes, R.W. and Hall, D.N.B. *Bull. AAS* **4** (1972) 389.
3. Ayres, T.R. *Ap. J.* **225** (1978) 665.
4. Ayres, T.R. and Testerman, L. *Ap. J.* **245** (1981) 1124.
5. Ayres, T.R. and Wiedemann, G.R. *Ap. J.* **338** (1989) 1033.
6. Ayres, T.R. *Solar Photosphere: Structure, Convection and Magnetic Fields,* J.O Stenflo (ed.) (1990) 23.
7. Solanki, S.K., Livingston, W.C. and Ayres, T.R. *Science* **263** (1994) 64.
8. Uitenbroek, H., Noyes, R.W. and Rabin, R.M. *Ap. J.* **432** (1994) L67.
9. Farmer, C.B. and Norton, R.H. *NASA Ref. Publ. 1224:* A High- Resolution Atlas of the Infrared Spectrum of the Sun and the Earth Atmosphere from Space, Vol 1, (1989)
10. Solanki,S. and Steiner, O. *Infrared Solar Physics: IAU Symp. 154,* Rabin, Jefferies, Lindsey (eds.) Kluwer Academic Publishers, (1994) 407.
11. Steiner, O. *Infrared Solar Physics: IAU Symp. 154,* Rabin, Jefferies, Lindsey (eds.) Kluwer Academic Publishers, (1994) 423.

DETECTION OF THE HI n = 20–19 SUBMILLIMETER LINE IN EMISSION AT THE SOLAR LIMB USING A POLARIZING FT SPECTROMETER.

T.A. CLARK
Physics and Astronomy Department
University of Calgary, 2500, University Drive, N.W. Calgary, AB T2N 1N4 Canada.

D.A. NAYLOR
Physics Department
University of Lethbridge, Lethbridge, AB, Canada.

and

G.R. DAVIS
Physics Department
University of Saskatchewan, Saskatoon, SK, Canada.

ABSTRACT

An enhancement of emission over that at the solar disk center has been detected near the solar limb at 29.622 cm^{-1}. This peak, at the position of the n = 20–19 Rydberg transition in hydrogen, has an intensity of up to 5% of the continuum intensity and line width of 0.07 cm^{-1}. This detection places a new lower limit upon the effective ionization level of hydrogen gas in the low chromosphere.

1. Introduction

The discovery 15 years ago of intense emission lines in the solar infra-red spectrum (Murcray et al., 1981[1]; Brault and Noyes, 1983[2]) and the subsequent identification of these as lines from n = 7–6 Rydberg transitions of MgI, SiI and AlI (Chang and Noyes, 1983[3], Chang, 1987[4]) opened up a new and powerful method for the study of atmospheric conditions in the Sun's high photosphere and low chromosphere. Many subsequent investigations, particularly of the strong limb brightening and high Zeeman sensitivity of these lines, have significantly extended this initial study. One particular advance, the measurement from the Space Shuttle of high quality infrared solar spectra by the ATMOS instrument (Farmer and Norton, 1989[5]) revealed a plethora of hitherto unknown solar features in both absorption and emission, from transitions from levels up to n = 10, from many solar atomic constituents. Emission line complexes from n = 12–11 through 16–15 have also been seen on far IR solar spectra taken with the University of Calgary balloon-borne solar telescope in which the dominant features were HI transitions (Boreiko and Clark, 1986[6]). HI lines transform from absorption to emission as the lower-n level increases (Boreiko et al., 1994[7]). This transformation occurs initially as emission cores superimposed upon broad absorption features at n = 7–6 and 9–7 in the ATMOS data, while n = 8–7 appears as an simple emission peak in data taken from mountain altitudes such as

Kitt Peak (Wallace, Livingston and Bernath, 1994[8]), Mauna Kea (Clark et al., 1994[9]) and Jungfraujoch (Farmer, unpublished research report, 1994).

Recent theoretical work has succeeded in explaining very successfully the main characteristics of both the MgI and HI lines at the lower n levels (MgI: Chang et al., 1991[10], Carlsson, Rutten and Shchukina, 1992[11]; HI: Carlsson and Rutten, 1992[12]) by invoking NLTE modeling in a radiative–equilibrium atmosphere without chromosphere. In the case of MgI, population inversion to produce emission occurs as a natural consequence of replenishment of upper levels from the next higher ionization state, while for HI lines it is driven by Balmer–continuum photoionization. Higher n–level HI lines are formed in the low chromosphere and are found to be sensitive to atmospheric characteristics, whereas MgI line predictions depend more strongly upon atomic parameters. As seen in Figure 3 of Carlsson and Rutten (1992)[12], observed line intensities for the higher-n transitions are not yet well fitted by theoretical calculations but lie between the predictions made by using two different atmospheric models in the calculations, namely the radiative–equilibrium model T5780 of Edvardsson et al., (1991)[13] and the empirical MACKKL model which includes a chromospheric temperature inversion (Maltby et al., 1986[14]). Better modeling must await the inclusion of the correct representation of inhomogeneities such as the chromospheric network and the cooler supergranular cells into these models.

2. Aim of the Present Experiment.

The present work is an extension of earlier far IR investigations into the submillimeter spectral range, in an attempt to measure and characterize higher level Rydberg lines in the Sun's spectrum, using a polarizing Fourier transform spectrometer (Naylor et al., 1994[15]) on the 15-m James Clerk Maxwell Telescope (JCMT). This wide-band modest-resolution spectrometer is better matched to the search for and measurement of these features than the higher-resolution heterodyne receivers, since these lines are predicted to be significantly Stark-broadened by protons. The spatial resolution of the JCMT (6–20″) is sufficient to explore limb brightening and possible enhancements over active regions and maybe even magnetic field measurements via the high Zeeman sensitivity of these long-wavelength lines.

3. Experimental Details

The Martin–Puplett interferometer was operated in a double input/single output configuration at the East Nasmyth focus of the JCMT. The JCMT facility detector, UKT14, cooled to 0.36K, was used for this work. For solar work, the second port of the interferometer viewed an ambient-temperature reference source. The instrument was operated in rapid-scan mode, a single spectrum being recorded to a resolving power of 75,000 once per minute. The polarizing beamsplitters for this spectral range were metal grids at 10μm spacing upon a 2.5μm Mylar substrate. Filters in UKT14 were used to limit the passband of the instrument to five different sub-millimeter atmospheric transmission windows. The present data were taken in the 350 μm

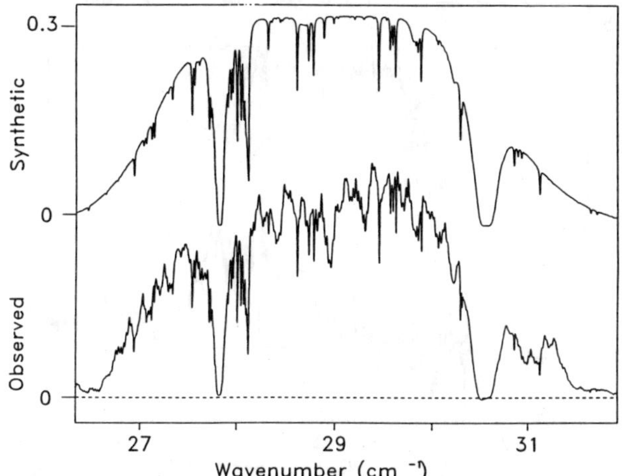

Figure 1: Solar sub-millimeter spectrum through the 350μm window, compared to a synthetic atmospheric transmission spectrum. The broad features are caused by H_2O and O_2 and the narrower lines by O_3. The measured solar spectrum shows additional structure as a result of channel effects in the detector dewar.

window between 320 and 370 μm (26.8 and 31.5 cm^{-1}).

4. Data

Figure 1 shows the measured spectrum from disk center through the 350 μm window, compared to a calculated atmospheric transmission spectrum. It can be readily seen that, while a significant amount of the structure in this spectrum is caused by atmospheric absorption by broad H_2O, O_2 and narrower O_3 lines, there is a great deal of remaining structure which is unrelated to atmospheric absorption. This is thought to be caused by channel–fringe effects in cavities within the UKT14 detector dewar. In principle, the effects of such fringes should be removable by ratio techniques but in practice, subtle differences in this pattern of effective detector sensitivity for different observing modes makes their removal difficult. Even the comparison of solar with equivalent lunar spectra does not result in cancellation of this fringe pattern, particularly when lunar spectra were taken at different times and air masses, since the broad-band shape of the window changes markedly with water vapor amounts above the site. Thus, the direct search for possible features on the solar disk has so far been unsuccessful.

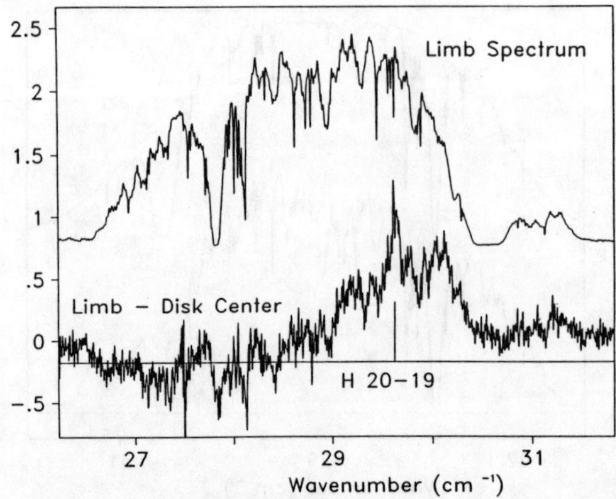

Figure 2: 350 μm limb spectrum of the Sun and the difference spectrum between limb and disk-center, showing excess emission at the position of the HI n = 20–19 transition.

5. Limb Spectra

As part of this search, a sequence of spectra were taken at different positions approaching the solar limb in the hope that solar lines, if present in the spectrum, would exhibit limb brightening equivalent to those at shorter wavelengths. Since these spectra were taken over a short period of time when air-mass and atmospheric conditions did not vary, a detailed comparison of these spectra could be attempted. Figure 2 shows a solar limb spectrum and the difference between it and a disk–center spectrum. It can be seen that a distinct feature appears in this difference spectrum at the position of the HI n=20–19 transition. Figure 3 shows a detail of this difference spectrum after spline fitting and minor smoothing. The peak emission is at the predicted frequency for the HI, n = 20-19 transition, 29.622 cm^{-1}, and has a line width of 0.07 cm^{-1} and an intensity of up to several % of the continuum intensity, with marked limb brightening. Thus, it appears that this high-level HI line is present, at least in the solar limb spectrum. There is some evidence of additional structure on the high wavenumber side of this peak. This could be emission from equivalent transitions in heavier elements, where their contributions will be blended at these long wavelengths. The limb brightening of the HI line in basic agreement with that seen in lower transition lines (e.g. Brault and Noyes, 1983 [2]). Generation of the correct limb-brightening curve for this line will require a deconvolution from the measured data of the broad and somewhat uncertain beam–pattern of the JCMT, determined by using solar limb scans (Clark et al., 1992[16]).

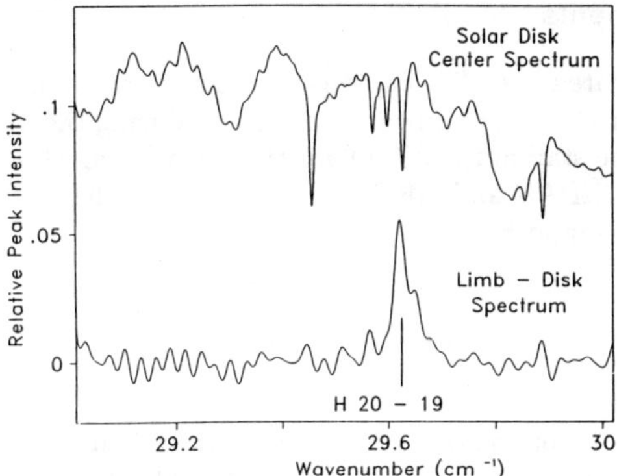

Figure 3: Upper graph: Portion of the near-limb spectrum, showing several narrow O_3 lines. Lower graph: Difference between limb and disk-center spectrum, showing the shape and structure of the HI n = 20–19 line in emission.

6. Conclusions

Recent calculations of conditions within the chromosphere by Kunc and Soon, (1992)[17] have placed limits upon the maximum principal quantum numbers of HI and the effective ionization level of hydrogen, as a consequence of line broadening and the lowering of the ionization energy of the hydrogen atoms. This maximum principal quantum number is predicted to be between 11 and 63 depending upon plasma temperature and density in the solar chromosphere. The present detection of a transition from n = 20 places a strong constraint upon this limit and therefore upon plasma conditions within the chromosphere at the source height of this emission line. Theoretical line formation models will need to be able to predict its characteristics along with those seen at lower wavelengths.

7. Future Observations

The commissioning of a new detector and more careful scheduling of solar and lunar observations to ensure that these spectra are taken under equivalent atmospheric conditions should significantly improve the quality of future spectra and permit a direct search for the HI n = 20-19 and higher n lines by comparison of solar with lunar spectra. Line intensity and width will be measured as a function of position towards the limb and attempts will be made to measure magnetic fields across active regions and chromospheric network using the high Zeeman sensitivity of these long wavelength lines.

8. Acknowledgements

The team is indebted to G.J. Tompkins for the construction, testing and operation of the electronics for these experiments, before and during the observing runs. It is a pleasure to thank the staff of the JCMT for their unstinting efforts on behalf of these measurements. TAC, DAN and GRD are supported by the NSERC of Canada, and are grateful for this support.

9. References

1. Murcray, F.J., Goldman, A., Murcray, F.H., Bradford, C.M. Murcray, D.G., Coffey, M.T. and Mankin, W.G. *Ap. J.* **247** (1981) L97.
2. Brault, J and Noyes, R.J. *Ap. J.* **269** (1983) L61.
3. Chang, E.S. and Noyes, R.W. *Ap. J.* **275** (1983) L11.
4. Chang, E.S. *Phys. Scripta* **35** (1987) 792.
5. Farmer, C.B. and Norton, R.H. *NASA Ref. Publ. 1224:* A High-Resolution Atlas of the Solar Spectrum of the Sun and the Earth Atmosphere from Space, Vol. 1, (1989).
6. Boreiko, R.T. and Clark, T.A. *Astron. Astrophys.***157** (1986) 353.
7. Boreiko, R.T., Clark, T.A., Naylor, D.A. and Busler, J.R. *IAU Symp. 154* Infrared Solar Physics, Rabin, Jefferies, Lindsey, eds. Kluwer Academic Publishers, (1994) 365.
8. Wallace, L., Livingston, W.C. and Bernath, P. *NSO Tech. Report 1994-01* (1994)
9. Clark, T.A. Naylor, D.A. and Tompkins, G.J. *8th Cambs. Workshop on Cool Stars, Stellar Systems and the Sun* A.S.P Conf. Series 64, (1993) 608.
10. Carlsson, M., Rutten, R.J. and Shchukina, N.G. *Astron. Astrophys.* **253** (1992) 567.
11. Chang, E.S., Avrett, E.H., Mauas, P.J., Noyes, R.W. and Loeser, R. *Ap. J.* **379** (1991) L79.
12. Carlsson, M. and Rutten, R.J. *Astron. Astrophys.* **259** (1992) L53.
13. Edvardsson, B., Gustafsson, B., Lambert, D.L., Nissen, P.E., Tomkin, J. and Andersen, J. *Astron. Astrophys.* 275 (1993) 101.
14. Maltby, P., Avrett, E.H., Carlsson, M., Kjeldseth-Moe, O., Kurucz, R.L. and Loeser, R. *Ap. J.* **306** (1986) 284.
15. Naylor, D.A., Clark, T.A. and Davis, G.R. *Proc. SPIE* Conf. on Astron. Instrumentation, (1994) in press.
16. Clark, T.A., Naylor, D.A., Tompkins, G.J. and Duncan, W.D. *Sol. Phys.* **140** (1992) 393.
17. Kunc, J.A. and Soon, W.H. *Ap. J.* **296** (1992) 364.

CLEAR : A CONCEPT FOR A CORONAGRAPH AND LOW EMISSIVITY ASTRONOMICAL REFLECTOR

JACQUES M. BECKERS
*National Solar Observatory/NOAO**
Sunspot, NM/Tucson, AZ

ABSTRACT

A number of efforts are underway to move into the next generation of solar telescopes. Among these are the Large Earth-Based Solar Telescope (LEST) and the upgrade of the Kitt Peak McMath-Pierce solar telescope to a 4 meter aperture. In this paper I outline a competing concept for a 4 meter class solar telescope aimed at observing both the solar disk and the solar corona at all wavelengths from 0.3 to 35 μm. Such a telescope has to be a single mirror reflector, using superpolished mirror technology, with stringent control of cleanliness of the primary mirror. Because of the similarities in the technical requirements, this telescope would, in addition to making an outstanding solar research facility, be a fine nighttime telescope for those research projects where low thermal background and/or low scattered light are required.

1. Introduction

Thw world's largest solar telescope, the 150 cm aperture McMath-Pierce facility on Kitt Peak, saw its first light over 30 years ago in 1962. It is still the only major solar facility capable of studying the sun at all optical wavelengths from 300 nm through 35 μm. Solar infrared astronomy has since then made major strides[1], and the lack of a modern, large aperture facility is a major handicap. The largest coronagraph in the US is the 40 cm John Evans Solar Facility at the Sacramento Peak Observatory. It saw its first light in 1953 and, by lack of a more modern facility, is still in use. Because it uses refractive optics, its wavelength range is limited to 300 nm through 2.5 μm. Other telescopes (e.g. the Sac Peak Vacuum Tower Telescope dedicated in 1969), principally aimed at high angular resolution, have been built since then, but all are limited to apertures well under 100 cm. They all are vacuum telescopes using glass windows or lenses, thus removing the longer wavelengths (> 2.5 μm) from their foci and adding scattered light making the telescopes unusable for coronal studies. In terms of modern, large aperture telescopes, solar astronomy has therefore fallen far behind the capabilities available to nighttime astronomy where well over a dozen 8 meter class telescopes are presently under construction[2].

* The National Optical Astronomy Observatory (NOAO) is operated under a cooperative agreement between the U.S. National Science Foundation (NSF) and the Association of Universities for Research in Astronomy (AURA).

Progress in solar astronomy is severely hampered by the absence of large, modern facilities capable of doing both high angular resolution and sensitive studies of physical processes of all observable layers in the solar atmosphere in all wavelengths available for such studies from the ground. Development of theoretical and analytical techniques in solar physics have out-paced our observational capabilities, and image restoration techniques warrant larger aperture telescopes to achieve angular resolutions needed to get down to the basic physical elements in the solar atmosphere. The Large Earth-Based Solar Telescope (LEST) is an attempt to achieve some of these goals. It failed to get support in the USA, partly because it was too limited in its goals. In this paper I will describe a proposal for the construction of a large aperture telescope aimed at the high angular resolution study of all solar layers over all optical wavelengths from 0.3 through 35 μm. Its unique optical properties makes this telescope also very attractive to those observations of the night sky where low scattered light and/or low IR background are desired.

2. Description of the CLEAR Concept

Smartt et al.[3,4] give a detailed description of the program underway at the Sacramento Peak site of the National Solar Observatory (NSO/SP) to develop reflecting solar coronagraphs. The solar coronagraph was invented by Bernard Lyot six decades ago. It includes many ingenious concepts to reduce the scattered light from the solar disk itself while observing the very faint solar corona outside the limb. Its main elements consists of : (i) a single optical element (the telescope's *"primary"*), of the highest quality as far as scattered light goes, which forms the solar image, (ii) an *occulting disk* in that image which blocks the bright radiation from the solar disk, and (iii) a circular aperture placed in the image of the entrance pupil formed by an optic following the occulting disk which blocks most of the light diffracted on the edge of the circular primary. The convention is to refer to this aperture as the *"Lyot stop."* Lyot used single refractive optical elements. The largest of such Lyot coronagraphs are the 52 cm aperture Russian coronagraphs and the 40 cm coronagraph of the J.W. Evans Solar Facility at NSO/SP. As is the case for nighttime telescopes, the availability of good, flawless optical glasses limits the aperture of refractive coronagraphs. Reflecting coronagraphs do not suffer such aperture limitations. In addition reflecting coronagraphs offer the advantage of achromatism and all wavelength coverage. The NSO/SP program by Smartt et al.[3,4] explores the limitations of using superpolished off-axis reflecting primaries rather than refracting primaries for this purpose. It has achieved apertures of 5 and 15 cm with performance comparable to the best of refractive coronagraphs and is presently developing a 50 to 60 cm aperture reflector.

The technical requirements placed on a reflecting solar coronagraph match almost all of the requirements one would place on a nighttime telescope aimed at low scattered light and low IR emissivity observations : (i) a single, low scattered light image forming element minimizing scattered light from astronomical objects and from IR radiation from the telescope environment, (ii) the absence of a secondary mirror and its supporting "spider", eliminating the IR radiation and the scattered and diffracted light from those components, and (iii) a system to keep the dust and other contamination of the primary to an absolute minimum. Existing coronagraphs have in fact been used for nighttime observations requiring low scattered light. The "Coronagraph and Low Emissivity Astronomical Reflector," or CLEAR, concept which is the subject of this paper proposes to use reflecting solar coronagraph experience and concepts to build a specialized astronomical facility which finds a broad range of application from solar to planetary to stellar to extra-solar planetary to extra-galactic research, working "around the clock" !

3. CLEAR Parameters

CLEAR at this moment is little more than a concept at the pre-Phase A definition level. Figure 1 is a sketch of the CLEAR layout. Table 1 lists the preliminary CLEAR parameters.

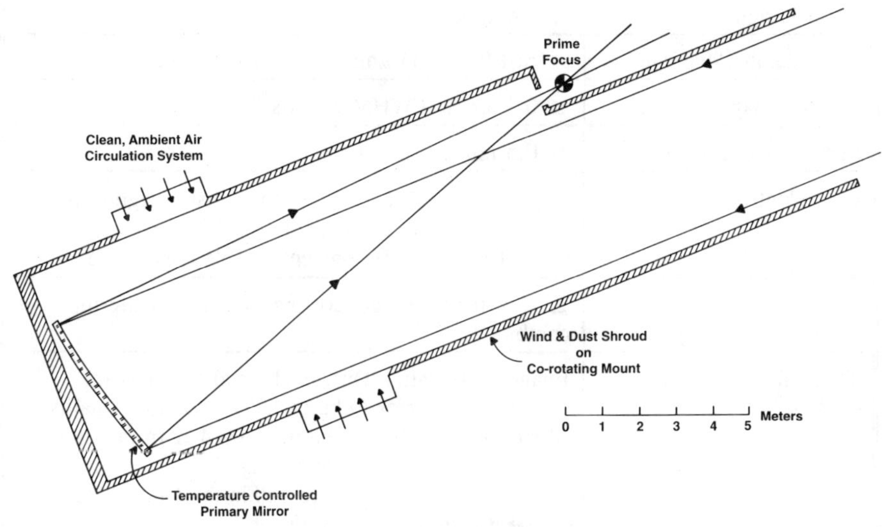

Figure 1 : Sketch of CLEAR Layout

The following comments accompany these parameters :

3.1 Optical Configuration

To minimize both the scattered light and the telescope emissivity it is essential to have an optical system with an absolute minimum of components in front of the primary image where occulting disks and cooled instrumentation can be used to minimize the scatter and emission beyond. CLEAR thus uses a single, superpolished, off-axis paraboloid, eliminating e.g. the conventional Cassegrain secondary mirror and its support which are common in nighttime telescopes and the vacuum window which are part of most modern solar telescopes. Such a window would both destroy the coronagraph properties and block the infrared radiation.

Table 1 : Preliminary CLEAR Parameters

Item :	Property :
Configuration	Single Superpolished Mirror giving Prime Focus
Mirror Diameter	200 to 400 cm
Mirror Shape	Off-Axis Paraboloid
Focal Ratio	F/3.75
Focal Length	750 to 1500 cm
Wavelength Coverage	at least 0.4 to 35 μm; 0.3 to 35 μm preferred
Optical Image Quality	$\leq$ 0.1 arcsec FWHM on-axis
Surface Microripple	$\leq$ 0.5 nm for spatial frequencies $\leq$ 1 cm^{-1}
Dust Control	to give $\leq$ 1 μ☼ ($\equiv 10^{-6}$ of solar disk) brightness at 1 arcmin above the solar limb Class 100 Dust Environment in Telescope Tube
IR Emissivity	$\leq$ 1 % for telescope, coating, dust and scattering combined
Thermal Control	Image Degradation Due to Thermal Warp of the Primary Mirror to be Included in Image Quality Mirror Temperature to be within 0.4 K from Ambient Air to Control Mirror Seeing Air Temperature anywhere within the Optical Path to be within 0.2 K of Ambient Air
Adaptive Optics	For $\lambda \geq 0.5$ μm Strehl Ratio $\geq$ 70 % Solar Disk using Feature Wavefront Sensing Solar Corona using Laser Guide Stars Nighttime using Natural or Laser Guide Stars
Pointing	Absolute : $\leq$ 2 arcsec at Nighttime; $\leq$ 5 arcsec in Daytime Offset Pointing : $\leq$ 0.1 arcsec over 1.0/0.5 degree at Nighttime/Daytime

3.2 *Telescope Diameter*

Light collection and angular resolution are important parameters for both solar and nighttime applications. In the era of 8 meter aperture nighttime telescopes[2] , a 200 to 400 cm aperture appears modest. On the other hand, superpolishing of large off-axis aspherical mirrors and the maintenance of the superpolished quality by means of mirror coatings remain among the main technical issues for further study.

Table 2 : Image Quality in Primary Image

Distance from Axis (arcsec)	Geometric Image Size (arcsec)	Comments
0	0	
900	5.0	Solar & Lunar Limb (point at disk center)
1800	10.0	2 $R_\odot$ (solar radii) Field-of-View
23	0.13	Jupiter Limb (point at jovian disk center)
1.5	.008	Approximate Radius of 0.5 μm Isoplanatic Patch Diffraction Limit of 400 cm Telescope is 0.025 arcsec
55	.31	Approximate Radius of 10 μm Isoplanatic Patch Diffraction Limit of 400 cm Telescope is 0.5 arcsec

3.3 *The Focal Ratio*

The choice of focal ratio is determined by both the requirement to handle the solar energy load at prime focus, the need for a good image quality field-of-view at the prime focus, and the cost of telescopes with large focal lengths. The f/3.75 focal ratio is a good compromise considering all of these factors. It is slower than the f/2.4 primary considered for the Large Earth-based Solar Telescope (LEST) which has a cooled occulting disk, as will be used for CLEAR, but one which will, unlike the CLEAR disk, "meltdown" in case of a cooling failure. The curvature of the f/3.75 and 15 meter focal length off-axis paraboloid is similar to the curvature of most of the on-axis mirrors used for the 8 to 10 meter class telescopes. The (geometric) image quality of the optical system, allowing for a distance between the parabola axis and the center of the mirror of 125 cm and 250 cm for the 200 and 400 cm respectively is given in Table 2. It refers to an image plane at right angles to the line connecting the mirror center and the prime focus. It should be noted that this is the image size at prime focus where the image occulting quality is determined. The final image quality will, of course, be better as determined by the optics following prime focus. Notable are : (i) the image quality at the edge of the solar occulting disk when pointing at the disk center is 5 arcseconds. For many applications this is acceptable. If better occulting of the solar disk is needed, it will be necessary to point at the solar limb, (ii) the image quality within the isoplanatic patch is good enough to maintain the adaptive optics corrected image quality even without secondary image correcting optics.

3.4 *Wavelength Coverage*

Being a fully reflecting system, the wavelength coverage will be determined by the coating applied to the primary mirror. The primary requirements for the coating are : (i) low scattered light, and (ii) low emissivity at IR wavelengths. The latter may imply the use of protected silver or gold coatings, the former needs investigation. Although UV wavelength coverage is desirable, it should not be allowed to compromise the coronagraphic and IR performance.

3.5 *Microripple*

As already pointed out by Smartt et al.[3], one of the disadvantages of a reflecting coronagraph is the need to reduce the surface microripple (high frequency surface imperfections, roughness, including scratches) by about a factor of three over that required for a refractive coronagraph. On the other hand the requirements on internal imperfections in the optics (bubbles, striae) are, of course, much relaxed. The prototype reflecting coronagraphs at NSO/SP have a microripple (σ) of 0.4 nm RMS. At that level the scattered light (= the light in the far wings of the point-spread-function, PSF) is indeed brought down to acceptable levels[3]. Similar levels of microripple are required for X-ray telescopes. The 120 cm and 84 cm diameter outer AXAF and ROSAT mirrors, for example, have a microripple of 0.25 to 0.8 nm RMS[5,6,7]. The total amount of scattered light can easily be estimated from the equation for the Strehl ratio SR for an otherwise perfect optical system :

$$SR = 1 - (4\pi \cdot \sigma/\lambda)^2 \equiv 1 - \Delta SR. \qquad (1)$$

$\Delta SR = (4\pi \cdot \sigma/\lambda)^2$ represents the amount of light distributed in the far wings of the PSF. The shape of this distribution depends on the power spectrum of the spatial variations in the microripple. For $\sigma = 0.4$ nm one obtains $\Delta SR = 10^{-4}$ and 2.5×10^{-7} for wavelengths λ of .5 μm and 10 μm respectively. The effects of microripple are therefore primarily noticeable at the shorter wavelengths where the NSO/SP tests were made.

3.6 *Dust Control*

Lyot made major attempts to prevent his coronagraph lenses to become contaminated by dust. Following Babinet's principle[8] it is possible to obtain a rough estimate for the tolerable amount of dust. Assuming that a (small) fraction ϵ of the primary mirror is occupied by dust, the wings of the PSF are enhanced by the PSF of the complimentary "mask" to the "dust mask"which contains an amount of energy equal to $\epsilon/(1-\epsilon) \approx \epsilon$ of the energy in the full aperture image. The distribution of the PSF wing enhancement is dependent on the dust particle sizes and on their spatial distribution. Things are thus similar to the case for the microripple/scratches effects. Assuming the same PSF wing distribution one obtains a requirement for ϵ of $\leq 10^{-4}$ if the microripple tolerance for 0.5 μm wavelength is taken as the reference. That implies an IR emissivity due to dust (and other mirror contaminants) of $\leq 10^{-4}$ as well. The assumption of equal spatial distribution for the microripple/scratches and dust is of course only a course approximation, so that this result should be taken as giving only an order of magnitude estimate.

It is important to note that the contribution of microripple and dust/contamination to the wings of the PSF increases with telescope diameter D in proportion to the aperture area, or D^2. The contribution by the diffraction on the edge of the primary mirror increases proportionally to its circumference, or D. For non-coronagraphic telescopes the diffraction by spiders supporting the secondaries increases proportional to D as well. Therefore for large aperture Lyot coronagraphs it becomes increasingly important to control the scatter on the optics in front of the primary focus as compared to the removal of the diffraction effects by the Lyot stop. It can be shown[2] that in the D = 8 meter case, the occulting disk - Lyot stop combination by itself does little to help the scattered light situation as far as most of it originates in the coronagraph itself (it may help in the instrumentation). The now common use of the term "coronagraph" for that combination in nighttime application[9,10] appears therefore inappropriate.

To control the dust in CLEAR it is proposed that the telescope be enclosed in a tube which has properties similar to a clean room. Like the enclosure of the Sloan Digital sky Survey (SDSS)[1112] this enclosure would have its own alt-azimuth drive system, separate from that of CLEAR, both to protect the CLEAR from wind baffling and from unwanted light contamination. This "co-rotating clean room" would include an air-filtration system on its sides to (i) maintain the cleanliness of the air in the lightpath, (ii) to maintain it at ambient temperature for seeing control, and (iii) to provide a slight overpressure in the tube to prevent external dust from entering the CLEAR system.

3.7 IR Emissivity

The primary mirror coating together with the atmosphere are likely to be the main contributors to the IR emissivity. The low ≤ 1 % goal is important not only for nighttime observations, it is also essential for solar corona observations. Table 3 gives the present estimates for the brightness of the solar aureole at 10 μm wavelengths for the different contributors.

3.8 *Adaptive Optics*

CLEAR should be equipped with adaptive optics. The adaptive mirror and wavefront reconstructor/controller can be identical for all applications, the wavefront sensor will be different. The CLEAR adaptive optics system should at least be designed for operation under medium seeing at visible wavelengths (0.5 μm). For medium seeing of 0.8 arcsec (r_0 = 12.5 cm) that implies an ≈ 1000 element system to achieve a Strehl Ratio of 70 %. In case high Strehl ratios are needed, as in the case of searching for extra-solar planetary systems[13], many more elements are needed as well as an amplitude adaptive optics system. Wavefront sensing for solar disk application should use extended object wavefront sensing on the solar granulation structure. The solar corona is faint enough to make use of sodium laser guide stars for wavefront sensing[14] possible. For planetary and stellar applications the object itself or laser guide stars can be used.

Table 3 : Contributing Sources to Solar Aureole at 10 μm

SOURCE	$R/R_\odot = 1.15$	$R/R_\odot = 1.5$	$R/R_\odot = 2.0$
Sky, Rayleigh Scattering	≪ 0.1 μ☼	≪ 0.1 μ☼	≪ 0.1 μ☼
Sky, Aerosol Scattering	≤ 2.7 μ☼	≤ 1.4 μ☼	≤ 0.9 μ☼
Optics, Micro-Ripple	< 0.1 μ☼	< 0.01 μ☼	< .001 μ☼
Optics Imperfections : Glass	0.1 μ☼	0.01 μ☼	< .001 μ☼
Optics Imperfections : Coating	≤ 2 μ☼ ?	≤ 2 μ☼ ?	≤ 2 μ☼ ?
Optics/Atmosphere, Emissivity	~ 20 μ☼	~ 20 μ☼	~ 20 μ☼
Residual Diffraction Effects	≤ 1 μ☼ ?	≤ 1 μ☼ ?	≤ 1 μ☼ ?
TOTAL	**≤ 26 μ☼**	**≤ 25 μ☼**	**≤ 24 μ☼**

3.9 *Thermal Control of Optics*

There are two reasons for thermal control of astronomical mirrors : (i) *mirror warping* due to thermal inhomogeneities inside the mirror and (ii) *mirror seeing* caused by a difference between the mirror temperature and the ambient air above the mirror. The importance of the former depends on the coefficient of thermal expansion (CTE) of the mirror material. By choosing a material with low CTE (e.g. ULE or Zerodur) rather large thermal inhomogeneities can be tolerated. Alternately, materials with high thermal conductivities (e.g. metals) are occasionally suggested because of their reduced thermal inhomogeneities. Mirror seeing is a major problem for astronomical mirrors, and is likely to be especially troublesome for solar telescopes because of the high solar irradiance of the mirrors. The rule of thumb is that each degree K the mirror temperature is above the ambient air temperature causes an image deterioration due to mırror seeing of 0.4 to 0.5 arcsec[15], except that there may be an lower limit of 0.5 K below which no mirror seeing occurs[16]. Tests on the Mc Math-Pierce telescope on Kitt Peak show mirror heating by solar radiation to be in the 1 to 2 K range. resulting in mirror seeing of 0.4 to 1 arcsec per mirror. It is therefore essential to control the CLEAR mirror temperature especially, but not only, for its daytime use. Various techniques have been explored to do so in nighttime telescopes, using either cooling from behind the mirror or air/liquid circulation within the mirror cavities. In case of the use of very low CTE glass mirrors like ULE the use of channeled mirrors[17,18,19] appears to find here a particularly good application.

3.10 Open Air versus Vacuum Telescopes

In the past three decades it has become custom to evacuate high angular resolution solar telescopes. This is a drastic way of eliminating internal telescope seeing. It goes, however, at the cost of the loss of most of the infrared part of the solar spectrum and of an increase in scattered light, both due to the window. CLEAR is not evacuated. One might justifiably ask if that does not compromise the telescope image quality as compared to vacuum telescopes. The answer is not clear yet. It is clear that vacuum telescopes remove mirror seeing, but thermal control of the mirrors will do the same. Direct heating of the air inside the telescope is negligible and, even if it were not, would be of even more concern in the large air column preceding the telescope. On the other hand, vacuum windows themselves are very hard to control thermally and optically. What other factors does evacuation reduce or eliminate ? The only one coming to mind as a problem in the CLEAR type concept is the heating of the solar occulting disk. Cooling of that disk and control of its image deteriorating effects are indeed of concern. Being out of the optical path helps as do the apparently small effects on image quality which localized heat sources produce[20]. However, even if it causes only little Schlieren activity in the optical path, it may be enough to destroy the coronagraphic properties of CLEAR.

3.11 Pointing and Tracking

Telescope pointing for solar applications is normally done with the solar limb as a reference. For nighttime applications pointing is more critical. State-of-the-art absolute pointing for alt-azimuth telescopes is at this moment 1 arcsec. More critical is offset pointing which holds the promise of knowing the absolute pointing to 0.1 arcsec or better if the absolute coordinates of Hipparchos stars (and future astrometric missions) are used as offset references.

Solar tracking is commonly done using the solar limb for an autoguider. Nighttime tracking using relatively bright offset stars works well provided that the telescope short time ($\leq$ 10 sec) "blind" tracking capability is good enough so that differential seeing motions between the object and the guide star can be time averaged-out to a sufficient degree.

3.12 Polarization

Off-axis paraboloids have a small amount of polarization, the amount depending on the coatings used. In the strawman CLEAR design considered here, the average angle of incidence is only 4.7^0 resulting in a polarization $\leq 1^0$ and a retardation $\leq \lambda/1500$. These polarization effects are constant with time and should be largely removable by calibration or by optical means following the prime focus.

3.13 Location

CLEAR should be located on a site with : (i) good to excellent daytime and nighttime seeing, (ii) good coronal skies, (iii) low sky emissivities (low PWV), (iv) low cloudiness and airline contrail activity, (iv) low sky pollution/dark sky and (v) preferably with an existing infrastructure. The NSO/Sacramento Peak & Apache Point Observatory complex provide for one possible location which satisfies these conditions quite well.

4. Technical Issues to be Addressed

A number of technical issues have to be addressed in determining if CLEAR is a viable concept for an advanced solar/stellar facility. Among the most prominent of these are :

A. The prospects of producing super-polished ($\leq$ 0.5 nm RMS), off-axis parabolic mirrors of the sizes of interest (200 - 400 cm diameter).

B. The selection of the best material for the primary mirror (e.g. Si, ULE, ZERODUR, Fused Silica).

C. The thermal control of the primary mirror when exposed to sunlight (to avoid mirror seeing & thermal distortion).

D. The prospects of obtaining low emissivity, low scattered light mirror coatings covering as wide a spectral range as possible.

E. The design of a system to provide an adequate dust/contamination control of the primary mirror while at the same time maintaining low internal seeing.

F. The design of the prime focus solar heat rejection device (for both external & inverse occulting).

G. The design and development of a solar disk, solar corona and nighttime adaptive optics system.

H. The development of the design of the entire system, including telescope, co-tracking clean room and instrument complement, taking into account a 24 hour operating environment when CLEAR will be subject to the full daily thermal cycling including sun-pointing.

The outcome of these studies will show whether the CLEAR concept is indeed a realistic one and whether it makes any sense to combine solar and nighttime use of CLEAR. In addition it will lead to a firmer definition of the telescope parameters and an estimate of its cost.

5. Scientific Applications of CLEAR

As currently envisaged, CLEAR will have a broad range of applications in many different areas of astrophysics. The following gives a broad summary.

5.1 Solar Physics

As currently envisaged, CLEAR will be the best, all purpose solar telescope ever constructed. It will do so while introducing minimal compromises as compared to special purpose solar telescopes. Specifically : (i) CLEAR will be able to study both the solar disk and the solar corona, (ii) CLEAR will have a broad wavelength coverage from the UV (0.35 μm ?) into the far infrared (35 μm), (iii) CLEAR's very large collecting area will result in improved signal-to-noise ratios and high sensitivity, (iv) CLEAR's very high angular resolution will be well beyond that available now (using adaptive optics and/or rapid guiding : ≤ 0.1 arcsec for all $\lambda \leq 1$ μm, Diffraction Limited for $\lambda > 1$ μm), and (v) CLEAR's telescope polarization is small, constant and hence largely correctable ($P \leq 1$ %; $\Delta\varphi \leq .25°$; depends on coating)

These properties make CLEAR particularly suitable for most solar physics observations, especially those related to :

- The dynamics of the small scale structure of active and quiet region magnetic fields and its relation to coronal heating.
- Coronal magnetic fields.
- The energetics of the chromospheric small scale structure (e.g. spicules).
- The magneto-hydrodynamics of sunspots.
- The small scale structure of solar flares.

5.2 *Solar-Stellar Physics*

Of particular interest to both solar and stellar physics is the study of solar like phenomena on other stars representing different stellar parameters and as well as different epochs in their activity cycles. Both CLEAR's large collecting area, high angular resolution and low scattered light/emissivity are of interest. Specific topics might be the study of :

- Stellar rotation and activity cycles from CaII H&K Line and HeI 1.083 μm observations.
- The spectral variations of sun-like stars in M67 to study activity cycle variations in stars of identical age.
- The imaging and spectroscopy of stellar envelopes and coronae.

5.3 *Solar Systems Studies*

Especially the coronagraph mode of CLEAR will have interesting applications, like the following :

- The observation of faint envelopes (Na-D; S^+) of planets and their satellites (e.g. the jupiter torus; the moon; Io; asteroids ?).
- The detection and study of asteroid companions (243 Ida types)
- The study of ion tails/magnetospheres of planets (like Mercury ?)
- The observation of comet structure

5.4 *Search for Extra-Solar Planets*

CLEAR is a true stellar coronagraph, rather than just a coronagraphic instrument. In addition to having very low scattered light it has a very low thermal background, making it ideal for IR coronagraphic observations. Both this low scattered light and low thermal background will result in quality observations of faint stellar envelopes (β PIC type) and Faint Stellar Companions (brown dwarfs & planets).

Table 4 is taken from an earlier paper[2] giving preliminary estimates[21] of the aureole/halo intensity around bright stars for 8 meter telescopes for 0.7 arcsec seeing (uncorrected by adaptive optics) but adapted to the 4 meter aperture CLEAR with 0.1 % emissivity due to dust (vs. 1 % in the earlier paper) and 0.5 nm RMS microripple (vs, 3 nm RMS on two mirrors in the earlier paper). From it one concludes that : (i) in the close proximity of the star it is essential to have both high order phase and amplitude adaptive optics to suppress the seeing halo around the star image as was already discussed by Angel[7]. In addition adaptive optics is essential, of course, to decrease the planet's image size and to enhance its central intensity and hence its contrast),

Table 4 : Halo/Aureole Intensity for 4 meter Aperture CLEAR Telescope
(magnitude per □" for 0.7" FWHM seeing (at 0.5 μm)
star with 0 magnitude/□" central intensity)

	Contributor	Dependence on D	Magnitude (per □") at λ = .5/2.0/10.0 μm		
			θ = 1"	θ = 10"	θ = 100"
1	Seeing	D^2	4.3/6.9/11.8	201/319/545	--/--/--
2	Scintillation	D^2	6.4/10.7/17.7	207/330/563	--/--/--
3	Diffraction on Telescope Aperture	D	8.0/7.1/6.0	15.5/14.6/13.5	23.0/22.1/21.0
4	Scattering on Aerosols for Good Coronal Sky	D^2	26.3/27.8/29.6	26.3/27.8/29.6	26.3/27.8/29.6
5	Scattering on Mirror Microripple	D^2	15.0/18/6/22.8	20.0/23.6/27.8	25.0/28.6/32.8
6	Scattering on Mirror Dust	D^2	13.0/13.6/14.3	18.0/18.6/19.3	23.0/23.6/24.3
7	Rayleigh Scattering of Sky (full moon)	D^2	18.7/23.4/25.0	18.7/23.4/25.0	18.7/23.4/25.0
8	Galactic Pole : Astronomical Background	D^2	21.5/18.5/--	21.5/18.5/--	21.5/18.5/--
9	Thermal Background Telescope & Sky (10 % total emissivity)	D^2	--/17/-3	--/17/-3	--/16/-3

(ii) after reduction of the seeing halo the aperture diffraction is the main contributor to the star halo also in the close proximity of the star unless most of it is removed by the Lyot stop, and (iii) after removal of both diffraction and seeing effects, both microripple and dust scattering compete as the main contributors. In case the dust and microripple control is not as good as it should be in a coronagraph, their contribution would would have competed with the diffraction contribution (without the Lyot stop), and would dominate the halo in case a Lyot stop is used.

5.5 *Galactic and Extra-Galactic Astrophysics*

As compared to the ≈ 15 telescopes with 8 meter class diameter now under construction[2], CLEAR will have a very modest collecting area and adaptive optics aided resolution. Its low scattered light and low emissivity make CLEAR, however, an outstanding special purpose facility for a wide variety of nighttime studies beyond those of specific solar-stellar interest.

For examples :

- CLEAR is ideal for the study of faint unresolved & diffuse objects and backgrounds in the vicinity of bright objects (e.g. Mira's & red supergiant envelopes; halos around galaxies; AGN's; QSO's; accretion disks; gravitational lensed objects).

- Because of its low thermal background CLEAR is very well suited for the study of IR diffuse objects at spatial scales smaller than the resolution of IRAS/ISO/SIRTF.

- Of special interest may be the imaging of high-z galaxies, QSO's and other distant objects as distorted by the gravitational fields associated with foreground clusters of galaxies and their associated dark matter.

5.6 *Detection of Space Debris*

Fraunhofer diffraction on aerosol particles is the main origin of the atmospheric aureole around the sun. Similarly Fraunhofer diffraction on space debris particles causes them to be much brighter than they are eg. by solar illumination in the twilight sky. CLEAR therefore is a powerful tool in the studies of small (≈ 2 mm) space debris particles[22]. The large aperture of CLEAN allows the discrimination between space debris and aerosol particles which are out of focus even when at 10 km height. Detection of these particles is of course restricted to a small area surrounding the sun. Yet, the abundant number of detections which are expected will allow the establishment of much improved statistics of particle numbers, size and orbital inclinations over extended periods of time, statistics which are of practical importance for the studies of the evolution of space pollution by human activities.

5.7 *CLEAR in the Context of the Human Environment Studies*

CLEAR is an important component of the National Solar Observatories Future Directions Plan (FDP) which aims at understanding the origins of the solar cycle, on the basis of tomographic imaging during the upcoming solar cycle of the solar convection zone by means of helioseismology (using GONG), and on developing this way a model of the solar activity which, one might expect, will have predictive power. This plan combines high quality basic astrophysical research of solar and stellar processes with research areas which are considered at this moment to be of particular strategic importance. Specifically these are the areas of *global change* and *space hazards*. Although CLEAR cannot be expected to be in operation at the start the next solar cycle in the late 90's, it will undoubtedly play a major role as the FDP research programs progress. Below are listed some of the CLEAR observations of particular interest.

In Solar Disk Mode :

- High angular resolution observations of solar surface vector magnetic fields and motions. These will result in estimates of the magnetic flux & magnetic topology changes in the photosphere, chromosphere and corona and of their related short term and long term EUV radiation outputs.

In Coronagraph Mode :

- Measurements of coronal magnetic fields (using IR coronal emission lines), their comparison with the magnetic field topologies and their relation with the Coronal Mass Ejections.
- Imaging of coronal mass ejections.

In Solar-Stellar Mode :

- Observation of stellar activity cycles (e.g. using the CaII H and K lines) and related luminosity changes aimed at establish their relationship.

In Stellar Mode :

- Study of stellar envelopes and their relation to stellar activity.

6. Conclusion and Acknowledgements

The CLEAR concept is at this moment in a very preliminary, pre-Phase A state. Although it looks very promising for a very wide range of applications, the technical and financial realities which will undoubtedly arise in future studies of the concept may restrict the range of its applications. Those studies are the essential next step.

This paper includes the result of discussions with and of the work by many colleagues. Especially the contributions by Ray Smartt, Serge Koutchmy, and Don Neidig are gratefully acknowledged.

7. References

1. D.M. Rabin, J.T. Jefferies and C. Lindsey (Editors), IAU Symposium No. 154 on *Infrared Solar Physics*, 1992, Kluwer Academic Publishers.

2. J.M. Beckers, Proceedings of Tokyo Conference on *Scientific and Engineering Frontiers for 8 to 10 m Telescopes - Instrumentation for Large Telescopes in the 21st Century*, eds. M. Iye and T. Nishimura (1995), submitted

3. R..N. Smartt, S.L. Koutchmy, S.A. Colley, R. Caron, R. Schwenn and S.R. Restaino, SPIE Proceedings **1236** on *Advanced Technology Telescopes IV*, ed. L.D. Barr (1990), 206.

4. R.N. Smartt, Proceedings 15th NSO/Sacramento Peak Observatory Workshop on *Infrared Tools for Solar Astrophysics*, eds. J. Kuhn and M. Penn (1995), submitted.

5. K. Beckstette, B. Aschenbach and M. Schmidt, SPIE Proceedings **982** on *X-Ray Instrumentation in Astronomy II*, (1988), 2.

6. L. van Speybroeck, P. Reid, D. Schwartz and J. Bilbro, SPIE Proceedings **1160** on *X-Ray/EUV Optics for Astronomy and Microscopy* (1989), 94.

7. P.B. Reid, T.E. Gordon and M.B. Magida, SPIE Proceedings **1618** on *Large Optics II*, (1991), 45.

8. M. Born and E. Wolf, *Principles of Optics*, 1964, 381.

9. M. Clampin, F. Paresce, G. De Marchi, M. Robberto, A. Ferrari and S. Marta, in *Progress in Telescope and Instrumentation Technologies* (ESO Conference and Workshop Proceedings No. 42), ed. M.-H. Ulrich (1992), 713 (see also SPIE Proceedings **2198**, 172 (1994).

10. M. Tamura, Proceedings of Tokyo Conference on *Scientific and Engineering Frontiers for 8 to 10 m Telescopes - Instrumentation for Large Telescopes in the 21st Century*, eds. M. Iye and T. Nishimura (1995), submitted

11. R.G. Kron, in *Progress in Telescope and Instrumentation Technologies* (ESO Conference and Workshop Proceedings No. 42), ed. M.-H. Ulrich (1992), 635.

12. C.L. Hull, S. Limmongkol and W.A. Siegmund, SPIE Proceedings **2199** on *Advanced Optical Telescopes V*, Ed. L. Stepp (1994), 1178.

13. J.R.P. Angel, *Nature* **368**, 1994, 203.

14. J.M. Beckers, *Annual Review of Astronomy and Astrophysics*, **31**, 1993, 13.

15. C.M. Lowne, *Monthly Notices Royal Astr. Soc.* (1979) 249.

16. L.D. Barr, J. Fox, G.A. Poczulp and C. Roddier, SPIE Proceedings **1236** on *Advanced Technology Telescopes IV*, ed. L.D. Barr (1990), 492.

17. L.D. Barr, J.M. Beckers, J.M. Pearson, T.W. Hobbs, J. Spangenberg-Jolley, ESO Conference on Very Large Telescopes and Their Instrumentation, ed. M.-H. Ulrich (1988) 595.

18. J. Spangenberg-Jolley, Th. Hobbs and G. Proulx, SPIE Proceedings **1236** on *Advanced Technology Telescopes IV*, ed. L.D. Barr (1990), 744.

19. L.D. Barr, J. Fox, D.M. Dryden and G.A. Poczulp, SPIE Proceedings **1236** on *Advanced Technology Telescopes IV*, ed. L.D. Barr (1990), 812.

20. J.M. Beckers, Proceedings SPIE **2199** on *Advanced Optical Telescopes V*, Ed. L. Stepp (1994) 478.

21. J.M. Beckers, (1995) in preparation.

22. D.F. Neidig, I.S. Kim, S. Koutchmy and R.N. Smartt, Proceedings 15th NSO/Sacramento Peak Observatory Workshop on *Infrared Tools for Solar Astrophysics*, eds. J. Kuhn and M. Penn (1995), submitted.

REFLECTING CORONAGRAPHS: PROSPECTS

Raymond N. Smartt
National Solar Observatory/Sacramento Peak *
Sunspot, NM 88349, USA

and

Serge Koutchmy
IAP - CNRS, Paris, F 75014, France

ABSTRACT

Contemporary coronal physics is severely handicapped by the fundamental limitations of conventional ground-based coronagraphic systems. Examples of these kinds of problems are discussed briefly. Advantages of working in the IR are stressed. Mirror-objective coronagraphs offer many performance advantages over conventional solar coronagraphs. Although reflecting coronagraphs were first proposed in the 1960's, the possibility of fully exploiting their potential has become feasible only in recent years with advances in super-polishing and low-scatter reflecting film technologies as well as new substrate materials. Key aspects of these technologies are discussed. In a joint program between NSO/SP and IAP, two prototype mirror-objective coronagraphs of 5cm and 15cm apertures have been constructed, These instruments are described briefly. The design of a 60-cm aperture coronagraph based on a Si-SiC super-polished objective mirror is also described. This instrument will have some advanced features with applications for studies of the solar disk as well as the corona and also nighttime studies. It will also have significant advantages for IR observations including high-precision polarimetry. This "research-quality" instrument is seen as a breadboard for much larger daytime-nighttime coronagraphic telescopes.

1. Introduction

Solar coronal physics has advanced to a point where many fundamental questions are posed, but observational answers are inaccessible due to inherent limitations of existing instrumentation. We now cite some typical examples that serve to illustrate the observational problems.

Observations with emission-line and white-light ground-based coronagraphs have allowed extensive studies of the morphology of the inner coronal region. Such observations, usually with only modest angular resolution, have established the overall structural characteristics of the corona as well as typical changes that occur over time scales of minutes, to hours, to days, as well as over time scales of the order of a solar cycle.[1] Emission-line images reveal the typical complex loop structures associated with the photospheric magnetic field distribution. But the structure usually has a rather different form associated with lines of different characteristic temperatures.

*Operated by the Association of Universities for Research in Astronomy (AURA) under cooperative agreement with the National Science Foundation.

While it is clear that this is due to variations in loop temperatures, the detailed differences are not well understood. Gross features observed in the K-corona are found also in emission-line images[2], which is to be expected if the structures are controlled by the magnetic-field distribution.[3] However, lack of high-angular-resolution observations precludes precise comparison and the picture remains unclear. The relationship of hot coronal material with cooler prominence material, and with cool material in general in the coronal environment, has not been well studied due to instrumental limitations, such as insufficient sensitivity, angular resolution and observing flexibility. Objects such as plasmoids are poorly observed, except perhaps in radio observations, because of inadequate resolution and sensitivity, and mechanisms responsible for condensation and flow of coronal plasma above active regions and around prominences are poorly understood. Short-time-scale dynamical events such as rapid realignments[4] and oscillations and/or waves[5] have been observed, but such observations have provided only very limited data.[6]

Post-flare loop systems have been well-studied. But it is only with improved-quality data that a phenomenon of loop interaction with partial magnetic reconnection becomes evident.[7] These events can have dimensions of only a few arc seconds, and critical studies will require an angular resolution far superior (< 0.5 arcsec) to that typically available from existing coronagraphs. Extremely faint (relative to the average coronal field) features are occasionally observed at the limit of detection of existing instrumentation.[8] These data suggest that a whole regime of faint events that possibly occur frequently in the corona are simply never observed. Flow velocities associated with coronal holes are of critical importance in solar wind studies. Such measurements are extremely difficult to obtain because of the low light flux and hence extremely noisy signals, requiring long integration times that result in low precision data.[8] Current estimates of the magnetic field derived from theoretical considerations and from radio work differ up to an order of magnitude. Such examples serve to illustrate the current inadequacy of ground-based coronal observational capability. Specifically, much greater photon flux and also much higher spatial and temporal resolution than is available from any existing instrumentation is required.

Space-based, white-light coronagraphs[9,10] have provided a large database of coronal transients, especially coronal mass ejections, but the important inner corona is not accessible, due to the vignetting characteristics of externally-occulted coronagraphs and to the overall design constraints that must allow for the imprecision of the pointing system. Recent X-ray coronal observations from Yohkoh[11] constitute a rich source of data that directly links the structure of surface-activity patterns with that of the more extended corona. Nevertheless, the angular resolution is far less than can be obtained from existing ground-based coronagraphs.

Recent stellar observations have shown the need for a nighttime coronagraph. Primarily resulting from IRAS observations, it has become apparent that many stellar systems are surrounded by large, faint, very extensive cloud systems. Although chopping techniques can be used to explore such systems, it is clear that a nighttime coronagraphic telescope is required to make definitive measurements of cloud and/or planetary circumstellar material.

In summary, progress in the field of coronal physics requires substantial improvements in the quality of the observational data, and hence, major advances in the capabilities of the instrumentation. If this can be fully realized, there is the prospect that many outstanding and fundamental problems can be critically studied by observational methods, such as the origin of coronal heating, mechanisms for accelerating the fast component of the solar wind in coronal holes, and the origin and behavior of CME's.

2. Key Requirements for a New Coronagraphic Telescope

a. White-light (achromatic) high-resolution capability. Over a large field-of-view of order of 10 arcmin^2, the spatial resolution should be at least 1 arcsec, with the capability of a much higher resolution over a substantially smaller field.

b. Transmission extending from the near UV to the IR, with achromatic properties.

c. Very low net instrumental polarization. To permit precise measurements of the magnetic field, very small polarization aberrations are desirable. Both circular polarization and linear polarization aberrations are of concern.

It is noted that for coronagraphs in general, the level of scattered light should be $\lesssim 10^{-6}$ at a radial distance of 30 arcsec above the limb. For a conventional (non-coronagraphic) telescope, this level can sometimes reach 10^{-2}!

3. Infrared Advantages

Near-IR coronal emission lines have some advantages compared with those at visible wavelengths. First, the 10,747 Å Fe XIII line is relatively strong. The two lines, 10,747 Å and 10,798 Å, as well as the 3,388 Å line, all Fe XIII transitions, together provide a useful coronal density diagnostic. The line intensity ratios, I(10,747)/I(10,798) and I(3,388)/I(10,747), are functions of the electron density that are independent of the temperature in the equilibrium range of the Fe XIII line. The upper levels are excited by electron collisions (as well as by radiation); hence the populations for the two lines will be dependent on the electron density. Further, the radiative de-excitation coefficients are known. Significant uncertainties in the measurements are present when the density is relatively high. However, it has been established that this technique is useful as an electron density diagnostic in the corona provided that sufficient accuracy in the line intensity measurements can be achieved.[12]

Studies of the coronal magnetic field are, in general, more easily carried out in the infrared than the visible, with increased Zeeman and linear-polarization sensitivity.

Coronal emission displays linear polarization when radiatively excited. This is due to resonance fluorescence from the anisotropic excitation by the incident photospheric radiation field. Therefore, in general the magnitude of the polarization increases with height. The fluorescence arises in the magnetic dipole transition of the ground terms of coronal ions. The intensities of emission lines depend on density, temperature (primarily through the ionization equilibrium) and to a lesser extent on field geometry. The polarization and intensity together contain information about the magnetic field direction. An interpretive routine can be used to obtain three-dimensional maps of the direction of the magnetic field corresponding to each Stokes vector (I, Q,U)[13,14] in a plane-of-the-sky analysis, which is thought to be a reasonable assumption. The asymptotic limit of polarization of the 5303 Å line of Fe XIV, neglecting collisions, is 0.43. For the 10747 Å line of Fe XIII, the limit is unity, since its transition ends in only one magnetic quantum sublevel. It is already 0.9 at a height of two solar radii.[15] Observations in 10747 Å therefore have the advantage of high intrinsic polarization while observations in 5303 Å have low polarization, but the instrumentation is more standard.

In general, the performance of ground-based coronagraphs improves with increasing wavelength, especially due to the reduction in sky brightness, as well as the reduction in fractional instrumental scatter. The sky brightness decreases as a function of wavelength. Rayleigh scattering is approximately $\propto \lambda^{-4}$, which means that the spectral brightness due to molecular scattering quickly becomes negligible into the infrared. The angular and spectral scattering characteristics of aerosols depend on the size, shape, internal structure and complex refractive indices of the individual particles, and on their large-scale spatial distribution. Measurements of particles in the upper atmosphere (upper troposphere and stratosphere) indicate diameters typically < 0.2 μm.[16] Such particles would likely have irregular shapes, but if hygroscopic, would tend to be spherical. In the limit of $p \equiv 2\pi a/\lambda \ll 1$, where a is the particle radius and λ is the incident wavelength, the scattering of a spherical particle with a real refractive index has the same spectral dependence as that of Rayleigh scattering. For $p \gg 1$, the aerosol scattering coefficient can be assumed to be roughly constant with wavelength, and for p in the intermediate range, the scattering coefficient can be assumed to be proportional to λ^{-n}, with n usually in the range 0.8 to 1.6 for a high-altitude site such as Sac Peak. Even though particles of size larger than about 0.5 μm can potentially cause significant scattering for observations in the near infrared ($\lambda \sim 1\mu$m), in practice their number density is extremely small under "normal" conditions at mountain sites, with minimal degradation of clear-sky conditions.[17]

Beyond the advantage of the reduced level of sky background, the scattering properties of a coronagraph objective improve as a function of wavelength. The relationship between the roughness of an optical surface and the resultant scattered radiation has been treated extensively in the literature.[18] In the case of an rms roughness, σ, where $\sigma \ll \lambda$, for radiation of wavelength, λ, and with a Gaussian distribution of roughness, the scattered component of reflectance at normal incidence, R_s, can be characterized by, $R_s \sim R_o(2k\sigma)^2$, where R_o is the reflectance of a completely smooth surface, and $k = 2\pi/\lambda$. With the availability of high-quantum-efficiency IR detector arrays, high-resolution imaging can be expected, especially since the seeing characteristics of the atmosphere also improve with wavelength - Freid's seeing parameter, r_o, varies as $\lambda^{1.2}$. However, some observational tests at NSO/SP at 4440 Å and 16,000 Å indicated $r_o \propto \lambda^{1.5}$. Some observations in the near-infrared are possible with a conventional coronagraph, taking advantage of the new IR arrays. But high-angular observations would be possible only with a conventional coronagraph that has a design optimized for near-infrared wavelengths. An equivalent reflecting coronagraph would have high intrinsic optical quality independent of wavelength.[19] It is noted that a coronagraph operating in the IR should have a proportionally-larger aperture than one in the visible to maintain the same angular-resolution capability.

4. Coronagraph Mirror Technology

Significant improvements in the quality of coronal observations can be achieved by the use of high-quality image intensifiers, to improve temporal resolution, and advanced CCD detectors, providing the possibility of substraction of on- and off-band images to reduce greatly the scattered-light contribution in the final image. A combination of image intensifier and CCD detection techniques can allow video-rate recording, with the advantage of frame selection on the basis of the quality of the instantaneous seeing. However, all such observations are still fundamentally limited by the use of lens objectives.

A mirror objective avoids the problems inherent in a singlet-lens objective coronagraphic design and large apertures are feasible. The primary image is achromatic, and with an all-mirror design, high-angular-resolution infrared and ultraviolet observations can be achieved. Mirror objectives for coronagraphs have been proposed[20,21] and successfully applied to small, externally-occulted, rocket- and balloon-borne coronagraphs.[22,23] Recently, extremely low-scatter mirror development has allowed the use of such mirrors as objectives for internally-occulted coronagraphs.[19,24] Measurements of the scattered light of several super-polished mirrors have been carried out at NSO/SP. For a 5-cm diameter, silicon super-polished (0.25 nm rms) objective mirror, the measured scattered light level was 10^{-6} B_{source} at an angular distance equivalent to 1.5 R_o from Sun center. This mirror has obvious imperfections and a higher-level performance is to be expected. Nevertheless, its performance is similar to that of a typical internally-occulted, lens-objective coronagraph. This is to be compared with the brightness of the 5303 Å (Fe XIV) coronal emission above active regions with typical values of $> 10^{-5}$ B_o (solar maximum) to $> 10^{-6}$ B_0 (solar minimum) at 1.3 R_o.

Glass and other optical substrates can now be polished with an extraordinarily low level of micro-roughness. Such "super-polished" surfaces, with roughness values in the range of a few Angstroms rms, can now be achieved routinely. In fact, surfaces with areas of 100 square centimeters, or larger, have been polished with a sub-Angstrom rms roughness. Moreover, techniques have been developed for producing reflective films, such as aluminum, that have virtually no intrinsic micro-roughness. The amount of scattered light produced by a high-quality optical mirror depends primarily on the residual microroughness of the substrate and on the reflective coating, apart from that due to surface defects, dust particles and other contaminants.

The term "microroughness" generally refers to roughness over surface scales in the micron range. Surface ripple where the scale is in the millimeter range (roughly from a fraction of a millimeter to a few millimeters) commonly occurs on polished optical surfaces. Such roughness contributes also to stray light, but mainly in the form of extremely low-angle scatter (at visible wavelengths), since in broad terms light is scattered/diffracted through an angle λ/b, where b is the period of some characteristic scale of roughness across the surface. For example, for $\lambda = 0.5\ \mu$m, and b = 0.1mm, a first order maximum occurs at an angle from the specular direction $\sim$ 17 arcmin. While both surface characteristics should be minimized for a solar coronagraph, micro-roughness is of primary concern. Larger-scale roughness is of critical importance for typical nighttime studies where a bright source is angularly close (a few arcsec or less) to some faint emission under study.

The Zernike phase-contrast test is suitable for measuring surface ripple in the sub-millimeter, millimeter or larger-scale ranges of large optics, but this test is rarely used, even though easily carried out. Its principle is conceptually simple. In an image plane where the specular and scattered components of the light are spatially separated, a filter is introduced that attentuates the direct light to balance the amplitudes of the two components, while also putting them (ideally) in a mutual state of phase quadrature, thus maximizing the contrast of the surface features.[23]

The scattered component of reflectance at normal incidence, R_s, was given in Section 3 as, $R_s \sim R_o(2k\sigma)^2$, where R_o is the reflectance of a completely smooth surface and $k = 2\pi/\lambda$. The corresponding scattered component of transmittance is, $T_s \sim T_o[|n_p - n_q|k\sigma]^2$, where T_o is the transmittance of a specular beam through a smooth surface that bounds media of reflective index, n_p and n_q. For a glass surface in air, $T_s \sim T_o\ (0.5k\sigma)^2$. It is noted that under these conditions the scattered components are proportional to σ^2 and λ^{-2}. Hence, for reflecting and transmitting surfaces with the same characteristic roughness, the reflected scattered component is roughly 16x that of the transmitted scattered component. But two surfaces of a singlet lens contribute to the scattering. Hence, neglecting bulk scattering in a lens, a super-polished mirror would have a similar performance to a super-polished singlet lens when the rms roughness of the mirror, $\sigma_m \sim \sigma_1/3$, where σ_1 is the roughness of the surfaces of the lens. IR observations clearly have the advantage that the relative surface roughness improves $\sim \lambda^{-2}$.

5. Mirror-Objective Coronagraph Designs

5.1. MAC I

A miniature reflecting coronagraph has been constructed at Sac Peak.[19] This Mirror Advanced Coronagraph (MAC I) is the first of two prototype reflecting coronagraphs developed at Sac Peak. The objective is a 5-cm super-polished, spherical silicon mirror with a focal length of 1-m, and a micro-roughness < 0.3nm rms. The optical system is simply off-axis reflection from the primary mirror to a secondary optical system that is a conventional Lyot coronagraph configuration, namely an occulting disk, field lens, Lyot stop, narrow-band filter and camera system. Both a photographic camera, as well as a video CCD camera together with a Varo Image Tube, have been used to record images of the green-line corona, the first coronal images obtained with a ground-based reflecting coronagraph. Much of the background in coronal images is due to the contribution of "flying dust" particles. The trace of these particles in the video image can be removed post facto by applying an algorithm to the digitized data that discriminates between stationary and moving components.[25] Due to its very small aperture and the off-axis configuration, MAC I does not produce high-angular resolution images. However, the large annular field-of-view of this coronagraph, coupled with the use of an image tube, allows its application as a sensitive prominence monitor. With a digital recording system, software can be developed to trigger an alert of the onset of a prominence eruption, of fundamental importance in the study of CME's.

5.2. MAC II

The successful development and operation of MAC I has led to the construction of a second, more advanced, prototype instrument, MAC II, based on a 15-cm diameter, super-polished mirror objective. In a joint agreement with the Institut d'Astrophysique (IAP), part of the construction has been carried out by IAP. Two super-polished objective mirrors of 2.25 m focal length have been prepared by IAP, one of fused silica and the other of zerodur, each with an rms roughness ~ 0.4 nm. The objective mount, designed and constructed at IAP, allows extremely accurate tilting of the mirror under remote control, required to center the solar image accurately relative to the secondary optical system. A concave annular field mirror, located near the primary focal plane, forms an image of the objective at the position of the Lyot stop. This field mirror functions as an inverse occulting disk (unlike the MAC I, the coronal field is reflected, while the solar image is transmitted through the central hole to a light trap). The remainder of the secondary optical system consists of a Lyot stop, collimating lens, filter and camera system.

The design of this instrument represents overall the preferred optical system for a reflecting coronagraph, except that an equivalent all-mirror system would preserve full achromaticity and allow UV and IR observations to the full extent of atmospheric transmittance, mirror reflectance and available detectors. However, it should be pointed out that the optimum f/number of the primary objective of a reflecting coronagraph is to some extent a function of the aperture – objectives with off-axis, aspherical surfaces would allow compact, high-angular-resolution designs, but existing technology for producing such surfaces might preclude large aperture ($\gtrsim 1m$) systems with relatively small f/numbers.

However, the successful performance of this mirror-objective coronagraph points to a new era of astronomical observational techniques, namely the development of very low-scattered-light level achromatic telescopes of large aperture with important applications in the near UV and IR.

5.3. MAC III

A third, research quality, instrument has been designed based on a 60-cm diameter Si-SiC objective. A small field mirror covers only the part of the corona under study. Different regions of the corona are then observed by rotation of this mirror around an axis passing through the center of the solar disk. The optical system has four mirrors with an overall length $\sim$ 7m, and an effective focal length of 17m in a f/34 beam. Spectral analyzing instruments, including a grating spectrograph, an LC Fabry-Perot etalon and a polarimeter, incorporating fast spectral chopping techniques, are planned for this instrument.

It will have more advanced technologies than those of MAC I and II. For example, the high-thermal-conductivity objective mirror will be actively cooled to minimize mirror seeing. Electrostatic dust control will be used, both for the air flow system within the tube and for the objective itself. An addressable LC pupil mask will be actively controlled to suppress light scattered at the objective surface. A passive system such as photochromic glass has been considered, but this material is not sufficiently sensitive for this application in the form that it is currently produced. Image stabilization via a fast mirror will be used. MAC III should be especially useful for IR studies.

5.4. MAC IV

A dual-purpose (solar and nighttime applications) coronagraph with an aperture of two or more meters, and the same basic design as that of MAC III, was envisaged as the last stage in this program to develop reflecting coronagraphs, [19] provided that the development of the smaller instruments proved successful. A general concept for such a large instrument is suggested in figure 1. The need for a large-aperture (meter class, or larger) coronagraph has been advocated, [19,26,27] and optical design considerations for a 2-m aperture instrument investigated.[28] A preliminary proposal for a 4-m coronagraphic telescope has also been presented.[29] In principle, a large coronagraph could be deployed in space, a lunar base having advantages for some critical coronagraphic observations.[30]

Designs for reflecting coronagraphs outlined above are off-axis, with the advantage of a clear entrance aperture, thus avoiding scattering and diffraction and thermal emissivity from a central obscuration in this plane. However, it has been pointed out that masking the pupil (Lyot stop) plane can be effective in eliminating stray radiation from a central obscuration; for IR work, a cooled mask would be used. An on-axis design would reduce the critical demands on mirror manufacture, but the direct solar light might cause excessive internal scatter in solar use which could prove to be an insurmountable problem. A tradeoff study will probably be needed to compare the merits of on- or off-axis designs. In either case, since the secondary (field) mirror is located at the primary focal plane and because of the technical difficulties of producing super-polished surfaces, any design is likely to be a relatively slow system.

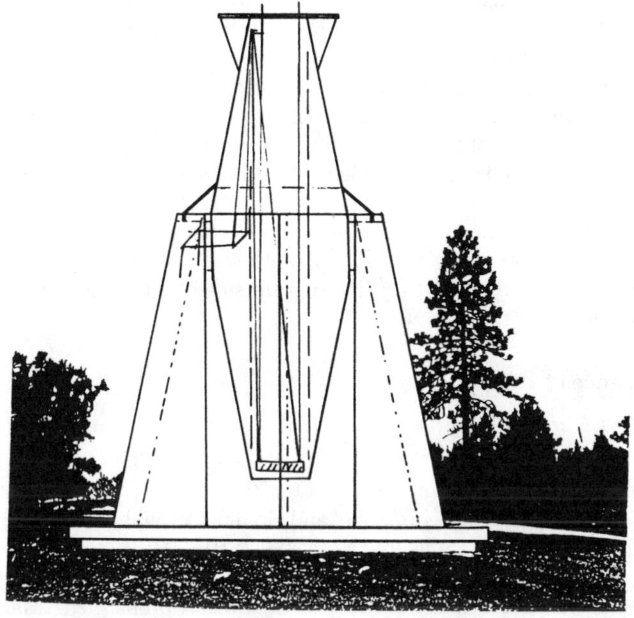

Figure 1: Concept drawing for a large-aperture coronagraph, shown in a stored position.

Although apparently technically feasible, even a large ground-based instrument presents special engineering challenges. Super-polishing of optics to the level of smoothness required, has been accomplished so far only on comparatively small optical components. As pointed out in Section 4, the finish on the objective would need to be such that it gives minimal scatter for both low- and higher-angle scatter to satisfy its dual use applications. The surface of the objective mirror must be maintained substantially dust-free under operating conditions. Techniques devised for MAC III should be effective here. Also, cleaning optical surfaces by Eximer lasers now appears extremely promising.[31] For solar use, special fail-safe engineering procedures will be required to manage the large energy contained in the primary image of the solar disk. Especially for many nighttime studies where the fields of interest are only a few arc seconds, an adaptive optics system is essential. All such aspects of the design and construction of a large coronagraph will need careful analysis and presumably some development studies, but it is anticipated that at least some of this work can be accomplished through the development of MAC III.

6. Conclusion

There is a pressing need for major improvements in the quality of coronal observations. Further independent work points to the need for low-scattered-light coronagraphic-type telescopes for many planetary, stellar, galactic and extragalactic studies. Both of these requirements can be met by the development of reflecting coronagraphs. A small reflecting coronagraph (MAC I) developed at NSO/SP has been used to obtain images of the solar emission corona in the green line, thus establishing the validity of this reflecting coronagraph technology for ground-based applications. The development of the second prototype instrument, and the design of a third, research-quality instrument are viewed as prerequisites for the eventual development of a much larger instrument suitable for both solar and nighttime studies. The work accomplished so far in the NSO/SP program suggests that such a major instrument is technically feasible.

7. References

1. Altrock, R.C.: 1988, "Solar and Stellar Coronal Structure and Dynamics", ed. R.C. Altrock (Sunspot, NM: NSO/SP), 288.
2. Koutchmy, S.: 1988, in "Solar and Stellar Coronal Structure and Dynamics", ed. R.C. Altrock (Sunspot, NM: NSO/SP), 208.
3. Querfeld, C.W. and Smartt, R.N.: 1984, *Solar Phys.*, **2**, 131.
4. Dunn, R.B.: 1971, in "Physics of the Solar Corona", ed. Macris (Dordrecht-Holland: Reidel), 114.
5. Pasachoff, J.M. and Ladd, E.F.: 1987, *Solar Phys.*, **109**, 365.
6. Tsubaki, T.: 1988, in "Solar and Stellar Coronal Structure and Dynamics", ed. R.C. Altrock (Sunspot, NM: NSO/SP), 140.
7. Smartt, R.N. and Zhang, Z.: 1987, in "Theoretical Problems in High-Resolution Solar Physics: Workshop Proceedings (2nd)", ed. G. Athay (Boulder, CO: NASA), 129.
8. Smartt, R.N. and Zirker, J.B.: 1981, in "Solar Terrestiral Physics, Proceedings of Second Indo-US Workshop on Solar Terrestrial Physics", ed. M.R. Kundu, B. Biswas, B.M. Reddy, S. Ramadurai (New Delhi: CSIR), 269.
9. MacQueen, R.M., Gosling, J.T., Hildner, E., Munro, R.H., Poland, A.I., and Ross, C.L.: 1974, *S.P.I.E.*, **44**, 207.
10. Sheeley, N.R., Jr., Michels, D.J., Howard, R.A. and Koomen, M.J.: 1980, *Astrophys. J.*, **237**, L99.
11. Ogawara, Y., Takano, T., Kato, T., Kosugi, T., Tsuneta, S., Watanabe, T., Kondo, I., and Uchida, Y.: 1991, *Solar Phys.*, **136**, 1 (see also complete issue of *Solar Phys.*, **136/1**).

12. Noëns, J.C., Pageault, J., Ratier, G.: 1984, *Solar Phys.*, **94**, 117.
13. Querfeld, C.W.: 1982, *Ap. J.*, **255**, 764.
14. Querfeld, C.W. and Smartt, R.N.: 1984, *Solar Phys.*, **91**, 299.
15. House, L.L.: 1977, *Astrophys. J.*, **214**, 632.
16. Bigg, E.K.: 1976, *J. Atmos. Sci.*, **33**, 1080.
17. Halthore, R.N.: 1992 (private communication).
18. Elson, J.M., Bennett, H.E., and Bennett, J.M.: 1979, in R.R. Shanon and J.C. Wyatt (eds.), "Applied Optics and Optical Engineering 7", Academic Press, New York, p. 191.
19. Smartt, R.N., Koutchmy, S.L., Colley, S.A., Caron, R., Schwenn, R., and Restaino, R.R.: 1990, *S.P.I.E.*, **1236**, 206.
20. Newkirk, G., and Bohlin, D.: 1963, *Appl. Opt.*, **2**, 131.
21. Zirin, H. and Newkirk, G.Jr.: 1963, *Appl. Opt.*, **2**, 977.
22. Kohl, J.L., Reeves, E.M., and Kirkham, B.: 1978, in K.A. van der Hucht and G. Vaiana (eds.), "New Instrumentation for Space Astronomy", Pergamon, New York, p. 91.
23. Smartt, R.N.: 1979, *SPIE*, **190**, 58.
24. Epple, A. and Schwenn, R. (these proceedings).
25. Koutchmy, S., Colley, S., Smartt, R.N., Nitschelm, C., and Zimmermann, J.P.: 1990, *SPIE*, **1235**, 849.
26. Tsubaki, T. (these proceedings).
27. Smartt, R.N. and Koutchmy, S.: 1992, in M. Giampapa and J. Bookbinder (eds.), "Seventh Cambridge Workshop on Cool Stars, Stellar Systems and the Sun", ASP Conference Series Vol. 26, p. 660.
28. Cross, E.W.: 1992, *SPIE*, **CR41**, 225.
29. Beckers, J.M. (these proceedings).
30. Smartt, R.N.: 1992, in "Eng. Construction and Operations in Space III", eds. W.Z. Sadeh, S. Sture, and R.J. Miller (N.Y.: A.S.C.E.), p. 1890.
31. Balick, B.: (private communication).

4-METER UPGRADE OF THE MCMATH-PIERCE TELESCOPE AND THE DETECTION OF SPACE DEBRIS

WILLIAM LIVINGSTON
National Solar Observatory
P.O. Box 26732, Tucson, AZ 85726

and

D.K. Lynch
Aerospace Corp., P.O. Box 92957, Los Angeles, CA 90009-2957

ABSTRACT

We review the IR criteria for a solar telescope, including response to 40μm, and show how the McMath-Pierce telescope could be improved by an increase of aperture and modification to the tunnel baffling. Ground-based solar telescopes might be used for the detection of space debris. Atmospheric calculations show that the optimum wavelength for the daytime sensing of space debris lies around 2.4μm, but for objects up to 30 cm in size, the contrast will be less than $10-1$. We conclude that a non-coronagraph type telescope is inadequate for the task.

1. Introduction

The existing McMath-Pierce telescope is well suited for solar observations in the infrared for the following reasons:

1. All reflecting system - including spectrograph with large grating that functions to 14μm.
2. Moderate aperture of 1.5 m - diffraction limit of 0.25 arcsec at 1.5μm; 0.8 arcsec at 5μm; 2 arcsec at 12μm.
3. Unobstructed aperture - lower thermal background than with a Cassegrain arrangement.
4. Long focal length - again low thermal background because solar flux at mirror surfaces is modest.

The facility could be improved for IR work by an increase to 40meter aperture and by attention to telescope seeing problems.[2] A conversion to 4-meter would lead to diffraction limit of 0.75 arcsec at 12μm. Mirror seeing could be eliminated by use of a high thermal conductivity mirror substrate, such as silicon carbide, together with liquid cooling to track the ambient air temperature. Tunnel seeing could be ameliorated by baffling the upper portions of the tunnel/exterior interface.

2. Wavelengths for the Solar IR

Figure 1 gives the calculated transmission of the atmosphere showing the nominal windows for IR work at a sea level site. Water vapor is the crucial absorber. In practice, however, IR observatories range in elevation from Kitt Peak at 2000 m, Sac Peak at 2800 m, the Jungfraujoch at 3580 m, to Mauna Kea at 4200 m.

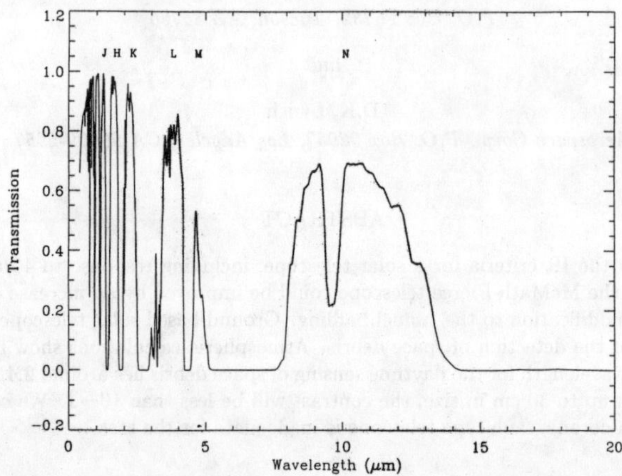

Figure 1: Atmospheric transmission for sea level with Sun at z = 30°.

Another factor is that cold, dry, air occasionally will invade from polar regions. Under the latter condition, atmospheric transmission as determined by water vapor is greatly increased and useful signals are available out to near 40μm, Fig. 2. Much of this domain is unexplored for solar observations.

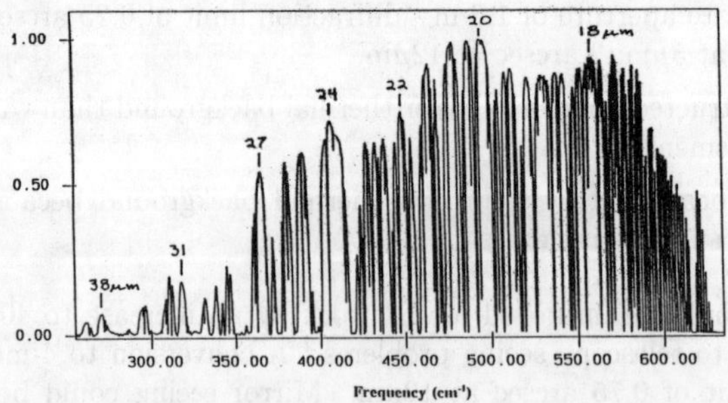

Figure 2: FTS spectrum from Jungfraujoch.[1]

3. Detection of Space Debris

Although not a solar physics issue, we have learned at this workshop the way an IR coronagraph could contribute to map and keep track of space debris. How would the McMath-Pierce perform for this problem?

Discrimination of a passive body in space against the sky background is a function of wavelength. As we go to longer wavelengths, Rayleigh scattering, which is proportional to λ^{-4}, falls off rapidly. For example, at 2μm sky brightness is 0.004, the value at 0.5μm.[4]

Figure 3 shows the theoretical noon day radiance of the clear sky for the Sun at the zenith and the observer looking at 30° zenith distance (solid line). There is a clear minimum near 2.5μm. To the short wavelength side of this minimum, the light is primarily scattered sunlight. Longward of 2.5 it is largely thermal emission from the atmosphere. In principle, this minimum is a good wavelength region to look for space debris during the day. Fig. 3 also shows the midnight spectrum for the same observer looking 30° from the zenith. Note the presence of scattered sunlight at short wavelengths for the daytime spectrum (solid line).

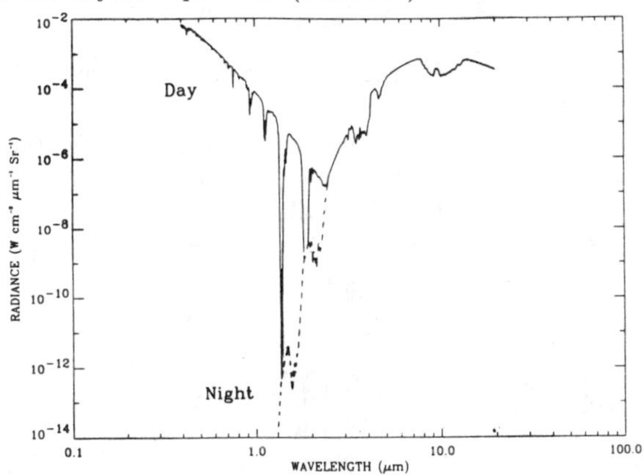

Figure 3: Day and night sky radiance

A telescope is also a scatterer/emitter. Dust and scratches on its mirrors scatter light and the surfaces also emit thermal radiation. The minimum light coming from any telescope is usually in the 2-4μm range, depending on the cleanliness of the mirror and the freshness of their coatings. A recently aluminized mirror will have a lower thermal emissivity than an old, tarnished one. A newly cleaned mirror is also less likely to have dust (both an emitter and scatterer) on its surface.

Just before the heliostat was recently recoated, the scattered light at 100 arcsec off the limb was 8.8 (10^{-4}). After aluminizing, it was 3.8 (10^{-4}, a factor of two improvement. For a comparison with the visible see Pierce.[3]

Figure 4 shows the theoretical noon day radiance of the clear sky as measured with a 3 arcsec pixel in the detector. The Sun is at the zenith and the observer looking at a 30° distance. Also shown are the radiances of two gray body spheres at an altitude of 1000 km, one 3 cm in radius, the other 30 cm in radius. The spheres are gray at all wavelengths with !abs - 0.5, and they are at an effective blackbody temperature of 5700 K.

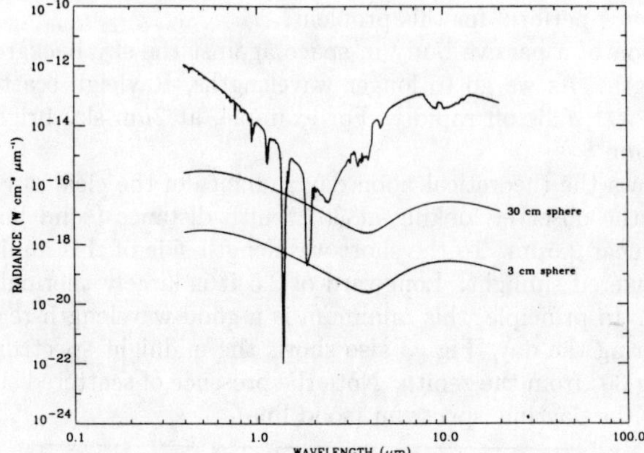

Figure 4: Sky and target radiance for a 3 arcsec pixel.

Figure 5 shows the ratio of the sphere brightness to the brightness of the atmosphere, again in a 3 arcsec pixel. There is an obvious maximum near 2.5μm, again suggesting that the contrast of the object will be highest here.

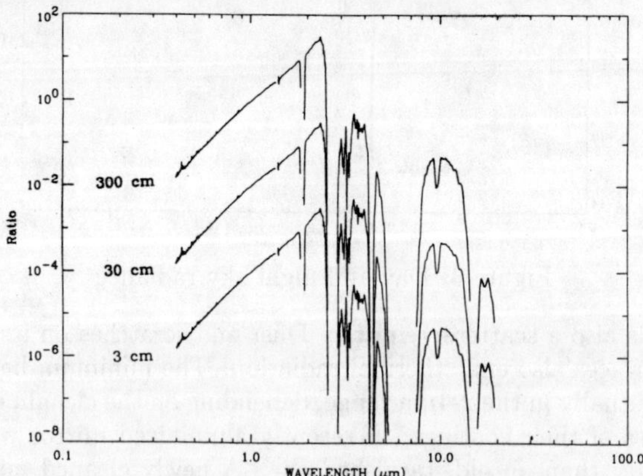

Figure 5: Target/sky radiance for different size spheres.

As the contrast for a 30 cm object is at best only around 10−1, we conclude that, within the above assumptions, a non-coronagraph-type telescope is inadequate for the detection of space debris.

4. Acknowledgments

S. Mazuk made the atmospheric radiance calculations using the MODTRAN code.

5. References

1. Farmer, C.B., Delbouille, L., Roland, G. & Servais, C.: 1994, *San Juan Capistrano Res. Inst. Tech. Rpt.*, **94-2**.
2. Livingston, W.: 1994, "Infrared Solar Physics", Rabin, D., Jefferies, J.T. and Lindsey, C. ed. Kluwer, Dordrecht: Kluwer.
3. Pierce, A.K.: 1991, *Solar Phys.*, **133**, 215.
4. Smartt, R.N., Koutchmy, S., & Noëns, J.-C.: 1994, "Infrared Solar Physics", Rabin, D., Jefferies, J.T. and Lindsey, C. ed. Kluwer, Dordrecht: Kluwer.

THE CORONAGRAPH AT THE DIFFRACTION LIMIT,

THE DETECTION OF FAINT SOURCES

CHRIST FTACLAS
Advanced Development, Hughes Danbury Optical Systems
100 Wooster Heights Road, Danbury CT 06804

ABSTRACT

At longer wavelengths, with better sites, improved telescope engineering and the use of adaptive optics, ground based optical performance can approach more closely the ideal diffraction limited case. As this limit is approached the stellar coronagraph becomes capable of very high background suppression. Realizing this enhanced performance, however, particularly in the detection of faint unresolved sources, requires careful design of the entire optical system which now must be considered as part of the coronagraphic instrument. Problems of faint source detection and the performance of coronagraphs are discussed within the constraints of ground based operation.

1. Introduction.

In developing any new instrument one strives always to explore some new region of observational parameter space. Observing in a new spectral window or in a new regime of spectral or spatial resolution is almost a guarantee of new and exciting science. The stellar coronagraph is designed to probe such a new window. That window is a part of the sky never before imaged, the region in the immediate vicinity of any bright source. Here direct, unaided observations are virtually impossible because the halo of diffracted and scattered light surrounding the source corrupts and elevates the local background. But it is here that important observations can be made on the structure of quasars, the birth and death of stars, and the existence, formation and evolution of other planetary systems.

Despite its central importance to astronomy, the circumstellar environment remains largely unprobed save for inferential methods. Direct imaging, however, must remain the ultimate goal. In the detection of extrasolar planets and protoplanetary systems, for example, where astrometry has often been invoked[1, 2] as the most robust, indirect probe of the circumstellar environment, direct imaging means that detection becomes possible on time scales considerably shorter than the orbital period of the system with a significant improvement in the ability to resolve multiple planet systems. Furthermore, imaging is the only direct probe of symmetrically distributed circumstellar matter at the present time.

Born from Lyot's[3] early considerations of the difficulties of observing the solar corona, the stellar coronagraph found limited application in the domain of unresolved sources. It has been the subject of renewed interest[4, 5, 6], however, following its use to discover a protoplanetary dust disk around β-Pictoris[7]. The stellar coronagraph shares one important characteristic with the solar coronagraph which is not common in the field of astronomical instrumentation. We traditionally think of an instrument as a device added to a telescope in order to make a given observation. In the case of the coronagraph, however, the telescope is an intimate part of the instrument itself.

The halo that dominates the circumstellar region consists of both ideal (i.e. unavoidable diffraction) and non-ideal (i.e. scatter) intensity contributions and in order to reduce the halo to manageable levels both must be reduced: the former by the coronagraph and the latter by appropriate design, manufacture and siting of the optics. Coronagraphic potential on the ground is necessarily limited by phase modulation of the incoming wave front by the atmosphere, yet that limit probably has not been reached in most telescopes because the rest of the optical system does not permit it. One exciting consequence of the current ongoing revolution in ground based astronomy is the possibility of extending coronagraphic performance to new and scientifically significant levels. As our understanding of site potential, locally induced seeing, and telescope system engineering improves we move closer to reaching limits set solely by the free atmosphere.

In addition to the dependence of coronagraphic performance on telescope optics and site quality there is also an optimal, sometimes counterintuitive, implementation of focal plane and data reduction strategies that must be implemented to exploit coronagraphic instruments fully. The following sections will describe the basic theory behind the operation of the stellar coronagraph and show how its performance has been dramatically improved. The special problems of making coronagraphic observations on the ground will then be addressed, followed by a discussion of the detection of faint unresolved sources using coronagraphs.

2. The Stellar Coronagraph.

In order to understand the action of the stellar coronagraph, it is best to consider it initially in the ideal domain of pure Fraunhofer aperture diffraction. In this limit the incoming wave front is flat, devoid of any phase irregularities and the imaging system also introduces no phase errors. In this case the field distribution in the focal plane can be written:

$$\mathcal{E}(p,q) = \mathcal{K} \int\int C(x,y) e^{-\frac{2\pi i}{\lambda}(px+qy)} dx dy \qquad (1)$$

where $\mathcal{K}$ is a normalization constant, p, q are (dimensionless) coordinates in the focal plane, λ is the wavelength in microns and $C(x, y)$ describes the shape of the telescope entrance pupil. With the exception of the wavelength dependence of the transform kernel, the integral in Eq. (1) can be recognized as the two dimensional Fourier transform of $C(x, y)$. If we rescale the focal plane coordinates by the wavelength, the identification is exact and the pupil plane and scaled focal plane amplitudes are Fourier transform pairs.

We can write this relation as:

$$\mathcal{E} = \widetilde{C'} \tag{2}$$

where the prime indicates scaled focal plane coordinates and the ~ indicates the Fourier transform operation.

Given the conjugate relation between focal and pupil plane amplitudes, it is most straightforward to consider the coronagraph as a Fourier filter. The stellar coronagraph requires that the stellar source be occulted in the focal plane and that this be followed by a reimaging of the pupil. Once we regard the point spread function as the Fourier spectrum of the pupil, it is clear that occultation in the focal plane is equivalent to filtering out the low spatial frequency content of the pupil. Reimaging the filtered pupil is equivalent to reconstructing the pupil only in its high spatial frequency content, that is, its edges.

It is tempting to view the occulting mask of the coronagraph as simply blocking light, but it is actually doing a spatial filtering of light diffracted from the slowly varying spatial frequency content of the entrance pupil. When this light is removed by the occulter the reimaged pupil contains only light that is diffracted by rapidly varying pupil features which are characteristically edges such as inner and outer aperture stops and secondary support structures (spiders). When the stellar source is occulted and the pupil reimaged, all edges in the pupil light up and its interior regions are dark. At this point more diffracted light can be rejected from the system by masking off these bright edges with a pupil plane (Lyot) stop.

When the beam is brought to a second focus following the Lyot stop, the central source appears with reduced intensity, particularly in its side lobe content. The side lobes are reduced even in regions of the focal plane that were not occulted, which makes it clear that the occulting and Lyot masks are doing something more complicated than simply blocking light. This combination of focal plane and pupil plane masks is the essential content of the stellar coronagraph and such systems are often called "doubly diffractive" in the optical literature.

The stellar coronagraph treats occulted sources differently from all other sources. Any in-field source whose image is occulted will not have its reimaged pupil image filtered and therefore suffers only the throughput reduction of the Lyot stop. Since the Lyot stop is necessarily undersized with respect to the pupil image, there is a throughput loss associated with its use. If the side lobe intensity is reduced by more than the throughput loss, however, there will be a net gain in visibility of faint regions around the central source. In a study for the Jet Propulsion Laboratory of diffraction control in space applications for extrasolar planet detection[8], we showed that a classical Lyot coronagraph could produce at most a diffraction side lobe reduction of order 100. That is, the diffractive side lobe intensity in the second focal plane would be about 1% of its value in the first focal plane with a throughput loss of about 20%.

In the ideal performance domain there is clearly a significant gain in visibility that can be achieved for background dominated observations using a basic Lyot coronagraph. Unfortunately it is still at least two orders of magnitude too low for the planet detection problem. The same study showed that the performance of the Lyot coronagraph could be significantly enhanced by modifying the occulting mask so that its transmission was tapered from a value of zero at the center of the field to unity at some radius away. The form of the taper, and its dimensions are disposable parameters but getting rid of the hard edge of the traditional occulting mask is crucial to enhancing the performance of the instrument.

For our tapered transmission occulting masks we find that several significant benefits accrue, including:

1. The tapered mask is more effective at leaving light only at the pupil edges making any Lyot stop more efficient (Figure 1.)
2. Imaging through the tapered mask is possible virtually up to the center of the field.
3. Centering the target star behind the tapered mask is significantly easier.
4. Elimination of the hard edge of the occulter eliminates annoying diffractive ringing in the reimaged focal plane.

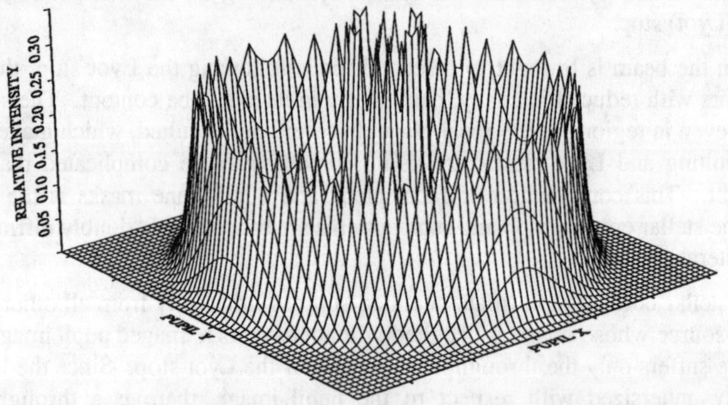

Figure 1. Model calculation for the reimaged pupil plane intensity following occultation of the central four diffraction rings of the Airy pattern by an appodized occulting mask. The input pupil plane was obscured with unit intensity. The very compact concentration of light at the edges of the reimaged pupil allows the Lyot stop to be very effective at getting diffracted light out of the system permitting diffraction reductions $> 10^3$.

Models of the optical propagation in a coronagraph showed that diffraction reductions in excess of 10^6 were possible using the tapered mask so that planet detection in space was enabled with a modest sized telescope[9, 10, 11]. Under HDOS Independent Research and Development funding a coronagraph breadboard laboratory was established and a demonstration instrument was built. We have already demonstrated diffraction reductions in excess of 10^3 and are limited primarily by our ability to work at the very low light levels required to align and characterize the system. Figure 2. shows images taken at the focal plane of the breadboard instrument before and after various Lyot stops are applied.

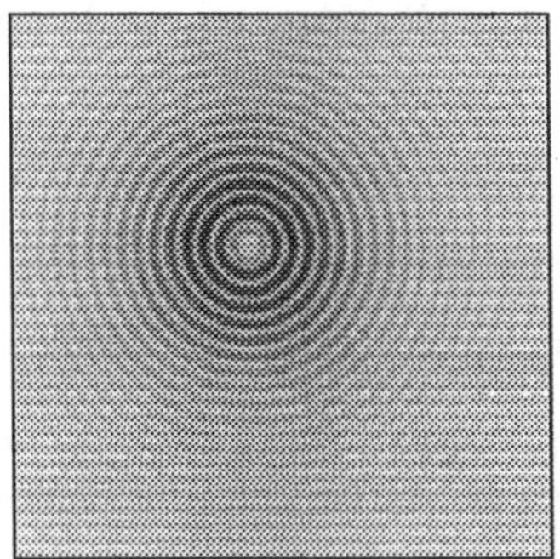

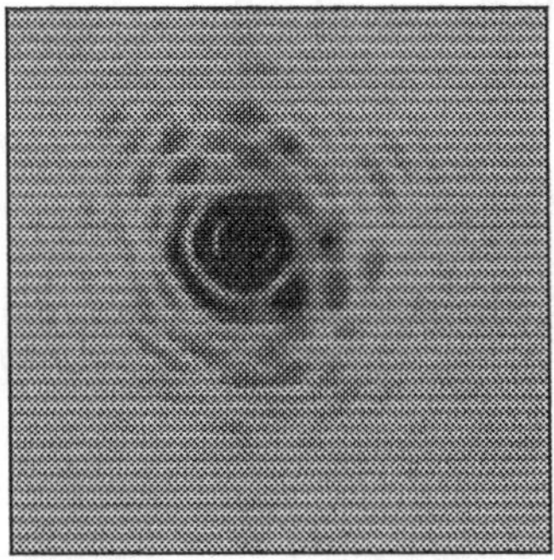

Figure 2. The negative images above were taken in the HDOS coronagraph laboratory. The image on the left has been occulted by a tapered transmission occulting mask whose 50% transmission point is at $4\frac{\lambda}{D}$ but no Lyot stop. Note that imaging is possible through the mask. In the figure on the right both an occulting mask and Lyot mask are in place completing the coronagraph. The difference in the exposure parameters of the two images is such that if the exposure on the left was 1 second, the one on the right would be 25 minutes. This is a diffraction reduction in excess of 10^3.

3. Scattered Light.

The considerations of the previous section were developed under the assumption of ideal wave fronts and perfect optical systems. The atmosphere is a significant departure from ideal and all real optical systems admit of fabrication errors at some level. In this section we consider the role of phase departures from ideal. The Fraunhofer diffraction integral in Eq. (1) now becomes:

$$\mathcal{E}(p,q) = \mathcal{K}\int\int C(x,y)e^{-\frac{2\pi i}{\lambda}(px+qy)}e^{2\pi i\sigma\widehat{F}(x,y)}dxdy \tag{3}$$

where $\widehat{F}(x,y)$ is a dimensionless function that describes the wave front phase variation in units of σ, the wave front rms. in waves. The function $\widehat{F}$ represents the deviation of the telescope wave front from ideal and can in general be written:

$$\widehat{F}(x,y) = \widehat{F}_A + \widehat{F}_M + \widehat{F}_L \tag{4}$$

The telescope wave front error contributions arise from the atmosphere ($\widehat{F}_A$), from manufacturing errors in the optics ($\widehat{F}_M$) and from alignment errors of the system ($\widehat{F}_L$). Maintaining a telescope in good alignment is crucial to optimizing performance. Since alignment can be easily optimized, in what follows we will consider only contributions due to turbulence and figure error.

Unfortunately the atmosphere at poor sites can produce large phase excursions and similarly one need not go far to find poorly figured optics. To illustrate the role of phase errors, however, it is instructive to consider scatter in the limit of small phase variations. In particular the case when $2\pi\sigma < 1$. In this limit the phase error term can be expanded to give:

$$\mathcal{E}(p,q) = \mathcal{K} \int\int C(x,y) e^{-\frac{2\pi i}{\lambda}(px+qy)} \left(1 - 2\pi i \sigma \widehat{F}(x,y)\right) dxdy \tag{5}$$

or:

$$\mathcal{E}(p,q) = \mathcal{E}_o + \mathcal{E}_{\widehat{F}} \tag{6}$$

where $\mathcal{E}_o$ is the ideal amplitude and $\mathcal{E}_{\widehat{F}}$ is the perturbation caused by phase errors. Note that because of the linearization of the integral the second term is proportional to the Fourier transform of the wave front Carrying through the perturbation expansion and applying conservation of energy the intensity can be written in a very simple form as[12, 13]:

$$I_T(\rho) = \left[1 - (2\pi\sigma)^2\right] I_o(\rho) + (2\pi)^2 PSD(\rho) \tag{7}$$

where I_T is the total observed intensity, I_o is the ideal intensity observed in the absence of phase errors and PSD is the wave front power spectral density with the normalization:

$$\sigma^2 = \int PSD(\rho) d^2\rho. \tag{8}$$

The dimensionless focal plane coordinate ρ is just $\frac{\theta}{\theta_o}$ where $\theta_o = \frac{\lambda}{D}$ with D the diameter of the telescope. With this definition ρ is also the spatial frequency in units of cycles per aperture. It therefore follows from Eq. (7) that scatter at field angle $\theta = \rho\theta_o$ arises from wave front errors of frequency ρ cycles per aperture. In other words the wave front should be thought of as consisting of a collection of phase gratings that create diffracted orders in the focal plane with an efficiency proportional to the square of their depth. The function $PSD(\rho)$ simply gives the distribution of scattering efficiency as function of spatial frequency.

Although the term "scatter" is almost universally used to describe the perturbed intensity term in Eq. (7) it is actually a diffractive redirection of light by wave front phase errors. In Eq. (7) it is evident that the energy that shows up in scatter is borrowed from the ideal diffraction pattern. The factor $\left[1-(2\pi\sigma)^2\right]$ that measures the fraction of total energy going into scatter is often called the Strehl factor of the wave front and measures the decrease in central intensity of the point spread function due to figure errors.

For our purposes there are two important properties of the intensity distribution described by Eq. (7). The first is that for all intents and purposes the stellar coronagraph has no effect on the scattered intensity. This occurs because the diffracted order caused by errors of ρ cycles per aperture falling at the field angle $\rho\frac{\lambda}{D}$ should be thought of as an image in its own right (a speckle). If it is to be diminished it must be occulted, but if it is occulted, so is any target at that location and there can be no net gain from the coronagraph. As is evident in the post coronagraph image in Figure 2, monochromatic scatter produces speckles in the focal plane. Reducing the intensity of these speckles is no different from reducing the intensity of any possible target at the same locations.

The second important property of scatter is that it has a different character for figure errors and for the atmosphere. This is true beyond the obvious differences in their time dependent behavior. Given the scintillation of the atmosphere, it is safe to assume that the PSD of the combined wave front is just the incoherent sum of the $PSD's$ due to figure errors and the atmosphere. Thus in the simplest linear model we can write that the intensity distribution in the focal plane can be written as:

$$I_T = I_D + I_F + I_A \tag{9}$$

where I_D is the diffractive intensity including Strehl losses, I_F is the scatter intensity due to figure error, and I_A is the scatter intensity due to the atmosphere. It is the balancing of all three of these factors that makes for good system design. The diffractive intensity contribution, as illustrated in the previous section, can be reduced by essentially arbitrary factors. The stellar coronagraph, as modified by our study, is a particularly low l way to achieve diffraction reduction but other approaches are possible[14, 15]. Any diffraction reduction necessarily requires some throughput loss so the reduction levels chosen should be such that the post coronagraphic diffractive component is a small fraction, say 10%, of the residual focal plane intensity. Reductions beyond this level only reduce throughput without net performance improvements.

It is always possible to reduce the contributors in Eq. (9.) to just the two scatter terms. The power spectrum describing figure errors can be approximated by[16]:

$$PSD_F = \frac{\sigma^2}{2\pi}(\gamma-2)\,\rho^{-\gamma} \tag{10}$$

where $2 \leq \gamma \leq 3$ depending on the manufacturing history of the optic and the spatial frequency domain. The power spectrum describing atmospheric phase errors can be written[17]:

$$PSD_A = 5.8 \times 10^{-4}\, b^{5/3}\, \rho^{-11/3} \tag{11}$$

where b is the diameter of the telescope in units of the Fried length[18] r_o. It follows that for typical optical systems, atmospheric scatter is considerably steeper than scatter from manufacturing errors. Setting the two power spectra of Eq. (10) and (11) equal we can find the point where the focal plane transitions from seeing dominated to figure error dominated as:

$$\log \rho^* = \frac{5}{11-3\gamma} \log b + K \tag{12}$$

where

$$K = \frac{3}{11-3\gamma}\left(\log \frac{2\pi}{\sigma^2 (\gamma-2)} - 3.24\right). \tag{13}$$

For the standard atmospheric model ρ^* is independent of wavelength since both power spectra scale quadratically with the wavelength. To illustrate the meaning of Eq. (12) consider a telescope with a manufacturing wave front error of $\frac{\lambda}{10}$, which is respectable for a large telescope, working in an atmosphere for which $b = \frac{D}{r_o} = 10$. This is equivalent to a 3 meter telescope in very good seeing in the visible. Assuming a typical value of γ=2.5 gives ρ^*=20 which implies that at 0.6μm the telescope side lobes are dominated by figure error for field angles in excess of 0.8 arc sec.. If the telescope wave front quality is improved to $\frac{\lambda}{20}$, the crossover point is moved out to ρ^*=67 or 2.8 arc sec. In the short IR where $b = 1$ is possible for a 3 meter telescope, even for a twentieth wave wavefront figure error dominates over atmospheric effects at ρ^*=2.5 or 0.4 arc seconds at a wavelength of 2.2μm.

Thus for the circumstellar problem the telescope figure quality must be specified in concert with the size of the field under investigation and the quality of the site. No mention has been made of adaptive optics since it has been assumed that wave front correction is capable of fixing both telescope and atmospheric phase errors out to the Nyquist frequency of the actuator system. For field angles that lay beyond that frequency the considerations given above apply.

4. Detecting Faint Sources.

The objective of the coronagraphic instrument is to reduce the background to such levels that faint objects in the circumstellar environment become detectable. For purposes of this discussion a faint source is one for which the bulk of the photons collected over the dimensions of its image arise from the background. Since background levels can be quite high in a stellar halo, it follows that were it to be located elsewhere such a source might not be considered faint. In monochromatic light the stellar halo is made up of the speckles of the wave front power spectrum. These speckles are evident in the post coronagraph image of Figure 2. Even for observations made with a broad bandpass, the scattered light intensity will retain some structure and a slope characteristic of the power spectrum. The one small advantage that accrues to ground based observations is that the turbulence of the atmosphere will somewhat smooth the speckle structure naturally making background removal easier.

If the concepts of the previous two sections are followed, telescope background levels can be reduced without significant throughput losses but there will always exist some performance edge of the system where source detectability is limited by the ability to remove or filter the remaining background. In considering extrasolar planet detection Brown and Burrows[13] noted that the ratio Φ of source counts to star background counts in the same pixels may set the ultimate limit to planet detection since fixed noise in the background subtraction process will ultimately dominate over shot noise. For a background count B we can write that the noise remaining after its removal is $\sqrt{B+\delta B^2}$ where the first term is the usual shot noise contribution and the second is the residual background after removal. Once δB becomes proportional to B, longer integration times no longer reduce the noise and there is no fainter source that can be detected.

We have performed numerical experiments with real scatter backgrounds[19] and have shown that background removal by filtering is well modeled by $\delta B \propto B$ so that the background subtraction for an idealized shot noise free background sets a fundamental limit on detection. This will be true if one assumes that the residual background is not sufficiently constant in time to permit characterization. For closely spaced observations the background can be considered constant at some level and removed. But it will not be absolutely constant and again the residual structure will limit source detection. In order to make the coronagraph really achieve deep sensitivity in the circumstellar halo, it is necessary to develop a strategy for coping with the residual background. We have argued for a filtering rather than image differencing technique because it leads to an image capture and analysis strategy that is robust against secular variations in the background.

Since it is the flux ratio Φ that may set the ultimate limit on detection the system should be designed to optimize this parameter. Formally Φ may be written:

$$\Phi(\rho) = \frac{EE(\rho)}{\pi\rho^2 B} \tag{14}$$

where $EE(\rho)$ is the encircled energy function of the telescope point spread function and, as above B, is the average background level. Clearly Φ will be optimized when the encircled energy is maximized (i.e. the Strehl is closer to unity) which obviously argues for better wave front quality. But encircled energy is also reduced by obscurations and secondary supports, so there are other considerations in designing a system. For a coronagraphic instrument this point is quite important. In principle the Lyot stop should mask off all pupil edges. This results in an increased loss due to secondary spiders but also results in a significant increase in telescope obscuration. As the obscuration of a telescope increases, the encircled energy of its point spread function decreases dramatically (Figure 3.) Thus gains made in improving Strehl through adaptive optics can be easily washed away by the effects of the Lyot mask on the point spread function. This argues strongly that an unobscured telescope is the optimal choice for coronagraphic observations.

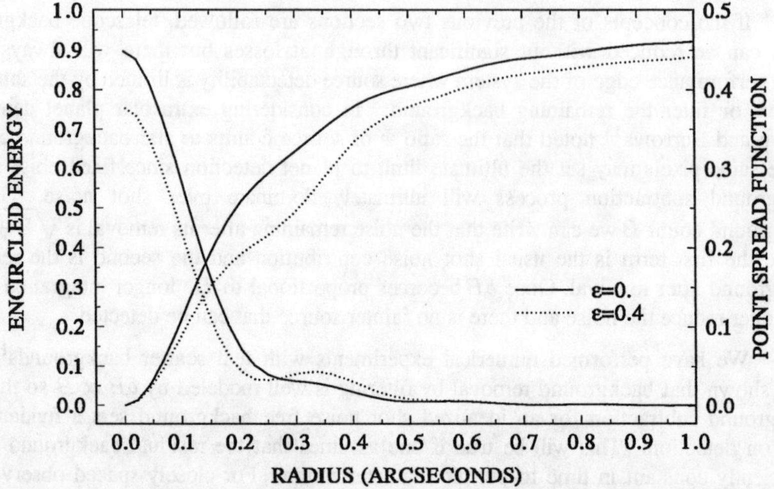

Figure 4. A comparision of the point spread function and encircled energy at 2.2 μm for unobscured and 40% obscured telescopes with 2 meter apertures. In both cases there is an assumed fabrication wavefront error of 0.06 μm. A value of $r_o = 1$ meter and a tip-tilt corrected system have been assumed. The point spread functions in both cases have been normalized to unity so the difference in collecting area, which would increase further the difference between obscured and unobscured systems, has not been included.

In Eq. (14) it is evident that Φ is a function of the radius ρ over which one chooses to calculate it. This is effectively a statement about pixel sizes. The quantity $\overline{I} = \frac{EE(\rho)}{\pi\rho^2}$ in Eq. (14) is just the average point spread function intensity over in a circle of radius ρ. Since the point spread function is assumed to have some sort of central peak, it follows that $\overline{I}$ peaks at the origin. That is, for a uniform background B, Φ is optimized when $\rho = 0$. In other words, vanishingly small pixels will be able to capture just the peak of the PSF giving an absolute maximum for Φ. This is obviously disastrous for the integration time since no photons are collected but it makes the point that pixel sizes can be adjusted to optimize Φ, and that optimizing Φ is not necessarily consistent with minimizing the integration time[9]. Assuming shot noise in the background dominates the observation, the integration time based solely on shot noise considerations will be:

$$t \propto \frac{1}{\phi_S(\rho)\Phi(\rho)} \tag{15}$$

where $\Phi(\rho)$ is as defined above and $\phi_S(\rho)$ is the rate at which photons are collected from the source. As the pixel size is decreases Φ increases but so does $\phi_S(\rho)$. As shown in Figure 5., initially the effect of the more favorably balanced photon mix dominates over the loss of source photons and the integration time starts to decrease despite the fact that fewer photons are collected in the smaller aperture. As the pixel size is decreased further, however, the improvement in $\Phi(\rho)$ is limited and the lack of photons begins to outweigh the mixture richness and the integration time increases monotonically to infinity.

Considerations like those in the paragraph above argue for the smallest pixel size that can reasonably be used for a given observation. That is, the smallest pixel size that permits shot noise to dominate read noise. This notion is not a popular one in astronomy where field of view is often a driver. This is evident, for example, in the instrument complement of the Hubble Space Telescope which under-samples the point spread function in the visible. But the circumstellar problem is by its very nature a narrow field one which must be properly instrumented to succeed. For small pixels, where small is defined as a sampling greater than three pixels per $\frac{\lambda}{D}$, one can always recover the image at coarser sampling and over-sampling the background helps enormously in its removal.

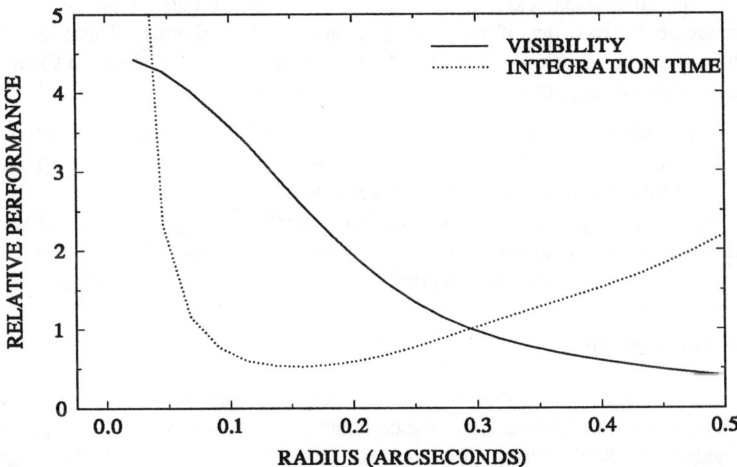

Figure 5. The relative values of integration time and visibility (Φ) are plotted for the unobscured point spread function of Figure 5. as a function of the radius of the aperture over which Φ, or integration time are calculated. Both functions are normalized to their value at the edge of the central Airy maximum. It is evident that as pixel size is varied the system can be tuned to optimize target visibility or integration time. Note that at a radius about 0.08 arc seconds the integration time to the same signal to noise level is identical to its value at the edge of the Airy maximum but that the source visibility is about four times greater.

5. Conclusions.

For several years we have used both laboratory experiments and computer simulations to study the problems of imaging within the circumstellar halo. The overwhelming conclusion of all these studies is the critical interdependence of the instrument, detector, telescope, site and data reduction approach required in order to work successfully in this region. The coronagraph cannot reach its performance potential in the absence of the required systems engineering and goal definition.

The question of goal definition is an important one. Most observatories try intentionally to be general purpose. They were not designed to address a specific type of observation and generally need to service a relatively broad user community. For circumstellar imaging, this is the least desirable way to begin since it leads to unhappy compromises. Typical of this is the discussions above on pixel size. A similar consideration is that for circumstellar imaging, an optical design that achieved decent image quality over a narrow field with an unobscured, unobstructed telescope, is preferable to one with perfect image quality over a larger field if it requires an obscuration and spiders. Similarly, adaptive optical systems are useful if they result in a net gain in Strehl with no residual excess figure error (i.e. scatter) outside the actuator Nyquist frequency, but disastrous if they produce such excess figure. These considerations illustrate the basic problems of circumstellar imaging and make the case for a dedicated facility optimized for addressing the circumstellar problem.

Interestingly, observatories designed to study the Sun have much in common with other kind of circumstellar imaging facility I have described. They are both sensitive to stray and scattered light, they both use variants of the coronagraph and both prefer no obstructions in the pupil. It is possible that a carefully designed, built and sited facility could satisfy the needs of both observing communities and provide outstanding observing efficiency to both scientific communities.

6. Acknowledgments.

I would like to acknowledge the many scientists and technologists who have contributed to my understanding of the circumstellar imaging problem including R. Terrile, S. Shaklan and S. Pravdo at JPL, and R. Hufnagel, R. Noll and A. Nonnenmacher at HDOS. In particular, it was E. Siebert at HDOS whose modification of the Lyot coronagraph was the breakthrough that enabled the possibility of direct detection of extrasolar planets with a modest telescope. I would also like to thank the staff at the National Solar Observatory for the kindness they showed me during my stay there.

References.

1. B. E. Burke *et al., TOPS: Toward Other Planetary Systems: report by the Solar System Exploration Division* (NASA, 1992).

2. G. Gatewood, *Astrophysical Journal*, **94** (1987) 213.
3. M. B. Lyot, *Monthly Notices of the Royal Astronomical Society* **99 (1939)** 579.
4. D. Jewett, this volume.
5. M. Clampin *et al*, *Astrophysics and Space Science* **212** (1994) 167.
6. F. Parece and C. Burrows, *Astrophysical Journal* **319** (1992) L23.
7. B. A. Smith and R. J. Terrile, *Science*, **226** (1984) 1421.
8. R. Terrile and C. Ftaclas, *SPIE Proceedings* **1113** (1989) 50.
9. C. Ftaclas *et al*, *Astrophysics and Space Science* **212** (1994) 441.
10. S. Pravdo *et al*, *SPIE Proceedings* **2199** (1994).
11. S. Pravdo *et al*, *Astronomy and Astrophysics* **212** (1994) 433.
12. C. Ftaclas *et al*, *Astrometric Imaging Telescope Project Report - Final Report for the AIT Metrology Definition Study* HDOS Project Report **PR B11-0348** (1992).
13. R. A. Brown and C. J. Burrows, *Icarus* **87** (1987) 484.
14. R. N. Bracewell and R. H. MacPhie, *Icarus* **38** (1979) 136.
15. J. B. Breckinridge, *et al*, *Optical Engineering* **23**, (1984) 816.
16. C. Ftaclas *et al*, *SPIE Proceedings* **1113** (1989) 56.
17. R.J. Noll, *Journal of the Optical Society of America* **66** (1976) 207.
18. D.L. Fried, *Journal of the Optical Society of America* **55** (1965) 1427.
19. C. Ftaclas *et al*, *SPIE Proceedings* **2198 (1994)** 1324.

2. G. Gatewood, Astronomical Journal 94 (1987) 213.

3. M. R. L[illegible], Monthly Notices of the Royal Astronomical Society 99 (1939) 579.

4. D. Jewitt, this volume.

5. A. Cameron [illegible], Astrophysics and Space Science 212 (1994) 107.

6. B. Pa[illegible] and C. Burrows, Astrophysical Journal 313 (1992) 123.

7. B. A. Smith and R. J. Terrile, Science 226 (1984) 1421.

8. R. Terrile and C. Fincher, SPIE Proceedings 1113 (1989) 54.

9. C. [illegible] et al., Astrophysics and Space Science 212 (1994) 463.

10. [illegible] et al., SPIE Proceedings [illegible] (1994).

11. S. [illegible] et al., Astronomy and Astrophysics 282 (1994) [illegible].

12. [illegible], Astronomical Infrared Telescope Technology Assessment - Final Report for the [illegible] Definition Study, HDOS Project Report FR 1111-2248 (1992).

13. E. M. [illegible] and C. J. Burrows, Icarus 87 (1990) [illegible].

14. R. [illegible] and R. H. [illegible], Icarus 18 (1973) [illegible].

15. W. B. [illegible] et al., [illegible] 75 (1975) [illegible].

16. [illegible] et al., SPIE Proceedings 1113 (1989) 30.

17. R. J. Noll, Journal of the Optical Society of America 66 (1976) 207.

18. D. L. Fried, Journal of the Optical Society of America 56 (1966) 1422.

19. C. F. [illegible], SPIE Proceedings 2198 (1994) 1137.

THEMIS IR CAPABILITIES

T. ROUDIER
Observatoire Midi-Pyrenees (Pic du Midi)
BP 136 65201 Bagneres de Bigorre, France

P. Mein
Observatoire de Paris, section de Meudon, F-92195 Meudon Cedex France
and

J. Rayrole
Observatoire de Paris, section de Meudon, F-92195 Meudon Cedex, France

1. Introduction

THEMIS is a French-Italian solar facility. More particularly, THEMIS is a 90 cm telescope, polarization-free, located on Tenerife Island (Spain) at the Teide observatory. The building is under construction at that site. THEMIS should be operated in 1996. The telescope (Fig. 1) feeds either a long predisperser and Echelle spectrograph, or narrow-band filter. Direct imaging will be also possible. The polarization analysis is completed in the first focus by an achromatic analyzer. The basic detector will be 20 CCD cameras (288x384). Spectro-polarimetry in 10 lines should be possible simultaneously.

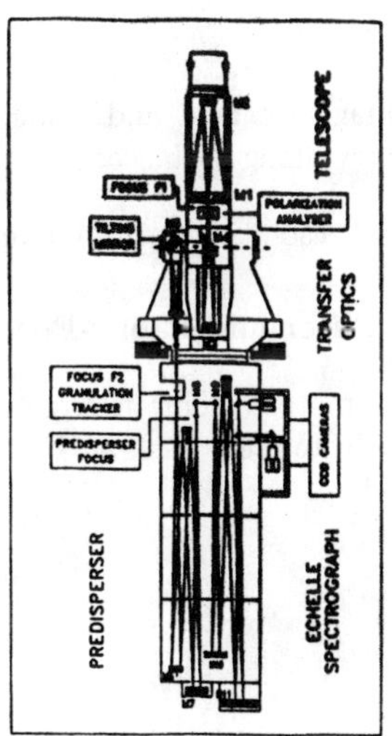

Figure 1: THEMIS Optical Scheme

2. Basic THEMIS Capabilities of Spectrographs

Several observing modes will be available. The main ones can be characterized as follows:

2.1. Spectra

- High spectral and spatial resolution; many spectral lines; best measurement of vector magnetic field and 4 Stokes parameter.

Difficulties: slow for large coverages over the disc; need of very good active optics to follow small structures.

2.2. MSDP

- Imaging spectroscopy, highest spatial resolution (no slit), tracking of very small structures (2D); simultaneous channels (same atmospheric effects).

Difficulties: limited spectral resolution(0.023: limited accuracy for transverse magnetic field (Zeeman).

2.3. Full disk

- Fastest spectroscopy (magnetic field and velocity field by barycenter of line profile); accurate geometry along each scan (diurnal motion).

Difficulties: Medium spatial resolution (pixel 0.5 arcsec), no line profile.

3. Universal Birefringent Filter and Fabry-Perot

The "Panoramic Monochromator" built in Arcetri is a servo-controlled Fabry-Perot interferometer mounted in tandem with an Universal Birefringent Filter. The detector is a 1024 x 1024 CCD.[1]

4. IR Capabilities

4.1. Spectral Range

The entrance window is in BK7 65mm (2.6 inches). The expected spectral range lies between 3500 Å to 2.5 micrometer. Up today the theoretical computation have been done up to 1.1 micrometer (see figure 2). The relative out- flux transmission at the exit of the Echelle spectrograph is optimized by combining 5 aluminum and 14 silver mirrors. The relative transmission is quite good in the range of 5000 Å to 1 micrometer where a great number of lines are interesting (see figure 3). Beyond 1.1 micrometer the transmission has not been computed but is expected, at least, to be good enough up to 2.5 micrometer.[1]

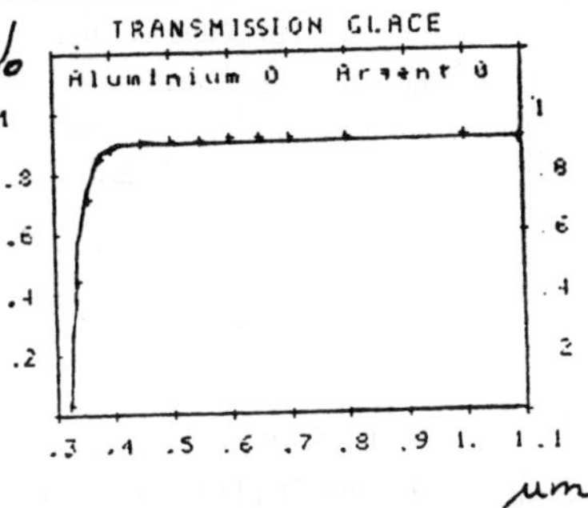

Figure 2:

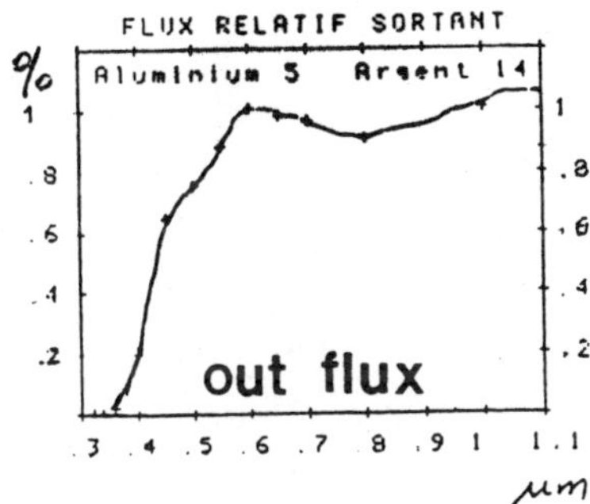

Figure 3:

So, the real IR capabilities of THEMIS are not yet known today, due in great part to the lens coating, which will limit the telescope transmission in the IR band. The IR noise of the telescope has also to be determined.

4.2. Analyzer and IR Detector

The polarization analyzer has been designed and optimized for the visible wavelength. It is not well adapted to IR measurement. However, by changing the "plates" temperature, it will be possible to switch over the He 10830 Å . line but with a passband of 300 Å only. The first generation of detectors have been optimized for the visible range of the spectrum and IR cameras have not yet been planned to be developed in this receptor phase. So, new analyzer and IR detectors are needed in the future as second generation of instrumentation.

5. CONCLUSION

THEMIS which has been designed for the visible part of the spectrum, has probably a great potentiality in IR range. However, a new generation of IR analyzer and receptor are required and to be developed. So, collaboration are welcome either in the international time facilities, or with French, Italian or Spanish teams, to test and use THEMIS in IR band.

6. References

1. J. Rayrole, P. Mein and Cavallini, "Solar Surface Magnetism", R.J. Rutten and C.J. Schrijver (eds.) Colloque NATO Soesterberg p. 507 (1994).
2. P. Mein, "How to use THEMIS for key-program", December 1993.

Coronagraph Stray Light Analyses

Bernard V. Jackson and Lance A. Lones
Univ. Calif. at San Diego

ABSTRACT

We analyze the possible stray light paths present at the front end of the SOHO C3 coronagraph in order to determine which are the most dominant. Using a combination of the Breault APART scattered light software code and our own scattering analysis, we determine a theoretical stray light level at the coronagraph image plane somewhat below that observed in laboratory tests. We find that reflective paths from the instrument walls and vanes determine the ultimate stray light level present in this instrument. We have begun the same analysis with the SWATH coronagraph design. Although generalizations are difficult from one instrument to another, we find that some ideas are basic to coronagraph design, especially concerning stray light from reflections within the instrument.

OBSERVATIONAL TECHNIQUES FOR HIGH-BACKGROUND MID-INFRARED (5 - 20 μm) ARRAY IMAGING

DANIEL Y. GEZARI
NASA/Goddard Space Flight Center
Infrared Astrophysics Branch, Code 685, Greenbelt, MD 20771

ABSTRACT

This paper gives an overview of the fundamental conditions encountered in high-background mid-infrared imaging, and summarizes the techniques used to acquire and process the image data, illustrated with examples from our 58 x 62 pixel Si:Ga infrared camera observing program. In broadband mid-infrared imaging observations, background radiation from the telescope and sky dominates the detected flux. Chopping is routinely employed to remove the large background signal and to flat-field the image data. Residual signal offsets caused by unbalanced optical paths often must be corrected for. Detector system non-linearities can produce fixed-pattern image defects, and images must be flat-fielded using an accurate array gain matrix. In some cases the defects are uncorrectable. Presently, slow ($\sim$ 1 Hz) chopping seems to be an essential part of the observing procedure, but unchopped observations can be successful. Although techniques exist which can correct for sky changes without chopping, experience suggests that chopping may always be required to achieve limiting sensitivity. The implications for the design of new large infrared-optimized telescopes are discussed.

1. Introduction

High-background imaging observations in the mid-infrared (5 - 20 μm) involve procedures which are significantly different from those used in the near-infrared, since a large thermal background signal from the sky, telescope and instrument dominates broadband measurements. In broadband mid-infrared imaging observations, the background radiation half-fills the detector "well" in the brief (30 *msec*) exposure time for each pixel. Typical bright infrared stars account for only about 1% of the full well, and the noise level in the background is about 0.1% of the well. The brightest mid-infrared stellar source (IRC+10216) is equal in the strength to the background level. Only a few other bright stars are noticeable in real time above the background. But broadband imaging of a very bright source like the Sun is actually a low-background observation since a cold 0.001 neutral density filter must be used to prevent detector saturation, resulting in a negligible contribution from the telescope and sky backgrounds compared to the detected flux of solar photons.

Small changes in the sky brightness on time scales comparable to the sampling, integration or calibration intervals can result in degraded images and reduced photometric accuracy. In principle, chopping is not required for background subtractraction or flat-fielding of high background array image data, and unchopped observations can be successful. However, in practice, image quality seems to break down on time scales of minutes when attempting staring (unchopped) observations under good weather conditions. Techniques exist which can be used to correct for sky changes and do background subtraction without chopping. However, experience suggests that chopping may always be required to achieve limiting sensitivity. This paper gives an introduction to the basic techniques of high-background (stellar) and low-background (solar) broadband mid-infrared imaging and summarizes the techniques used to acquire and process the image data, illustrated with examples from our 58 x 62 pixel gallium-doped silicon (Si:Ga) array camera observing program.

2. The Array Camera Instrument

A detailed description of our array camera optical and electronic design, detector characteristics, and operating strategy has been presented by Gezari, et al.[3] The camera uses a 58 x 62 pixel Si:Ga photoconductor array detector manufactured by Hughes/Santa Barbara Research Center (SBRC). The detector array is a hybrid device, assembled from a wafer of Si:Ga detector material (nominally sensitive between 5-17 μm), bump-bonded to a Hughes CRC-228 direct readout (DRO) integrated circuit multiplexer chip.[7] The array pixels are read out serially, although the switched FET multiplexer design allows them to be sampled in any order, or polled non-destructively with the penalty only of added read noise. The incoming images corresponding to the two positions of the telescope chopping secondary mirror are sorted by the data acquisition computer (synchronized with the chopper drive signal) and co-added into two algebraic arrays. Data are ignored while the chopper mirror is moving between end positions. A single 30 msec camera exposure is referred to as a "frame", and an integration consisting of a co-added group of frames is an "image". Two final images (source and sky), each representing typically 1 minute of integration time, are downloaded to the host computer and stored as an image pair.

The array camera noise equivalent flux density (1σ) with broadband ($\Delta\lambda/\lambda = 0.1$), transmission $\sim$ 80%) interference filters is NEFD $\sim$ 0.03 Jy/min$^{-1/2}$ pixel^{-1} (1 Jy = 1 Jansky = 10^{-26} Wm^{-2} Hz^{-1}). The NEFD expressed as noise equivalent brightness is NEB = 0.45 Jy arcsec^2 min$^{-1/2}$(1σ) since the point source flux is spread over 20-30 pixels by diffraction ($\sim$ 1 arcsec) and, to a lesser extent, by seeing (few $\sim$ 0.1 arcsec). The pixels can be "binned-up" (co-added) in data analysis to create a larger synthetic pixel or aperture and thus improve the signal/noise in the data, at the expense only of spatial resolution (as long as the source fills the aperture). For example, by binning the data into a 5 x 5 = 25 element synthetic pixel the signal/noise ratio in the data can be improved by a factor of 5. For many observational problems there is no significant loss of spatial information when binning-up pixels to the $\sim$1 arcsec seeing/diffraction limit of 3-meter telescopes of 10μm.

Several factors combine to make array camera observations more difficult than first impressions might suggest. NEFD *per pixel* numbers can seem deceptively good, but this is because the array pixels are small. Look at the noise equivalent brightness (NEFD arcsec^{-2}) numbers for a more realistic picture. NEFD is expressed as the 1σ noise level, but good images require $\sim 5\sigma$ results in most of the image, increasing required integration time by an order of magnitude or more (although this penalty is offset somewhat by routinely binning 3 x 3 pixels, thus recovering a factor of 3). Also, chopping doubles the elapsed observing time, and overhead must be allowed for calibration observations.

3. Sources of Mid-Infrared Background

The $10\mu m$ thermal background from the night sky and telescope optics in a T$\sim$270 Kelvin observatory environment, at the Cassegrain focal plane of a large ($\sim$3 meter) conventional telescope, is about 10^9 photons sec^{-1} $m^{-2}\mu m^{-1}$ $arcsec^{-2}$. But detector well capacity of infrared photoconductor arrays is typically only 10^5-10^6 electrons. However, our small detector pixels (0.26 arcsec on the sky), actual optical efficiency of the instrument, and low photoconductive gain at which our detector can be operated combine to reduce the number of electrons which actually accumulate in each detector well during the short (30 msec) integration by about four orders of magnitude, making operation of the large-format array feasible.

The infrared photometric bands corresponding to the principal atmospheric windows are the J($1.2\mu m$), H($1.6\mu m$), K($2.2\mu m$), L($3.5\mu m$), L'($3.8\mu m$), and M($4.8\mu m$) bands, the broad N($10\mu m$) band covering the atmospheric window between 7-$13\mu m$, and the Q($20\mu m$) band between 18-$24\mu m$. In recent years, observations in the broad N and Q bands have commonly been made with $\sim 1\mu m$ wide interference filters (such as the OCLI Corp. "Silicate Filter Set") rather than with $\sim 5\mu m$ wide filters covering the full atmospheric windows. The principal chemical components of the atmospheric absorption spectrum at mid-infrared wavelengths are the vibrational-rotational transitions of the molecules H_20 and CO_2. While the transmission of the windows at these wavelengths can be high ($\sim$ 95%) in dry weather conditions, there are deep absorption lines which can contribute considerable background if filters are not selected carefully to exclude them. A informative model analysis of atmospheric properties in the infrared between $1\mu m$ and 1mm at site altitudes of 4 km (for Mauna Kea), and 14 km (aircraft) and 40 km (balloons) is presented by Traub and Stier.[11]

The strength of the background flux due to the sky is a function of the emissivity and temperature of the atmosphere, with the emissivity dominated by the humidity dominated by altitude above sea level. Thus considerable improvements in mid-infrared observing conditions have been achieved by going to cold, dry, high altitude observing sites. For example, the $12\mu m$ thermal background expected at Kitt Peak (altitude 6000 feet) can be compared to Mauna Kea (altitude 14,000 feet). All other things being equal, differences in ambient temperature (290 K vs. 270 K) and atmospheric precipitable water vapor ($1\mu m$ vs. $\sim 3\mu m$ in good weather) would reduce the $12\mu m$ background at Mauna Kea by about a factor of 4 as compared to Kitt Peak.

"Sky noise" has two major components: 1) statistical noise (shot noise) in the incident photon field (the random variation in the photon flux rate, proportional to the square root of the number of detected photons per second), and 2) variations in the emissivity and temperature of sky passing overhead, which give rise to variations in the detected background level. Sky fluctuations can have high frequency structure and low frequency drift components, with $1/f$-like noise characteristics.

Factors which effect the detected instrumental thermal background include the optical throughput of the dewar/telescope optical system, cleanliness of the telescope optics, type of optical surface coatings of the telescope optics, environmental temperature, telescope and camera optical alignment, optical balance between chopper beams, design of the telescope secondary mirror assembly and spider structure, etc. Instrumental background contributions include the emissivity of telescope reflective optics, telescope spider, and the camera instrument itself, (due primarily to the dewar window). Scattered light into the camera from sources such as the daytime sky, the observatory dome, telescope structure, room lighting, etc. have been found to be negligible. The background contribution from the camera (other than from the dewar window) is unmeasurable in the operating configuration, and presents itself as a very small signal indistinguishable from dark current in a well designed system. However, filters mounted inside the dewar must be kept near liquid helium temperature to avoid significant broadband, large solid-angle thermal emission to the detector.

4. Processing of High Background Array Data

4.1. Background Subtraction

For background subtraction, "reference" images of blank sky (observed nearly simultaneously using the telescope chopping secondary mirror) are subtracted from "source" images. The results are flat-fielded by dividing each by a normalized blank sky image (measure of the relative instrumental gain pattern of camera, or the "gain matrix"). Any residual sky signal (background offset level after subtraction) can be removed by subtracting another blank sky image from each difference image (an alternative to nodding the telescope). After being aligned spatially, the flat-fielded images are averaged together to improve signal-to-noise and produce a final mosaic. The following is a brief summary of the discussion by Gezari.[2]

The basic astronomical image $I(x, y, t)$ is the difference between a source image S and a reference image R.

$$S = g(x, y, t)[O(\alpha, \delta) + B_s + B_t + B_i]$$

$$R = g(x, y, t)[B_s + B_t + B_i]$$

where $O(\alpha, \delta)$ is the object intensity distribution in the sky and $g(x, y, t)$ is the net electronic/optical gain distribution. B_s is the sky background, B_t is the telescope background and B_i is the instrument background (all with possible spatial and temporal dependencies). Ideally, subtracting the source and reference images yields $O(\alpha, \delta)$, modified by $g(x, y, t)$. But if the background contributions are not exactly equal in the source and reference images, a "residual sky" background level ΔB persists in the image.

$$I = S - R = g(x, y, t)[O(\alpha, \delta) + \Delta B]$$

The residual background can result from very small differences in the optical path through the telescope between the two chopper positions, asymmetrical views of the telescope structure, sky opacity changes, etc. This offset can present itself as a featureless, constant background level or as a more complicated intensity gradient across the image. It can be removed from an image by "nodding" the telescope so that the astronomical object is placed alternately in the S and R images, which are subsequently subtracted from each other in pairs.

Nodding is effective, but it introduces a considerable additional data management and analysis burden. Alternately, a single image of nearby blank sky can be subtracted from a number of different object images. The final image co-added from several such aligned and corrected images would not be limited by the noise of the single blank sky image, provided that the telescope was repositioned slightly for each integration (to put the object at a slightly different position on the array). When the images are aligned for averaging, the position of the common sky image becomes shifted in the aligned stack, and the noise is randomized in the average.

4.2. Flat Fielding

Since most of the detected flux is due to background in broadband stellar observations, the detector is always operating at nearly the same level in the detector "well" regardless of source strength for all but the brightest sources and detector linearity would not seem to be a major concern. But very subtle detector problems become magnified when the background is subtracted and the 0.01 - 1% of the well contributed by the source flux is expanded to full contrast range in the final image or photometry. In this sense, high background imaging is much less forgiving than low-background (where most of the detected signal is due to source photons). Considerable attention must therefore be paid to the design of a stable detector electronics system.

If any image defects persist after initial background subtraction, they can be further corrected for by dividing the image by a normalized reference image R_o of nearby blank sky, the observationally generated map of net camera response described by $g(x, y, t)$. The reference image used for the division can be the direct sky image obtained simultaneously with the chopper, or a sky frame taken separately under the gain and background conditions as similar as possible. In principle, the same reference image can be used to flat-field many different data images. However, the background flux and electronic gain of the detector system have to be quite stable on the time scale of the observation. Flat-fielding techniques have also been discussed by Hoffmann, et al.[8] and McCaughrean.[10]

Fixed-pattern image defects commonly appear in the image when processing is attempted with a reference image obtained at a different background intensity. If

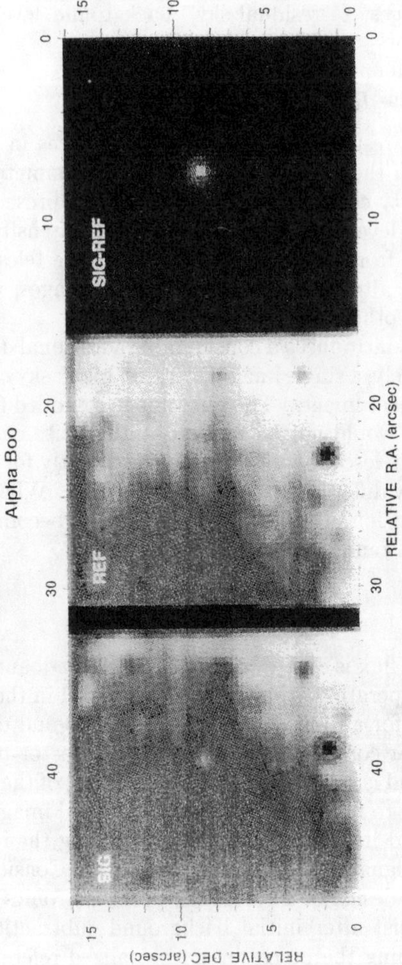

Figure 1: Individual source (SIG) and reference (REF) images of the bright star αBoo, and the final sky-subtracted image (SIG-REF) obtained by taking the difference between the two. αBoo can barely be seen above the large background level in the SIG image. Array gain defects up to 25% of the mean can be seen resulting from defective bumb-bonds on the detector chip. The gain variations among all of the pixels is 15% (1σ). Despite these gain non-uniformities, the sky-subtracted images can be flat-fielded to better than 0.01% (1σ) of the raw mean.

the source and sky images are not too different, a simple level matching process (adjusting offset and gain factors) can be used to generate a matching reference image. However, large background differences combined with small non-linear detector effects (discussed below) can result in two images which can not be successfully matched. Figure 1 shows individual source and reference images of the bright star αBoo, and the final sky-subtracted image obtained by taking the difference between the two. αBoo can barely be seen above the large background level in the SIG image. The REF of blank sky image depicts the gain matrix of the detector system response. A gradient of about 30% can be seen across the field due to gain differences among the array pixels. Gain differences resulting from defective bumb-bonds on the detector chip appear as dark spots. Despite these gain nonuniformities, the sky-subtracted images can be flat-fielded to better than 0.01% (1σ) of the raw mean. Slight differences in the SIG and REF images shown here are due to the auto-ranging color display; on the same brightness scale the images would appear essentially identical. If the slight differences between the images seen in this display were real (due to sky background changes, etc.) the final sky-subtracted image would be useless.

4.3. Photometric Calibration

Images of a standard star must be obtained at least once during the observation of a source for the purposes of absolute flux calibration, and are required more frequently (typically each hour) to determine an atmospheric extinction correction at each wavelength. A dark frame (cold shutter) is useful for instrumental calibration purposes but is not required for reduction of differential (chopped) data if the image contains blank sky, even in the single-sampling mode. The calibration star images must have signal/noise characteristics comparable to each of the source images to be calibrated, but are usually bright sources requiring shorter exposure times.

The experimental uncertainty of flux measurements within a single image (relative photometry) is estimated to be 2% of the peak, limited by the effectiveness of flat-fielding procedures. The absolute calibration uncertainty is estimated at 10%, due primarily to the variability of atmospheric transmission during the observations and the reliability of interpolating the calibration star flux data to our wavelengths.

5. Mid-Infrared Observational Considerations

5.1. Infrared Seeing

As summarized by Livingston, et al[9], the nominal relationship between angular seeing size Θ_λ in the infrared and wavelength λ for Kolmogorov turbulence is

$$\Theta_\lambda = \Theta_{vis}(\lambda/\lambda_{vis})^{-1/5}$$

which at 12.4μm is $\Theta_\lambda/\Theta_{vis} = (12.4/0.5)^{-1/5} \sim 0.5$, or a predicted factor of two

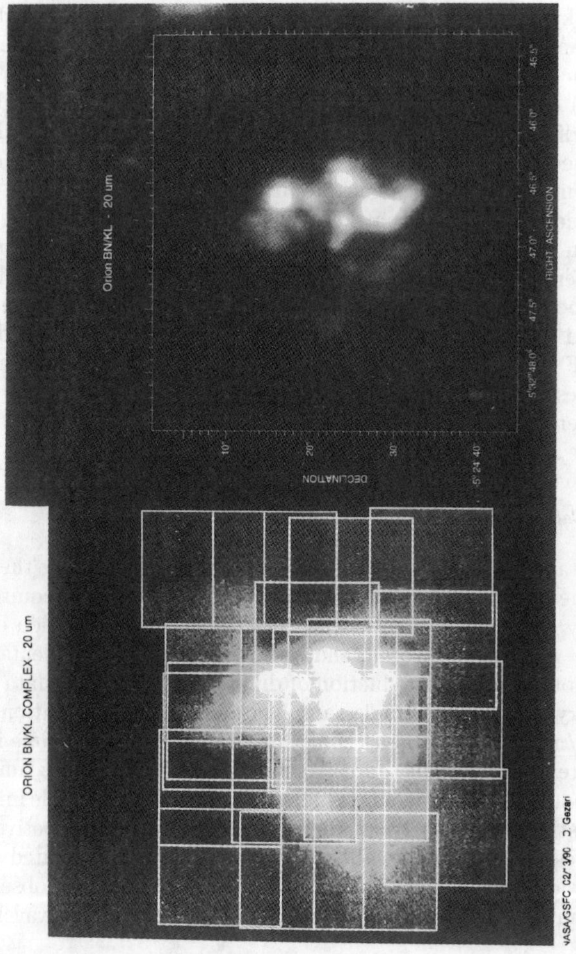

Figure 2: An example of a final, calibrated astronomical image, the $20\mu m$ continuum mosaic of the Orion BN/KL complex obtained with the 58 x 62 array camera[3] at the 3-m NASA/IRTF Telescope at Mauna Kea. The mosaic was assembled from 23 overlapping 1 minute integration frames (15 x 16 arcsec field-of-view, pixel size 0.26 arcsec) which were aligned, matched and co-added to make up the final mosaic using our MOSAIC software package[6].

better seeing at 12.4μm. However, Bester, et al.[1] suggested that in periods of good seeing the atmosphere may behave according to a random walk model, and seeing would be inversely proportional to wavelength, predicting a factor of 25 improvement at 12.4μm. In fact, our experience with 3-meter telescopes is that infrared seeing seems to be a stronger function of wavelength than $\lambda^{-1/5}$, more like $\lambda^{-1/2}$ (seeing at 5μm estimated roughly 3 times better than 0.5μm, and at 12.4μm roughly 5 times better than at 0.5μm. However, a systematic mid-infrared seeing study has not been attempted.

The number of speckles[4] in an infrared image is inversely proportional to wavelength, and so to first order infrared seeing should scale with wavelength. However, the index of refraction is also wavelength dependent. Observationally, in most 3-meter class telescopes, there are only a couple of speckles in the image at 12.4μm. This results in almost negligible ($\sim$ 0.1 arcsec) seeing (compared to the diffraction limited resolution of 1.0 arcsec at 12.4μm), which is primarily a slight image "wobble" rather than "blooming" (a more significant effect at near-infrared wavelengths). For this reason, simple real-time tip-tilt image sharpening can make a significant improvement in image resolution and observational sensitivity for infrared point sources. This procedure is most effective near 5μm where seeing is more significant, yet the seeing effect is primarily wobble.

5.2. Specific Requirements for Solar Imaging

Solar imaging with an array camera at mid-infrared (5-20μm) wavelengths presents several unique observational conditions and requirements. Not the least of these is a source signal so large that the detector would be saturated by orders of magnitude without using a 0.001 neutral density filter, making the mid-infrared thermal background from the telescope optics and sky negligible. Other difficulties include 1) obtaining suitable calibration images for flat-fielding using such a bright source with extensive surface structure, 2) precise pointing, guiding and registration of images on an extended object using an instrument capable of very high spatial resolution ($\sim$ 1 *arcsec*) and relative astrometric precision ($\sim$ 0.1 *arcsec*), and 3) assembling large mosaic images from individual exposures of low contrast infrared surface features which may not have visible counterparts.

Broadband solar image data are obtained without chopping and are flat-fielded by dividing them each by a featureless reference image. This image must be the same brightness as source region observed. To obtain this image, a small region of quiet Sun (at the same radial distance from center disk as the feature being observed) is scanned continuously over several array fields of view during the exposure to smear out any solar surface contrast features.

An example of a 12.4μm solar image is shown in Fig. 3. At these wavelengths the observed brightness is essentially proportional to temperature, providing an unambiguous thermal map of the region. The coolest part of the sunspot umbra is about 20% cooler (darker) than the surrounding quiet Sun, while the penumbra is about 3%

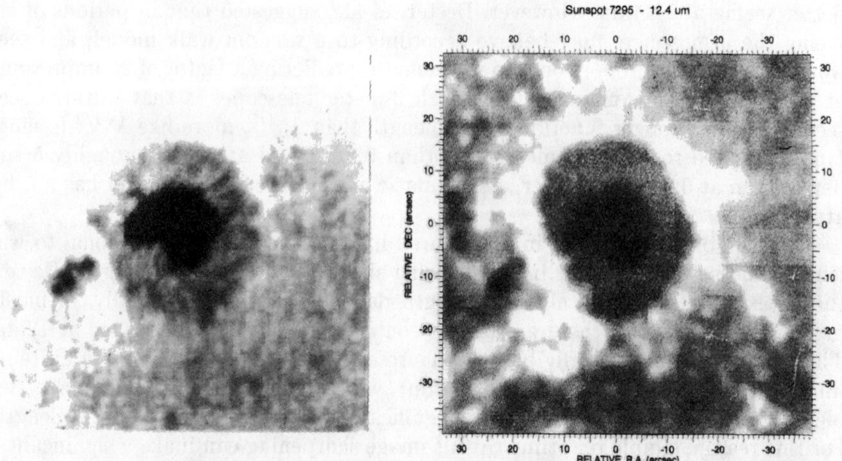

Figure 3: *Right*: 12.4μm images of Sunspot 7295 made with our 58 x 62 array camera[5] at the McMath-Pierce Solar Telescope using a 0.001 neutral density filter. The field of view is about 1 *arcmin*. Temperature contrast of about 25% is observed between the umbra and the disk, and the temperature resolution in about 10 Kelvin. The bright spots surrounding the penumbra are regions hotter that the mean quiet Sun which are not evident in the visible light image (shown at left), and may indicate luminosity from below the sunspot emerging at the surface around it.

cooler. Since the umbra is brighter in the mid-infrared than in the visible, scattered light contamination is much less of a problem at infrared wavelengths. This kind of source contrast represents a much larger fraction of the detector well capacity, as compared to less than 1% of the well for stellar sources in the high-background case.

While sky background emission level changes are reduced by a factor of 1000 when a neutral density filter is used for broadband solar imaging, atmospheric transmission changes have their full effect on the detected source flux (Fig. 4). Because of this, the solar signal detected can fluctuate considerably in poor weather conditions, making some kind of level matching necessary if images are to be co-added. The level matching corrections are typically less than 10% in both gain factor and offset level. Examples of these effects have been obtained in "movies" of 4.8 and 12.4μm mid-infrared solar surface structure[5] (Fig. 4), where the high source brightness made it possible to make a sequence of very short (30 *msec*) exposures of diffraction-limited source detail in the presence of atmospheric seeing. These short exposures freeze seeing motion. Seeing effects are obvious in images taken at the McMath-Pierce solar telescope at 4.8μm (diffraction limit $\Theta = 0.6$ *arcsec*) when short exposures are displayed as a time-lapse movie, but no seeing is evident at 12.4μm ($\Theta = 1.6$ *arcsec*), although the $\lambda^{-1/5}$ dependence would predict ~ 1 *arcsec* seeing, which should have been quite apparent.

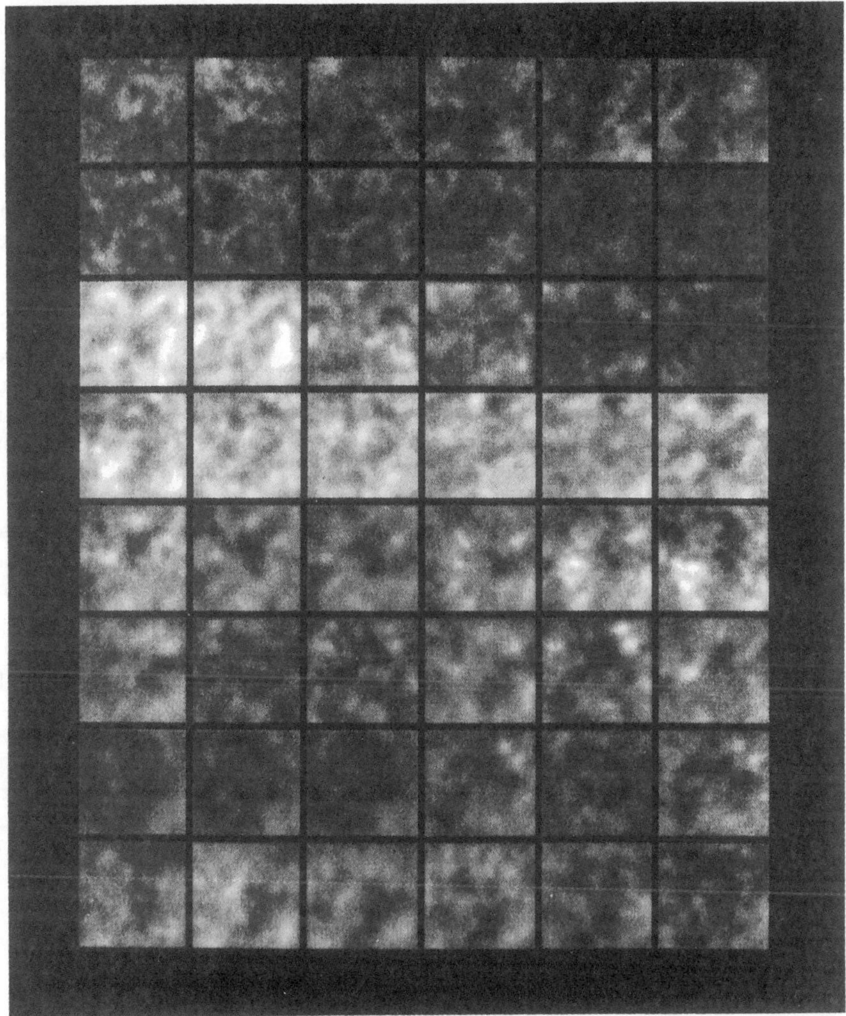

Figure 4: 12.4μm time-lapse images[5] taken at 40 *sec* intervals of a quiet region near Sun disk-center made at the McMath-Pierce Solar Telescope using a 0.001 neutral density filter. The field of view is 23 *arcsec*. Brightness changes from frame to frame are due to changes in atmospheric transmission (a long-term linear drift component due to changing airmass has been removed for clarity). Point-like ($\sim$ 2 *arcsec*) bright features are seen about 10 *arcsec* apart with thermal contrast of about 100 Kelvin. This sequence of short exposures shows random rather than convective behavior of the bright features, which appear to fade on a $\sim$ 1 *min* time-scale (about an order of magnitude faster than visible light granulation.

5.3. Chopping and Next-Generation Telescopes

The question of whether chopping is required for high-background array observations is of considerable current interest, particularly because of the cost and engineering impact of developing chopping optics for large, next-generation telescopes. Mid-infrared array imaging can be done without chopping under very stable weather conditions. The images shown in Fig. 5 were obtained on two nights with significantly different weather conditions. Each set of images represents an elapsed time of 8 minutes. In array imaging of brighter sources under high background conditions can be done for periods of at least 10 minutes without real-time chopping or special post-processing procedures.

Nodding the telescope (moving the telescope alternately between two positions in the sky) is effective for sky subtraction and eliminates the problem of offsets caused by imbalances in the optical paths in chopping, but has real limitations since the telescope can not generally be repositioned with the high (sub-arcsecond) relative astrometric precision of the array, especially when observing faint sources, thus compromising image quality. Nodding can not be done very quickly, and may not be effective in minimizing higher frequency components of $1/f$ sky noise.

Sky background subtraction can be done mathematically if some part of an image is known to contain blank sky, and an accurate net system gain matrix exists for that image. The two could be used to create a synthetic reference frame for background subtraction. Sky background subtraction can also be done by "dithering" (moving the telescope slightly to shift the position of the source on the array), subtracting the shifted images, and deconvolving the source structure (a process similar to nodding). In practice, sky subtraction using these alternate methods can be rather cumbersome.

Because of the sensitivity of the observations to environmental factors, slow (~ 1 Hz) chopping against blank sky provides real-time sky subtraction and gain matrix data at a reasonable cost in observing time. Fast chopping may not always be required for sensitive high-background array observations, depending on weather conditions. But to achieve limiting sensitivity for the faintest sources, chopping may always be required. Large infrared-optimized telescopes should be designed with the minimum capability of efficiently switching sky positions (nodding) by about an arcminute with good reproducibility (~ 0.1 arcsec) on a time scale of a few seconds.

6. Acknowledgments

We are grateful to Mary Hewitt of Hughes/Santa Barbara Research Center (SBRC) for her expert guidance during the development of the DRO array camera. We acknowledge stimulating discussions with Dick Joyce, Harvey Moseley, and Frank Varosi. We thank Michael Hauser at NASA/Goddard for his on-going support of this program. This research is funded by NASA/OSSA (RTOP 188-44-23-08).

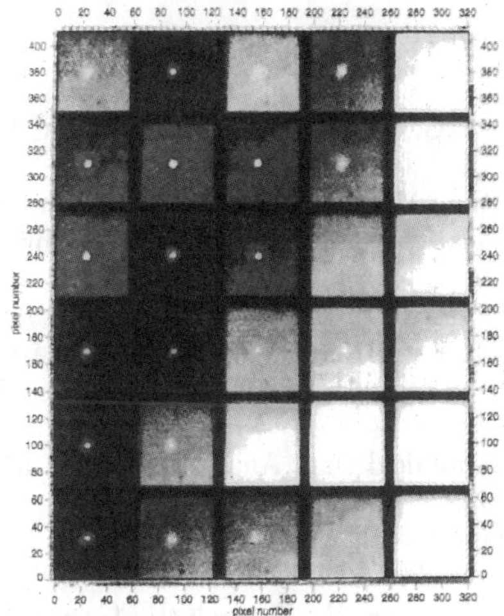

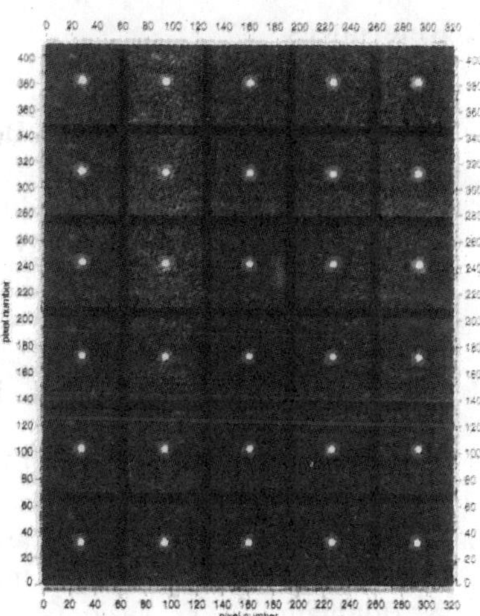

Figure 5: Two 12.4μm image sets taken with the 58 x 62 array camera on the 3.0 meter NASA/Infrared Telescope Facility (IRTF) at Mauna Kea. Each set was acquired without chopping during 8 minutes of elapsed time, illustrating the success of unchopped high-background array imaging under differing weather conditions. Each image is a 0.2 second exposure, taken at intervals of 15 seconds. A single sky image was subtracted from all 30 of the source images for background subtraction. (*Left*): 30 unchopped images of W Hya acquired under less than ideal weather conditions (humidity = 30%, traces of thin cirrus clouds). (*Right*): 30 images of αBoo obtained with the same experimental approach as the image set at left, but under excellent weather conditions (humidity = 4%).

7. References

1. Bester, M., Danchi, W.C., Degiacomi, C.G., Greenhill, L.J., Townes, C.H.: 1992, *Ap. J.* **392**, 357.
2. Gezari, D.Y.: 1995, Proceedings from the IAU Symposium #167: "New Directions in Array Technology and Applications", ed. A.G.D. Phillip, Kluwer Acad. Pub., Dordrecht, (in press).
3. Gezari, D.Y., Folz, W.C., Woods, L.A., Varosi, F.: 1992, *P.A.S.P.*, **104**, 191.
4. Gezari, D.Y., Labeyrie, A. & Stachnik, R.V.: 1972, *Ap. J. Letters*, **173**, L1.
5. Gezari, D.Y., Livingston, W., Kopp, G., and Varosi, F.: 1993, *BAAS*, Phoenix AAS Meeting (abstract).
6. Varosi, F. and Gezari, D.: 1993, "Astronomical Data Analysis Software and Systems II", *A.S.P Conf. Series*, **52**, 393.
7. Hoffman, A.W.: 1987, "Infrared Astronomy with Arrays" (Proceedings, Hilo Detector Workshop 3/87), Univ. of Hawaii, 29.
8. Hoffmann, W. F., Fazio, G. G., Tresch-Fienberg, R., Deutsch, L., Gezari, D. Y., Lamb, G. M., Shu, P., and McCreight, C.: 1987, "Infrared Astronomy with Arrays" (Proceedings, Hilo Detector Workshop 3/87), Univ. of Hawaii, 241.
9. Livingston, W., Kopp, G. Gezari, D.Y., and Varosi, F.: 1993, Proceedings from the IAU Symposium #158, "High Resolution Imaging", Sydney, January 1993.
10. McCaughrean, M.J.: 1988, Ph.D. Thesis, University of Edinburgh.
11. Traub, W.A. and Stier, M.T.: 1976, *Appl. Opt.*, **15**, 364.

INFRARED CAPABILITIES OF THE LARGE EARTH-BASED SOLAR TELESCOPE (LEST)

CHRISTOPH U. KELLER
National Solar Observatory, P.O.Box 26732
Tucson, AZ 85726-6732, USA

ABSTRACT

The Large Earth-based Solar Telescope (LEST) will be an instrument excellently suited for studies in the near-infrared up to 2.7 μm. The longer wavelength limit is due to the fused silica entrance window. The large 2.4-m aperture of LEST coupled with adaptive optics provides unprecedented spatial resolution. The post-focus instrumentation for the infrared part of the spectrum will consist of 1) detector and image acquisition systems with several detectors running in parallel and at up to 50 frames/sec, 2) an Echelle spectrograph containing a slit-jaw monitor and an independent spatial scanning system, 3) a Fabry-Pérot universal filter for the 0.5–2.0 μm range with a spectral resolution of 300,000, and 4) an imaging Stokes polarimeter covering the 1.0–2.7 μm range. All post-focus instruments support the telescope's full field of view of $2'$.

1. Introduction

The Large Earth-based Solar Telescope (LEST) is designed by the LEST Foundation, of which Germany, Israel, Italy, Norway, Spain, Sweden, Switzerland, and the USA are members. It is a 2.4-m aperture Gregorian-type telescope that is filled with Helium to optimize internal seeing[1]. The design is 'polarization-free', i.e. all the optical elements in front of the polarization analysis unit are axially symmetric. Adaptive optics will provide near diffraction-limited spatial resolution. A field of view of $2'$ at an image scale of 0.87 mm/arcsec will be available in the final focus.

The entrance window consists of fused silica, which limits the wavelength range to 0.3–2.7 μm. While the window transmits the solar spectrum down to the ultraviolet cut-off of the Earth's atmosphere, the carbon monoxide lines at 4.8 μm and the Mg I lines at 12 μm will not be accessible. This disadvantage is compensated by excellent performance in the visible and the near-infrared spectrum, which cannot be achieved any other way with present technology.

The core telescope has been designed in detail. Modern post-focus instruments are currently being designed. The latter work is being supervised by the LEST Instrument Advisory Group, consisting of Alberto Righini (chair, Fabry-Pérot filter), Wolfgang Schmidt (spectrograph), Manolo Collados (detector and data acquisition systems), and Christoph Keller (polarimeter). These design studies are of general interest for improving the capabilities of existing telescopes and will be discussed in more details.

2. Detector and data acquisition systems

The detector system will be the most often used post-focus instrument and therefore needs to cover as much as possible of the broad range of observational needs. The near-infrared detectors will have at least 1000×1000 pixels. The quantum efficiency needs to be as high as possible to make efficient use of the large aperture of the telescope. Digital read-out will be performed with 12-bit accuracy or better. A variable read-out speed will allow the user to make a trade-off between read-out speed and read-out noise. Up to 10 full frames per second will be read out while up to 50 frames per second (not necessarily full frames) may be recorded in a high-speed mode. The detector systems attach to spectrographs and filters or may be used on their own.

Flexible real-time data processing and storage capabilities will be provided that can handle up to 10 detectors operating in parallel. The processing capabilities will include real-time frame selection for all attached detectors and real-time displays of all attached detectors, including their data products, such as normalized Stokes parameters of the polarimetry system, summed images, etc.

3. Spectrograph

The spectrograph is designed for high-spectral-resolution spectroscopy at a resolution of about 500,000. The spatial resolution will be $0\farcs1$, which is accomplished by a *reduction* of the image scale by a factor of two in front of the entrance slit. It is anticipated that most studies using the spectrograph would not demand higher spatial resolution and therefore would profit from a higher photon flux per detector element.

The main spectrograph will be situated in the vertical pit in the center of the rotating instrumentation table and will not be evacuated. Vertically mounted, non-evacuated spectrographs that extend below the observing floor have been used with great success and achieve high stability (e.g. at the German Vacuum Tower Telescope on Tenerife and at the Kitt Peak Vacuum Telescope). The optical system consists of an 8-m-class Echelle-type spectrograph with a predisperser and/or interference filters to separate overlapping orders.

In front of the entrance slit, a fast scanning system will provide spatial scanning independent of the telescope pointing capabilities. A slit-jaw system features broad and narrow-band imaging capabilities. It will be possible to simultaneously operate the spectrograph and the tunable filter. While the tunable filter may provide high-temporal-resolution observations by constantly looking at the same area, the spectrograph may investigate the same region with its independent scanning system. To date it is not yet clear whether there will be separate spectrographs for the visible and the infrared, or a single system that covers the complete wavelength range.

4. Tunable filter

The universal tunable filter for LEST has been intensively examined in the design study by Darvann and Owner-Petersen[2]. It will consist of three separate double and triple Fabry-Pérot systems for three overlapping wavelength bands. In the near infrared the 0.5–1.3 and 1.2–2.0 μm systems will be of interest. While the spectrograph is trimmed towards high spectral resolution, the filter system will offer diffraction limited imaging capabilities at the expense of a lower spectral resolution of about 300,000.

Wedged interferometer plates will be used to avoid ghost images. The spectral stray-light will be less than 0.002 of the peak transmitted intensity. The spectral stability should be better than the width of the spectral bandpass over two hours. The wavelength scanning speed will cover 10 wavelength points per second. Fixed interference prefilters of 10 to 40 Å FWHM will be used to reject off-band transmission of the Fabry-Pérot elements. The overall transmission of the systems will be about 20%, which is almost an order of magnitude higher than systems that use Lyot filters as blocking filters.

5. Vector polarimeter

The LEST vector polarimeter for the near infrared will cover the 1.0–2.7 μm region and will attach to the spectrograph and the tunable filter. It should have a sensitivity of 1×10^{-4} in units of the average continuum intensity. The off-diagonal elements of the Mueller matrix describing the optics between the sensor and the Sun need to be determined to an accuracy better than 1×10^{-3}. The instrument should simultaneously cover at least 1000 Å in wavelength.

The entrance window is the dominant contributor to instrumental polarization. It should be possible to calibrate its effects on the polarization measurements[3] to a precision of about 5×10^{-5}. The polarization analysis should be done at the secondary focus, where the optical elements are still axisymmetric.

In the visible part of the spectrum, a ZIMPOL II-type[4] polarimeter with photo-elastic modulators is foreseen. The latter modulate the intensity signal as a function of the incoming Stokes vector at up to 100 kHz. While CCDs may be modified to demodulate light at that frequency, this cannot be done with currently available infrared array detectors. It is presently not clear how the vector polarimeter in the infrared should be designed. There are several possibilities, each with its own particular advantages and disadvantages.

Optical demodulation with electro-optical modulators in front of regular infrared array detectors is one of the possibilities. This is a technically feasible approach that would use the same modulator package as in the visible. A loss of efficiency by a factor of three must be taken into account due to the need for three demodulating systems that each work with one of the polarized Stokes components.

A near-infrared magnetograph (NIM)-type slow modulation scheme[5] employing liquid crystals is another approach. Unfortunately it is not possible to use a polarizing beam-splitter system at the secondary focus of the telescope, which would drastically reduce the sensitivity to seeing. The time constant of seeing is greater in the infrared than in the visible. Slow modulation is therefore not as detrimental as in the visible part of the spectrum. To reach the sensitivity and accuracy specified above, a switching frequency of the liquid crystal modulators of at least 50 Hz would be required, which is faster than currently available modulators that work over a large wavelength range. Two liquid crystal retarders at 45 degrees would allow the measurement of the complete Stokes vector. For this approach a different modulator package at the secondary focus would be needed.

Yet another approach consists in putting the polarization analyzer somewhere other than the secondary focus. This would make it possible to use a polarizing beam-splitter system with variable retarders to both reduce the effects of seeing as well as the pixel-to-pixel gain variations. Instrumental polarization due to the optical elements behind the secondary focus can be calibrated with polarizers at the secondary focus.

6. Discussion

LEST is a general purpose solar telescope for the visible and the near infrared parts of the optical spectrum. Its adaptive-optics system will feature diffraction limited resolution in the near infrared that is about three times finer than the spatial resolution achieved with any other solar telescope in operation or under construction. The large diameter also provides a high photon flux. LEST is therefore ideally suited for high-spatial-and/or-spectral-resolution, near-infrared studies of the Sun.

The post-focus instruments provide a variety of complementary observing possibilities. For high-spatial-resolution studies, the Fabry-Pérot tunable filters will be employed, while for high-spectral-resolution studies, the spectrograph will be the preferred instrument. A series of infrared focal-plane arrays connected to a versatile data acquisition system may be used on both post-focus instruments. A satisfactory design for an infrared polarimeter does not yet exist.

7. Acknowledgments

This work was supported by the Swiss National Science Foundation under grant No. 8220-037202, which is gratefully acknowledged. The National Solar Observatory is part of the National Optical Astronomy Observatories, which are operated by the Association of Universities for Research in Astronomy, Inc. (AURA) under cooperative agreement with the National Science Foundation.

8. References

1. O. Engvold and T. Andersen, *Status of the Design of the Large Earth-based Solar Telescope* (LEST Foundation, 1990).
2. T. Darvann and M. Owner-Petersen, *LEST Technical Report* **57** (1994).

3. C. U. Keller, in *Solar Surface Magnetism*, eds. R. J. Rutten and C. J. Schrijver (Kluwer, Dordrecht, 1994), p. 43.
4. J. O. Stenflo, C. U. Keller, and H. P. Povel, *LEST Technical Report* **54** (1992).
5. D. M. Rabin, D. Jaksha, C. Plymate, J. Wagner, and K. Iwata, in *Solar Polarimetry, Proc. of the Eleventh Sacramento Peak Summer Workshop*, ed. L. J. November (Sunspot, New Mexico, 1991), p. 361.

SOLAR PHYSICS FROM GROUND-BASED INFRARED AND SPACE OBSERVATIONS

JEAN-CLAUDE VIAL
Institut d'Astrophysique Spatiale, Université PARIS XI
Bât. 121. 91405 Orsay Cédex, France

ABSTRACT

We discuss the specific properties of infrared (IR) ground-based and space observations, showing how they complement each other. We shortly review the IR instrumentation and the ongoing solar space programs, with a special emphasis on SOHO. We propose a few targets for simultaneous IR and space observations: structuring of the solar atmosphere (the IR as a tracer of the very cool material), oscillations, sunspots (the IR for magnetic field measurements), coronal diagnostic. Other miscellaneous targets are also mentioned. We stress the importance of the immediate possibilities offered by joint IR/SOHO observations, with a special emphasis on the He I 10830 Å line. Finally, a large mirror coronagraph would open up new possibilities, especially in magnetography, in conjunction with space instrumentation.

1. Specific Properties of the Two Types of Observations

Infrared (IR) ground-based observations (GBO) have advantages which are inherent to GBO: large diameters allow for a spatial resolution which is not diffraction-limited - of course the seeing is a limiting factor which can be overcome with very short exposure times and/or adaptive optics; the large collecting power allows for more photons which implies a better spectral (and spatial) resolution and polarimetric capabilities.

There are also advantages specific to the IR: it give access to the cool material (molecular lines); the continuum is free of absorption lines and is formed from the photosphere to the low chromosphere; some lines may be formed even higher (He I 10830 Å); for some forbidden lines formed in the corona the Signal to Noise ratio may be better than in the optical domain; the Zeeman splitting being proportional to λ^2, in the IR, it is well separated from the Doppler broadening which makes it possible to measure directly the modulus of the magnetic field without any polarimetric device.

Space observations, on the other hand, give access to the whole electromagnetic spectrum, although most instruments have focused on the UV, EUV, and X-ray radiation emitted by hot regions. Consequently, the IR and sub-mm regions are still largely unexplored on the Sun except for integrated-disk ATMOS experiments.[14] However, space observations suffer severe constraints: small apertures limit the spatial resolution (even free of atmospheric problem) and the number of photons. Such a situation is worse in the IR and sub-mm where a diffraction-free instrumentation requires still larger apertures.

In view of the above specificities of IR and space observations, it is easy to see how they complement each other.

2. Complementary Diagnostic tools

First, UV and IR lies and continuua are emitted by regions of different altitudes and temperatures. Second, the IR may give access to parameters which are (until now) not accessed from space, e.g. the coronal magnetic field. These two aspects are very important ones for the studies of the solar atmosphere.

3. Present IR Instrumentation and Observable Wavelengths

A lot of recent IR data have been obtained with the McMath-Pierce telescope of Kitt Peak[30], the VTT of NSO/SP[20], and also during the 1991 eclipse.[23] They rely on the use of mirrors and the important progress made in IR detectors which now allow for useful imaging above 1μ. They are now limited by the size of apertures (1.5 m if the McMath) and the atmospheric windows. The available wavelengths can be summarized as follows:

The intensity of the continuum (bound-free and free-free of H and H^-) is about everywhere proportional to the temperature of the atmosphere. The region at 1.63μ corresponds to the minimum of opacity of the solar atmosphere which means that this continuum probes the deepest photosphere. Other windows are located around 5, 10, 30μ and above $300\mu$to the mm region which is emitted in the chromosphere.

As far as lines are concerned, they are also formed in a wide range of temperatures (e.g. HPβ at 1.28μ, formed in the chromosphere). The combination of a valuable Landé factor and the $(\text{wavelength})^2$ variation is very favorable for atomic Mg I 12μ, Fe I 1.56μ. The He I lines (1083 nm, 10.88μ) are interesting tracers of the high chromosphere and prominences.

Molecular lines exist only at low temperatures (e.g. OH 11065μ, FeH 1μ). CO has a special status since its bands at 2.32μ and 4.7μ evidence the existence of very low temperatures in the "chromosphere" which are explained by the cooling role of this molecule.[3] For a complete view, I refer the reader to papers of Jefferies (p. 1), Falchi et al (p. 113), Livingston (p. 589) in "Infrared Solar Physics", 1994, Symposium 154 of IAU, eds. Rabin, Jefferies and Lindsey, Kluwer Acad. Publ. and also to this workshop.

4. Present Space Instrumentation

Flying instruments on YOHKOH, NIXT, CORONAS are mainly soft X-ray or EUV full sun images of the corona which select either a wide range of (high) temperature contributions or a unique temperature. The spectroscopic capabilities on YOHKOH are devoted to the study of high energy flares. The approved Explorer mission TRACE also includes high resolution imagers, mainly of the transition region.

As for the SOHO mission to be launched in September 1995, the scientific package allows for helioseismology, corona and solar wind studies. Its location at the Lagrangian point L_1 between the Sun and the Earth gives the capability to observe 24 hours a day. Another major advantage will be the possibility to control the instruments in near real time about 12 hours/day. The SOHO coronal package offers interesting possibilities for joint observations with IR instruments and is discussed below. More details about the mission and the instruments can be found in Domingo and Poland[13] and a special issue of Solar Physics to be published in 1995.

The coronal package consists in two sets of experiments:

- instruments looking at the solar disc (EIT, SUMER and CDS)
- instruments observing off the limb (UVCS and LASCO coronagraphs, SWAN).

Some instruments have XUV (CDS) and EUV (SUMER) capabilities, others (UVCS and LASCO) have spectroscopic and imaging possibilities in the UV and the visible; EIT is the XUV "eye" of SOHO.

The Extreme ultraviolet Imager Telescope (EIT) provides an instantaneous full field-of-view of the solar disc (42' x 42' with 2.5 arcsec pixels) in a few coronal lines (Fe IX 171 Å, Fe XII 195 Å, Fe XV 284 Å) and the low transition region (He II 304 Å) line.

The two spectrometer instruments (CDS and SUMER) cover together the 170-1600 Å spectral range i.e. a temperature range from 5000° K (low chromosphere) up to several $10^{6\circ}$ (hot corona) (see Figure 1).

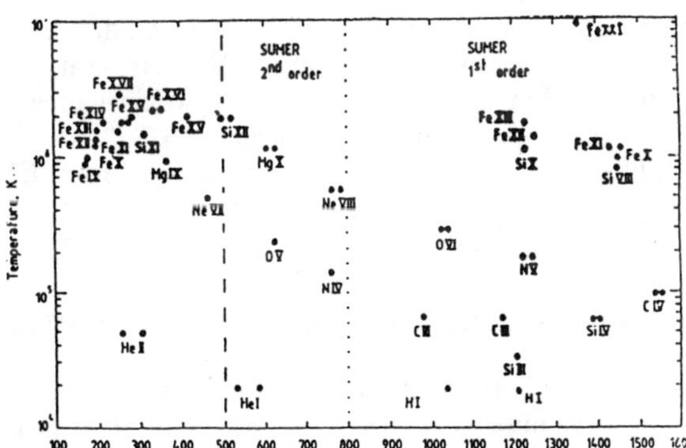

Figure 1: Temperatures of formation of SUMER, EIT and CDS lines (from the Blue and Red Books).

CDS consists in a grazing incidence telescope and two spectrometric channels; a grazing incidence channel or GIS (the 170-800 Å range in several bands) and a normal incidence channel or NIS recording two bands (310-380 Å, and 517-633 Å).

SUMER has a good spatial (1.5 arcsec) and spectral (1 km s^{-1}) resolution in the 500-1600 Å range: its instantaneous spatial range is 300", and spectral range 40 Å in first order and 20 Å in second order. The total field of view (fov) of the three instruments is about 3*3 solar radii.

Above the limb, UVCS also uses spectrometric techniques to diagnose the coronal plasma. Moreover, the Doppler dimming of resonance lines due to radial flows results in a decrease of the intensity of lines, which allows for the measurement of outflow velocities.

Its instantaneous field of view is 42' x 12", but the 1.2 to IO solar radii distance can be scanned radially. The spectrometer has a spectral resolution varying from 0.07 to 0.2 Å for H I Lα, O VI, Si XII and Mg X lines. The coverage of the whole corona is performed by rotating the entire instrument.

LASCO is a combination of 3 coronagraphs:

- C1 scans the 1.1 - 3 solar radii region with a 5.6" resolution, and uses a tunable Fabry-Perot interferometer to scan green, red and yellow lines of Fe XIV, Fe X, Ca XV, Na and Hα.
- C2 scans the 1.5 - 6.0 solar radii region with 11.4" pixels.
- C3 reaches high distances (3.0 - 30 solar radii) with 56.0" pixels.
- C2 and C3 record white-light (Thomson) emission which provides information about the large scale structures and density.

The SOHO instrumentation performs detailed diagnostics of cool to hot plasma from the low chromosphere to the corona to the solar wind.[34] They concern a wide range of parameters such as Emission Measure (EM), temperature and density (line ratio technique), flows, filling factor, etc. In Table 1, we give a few examples of these diagnostics taken in the "Blue" and "Red" Books of CDS.[16,35].

Everywhere a filing factor ($< n_e^2 > / n_e^2$) can be derived from EM and line ratios. However, these important parameters never include any magnetic information on the corona which is nevertheless a region of low β!

5. A Few Targets for Simultaneous and Space Observations

We now propose a few observing programs implying existing IR instruments and existing or planned space missions. These programs could serve as bedtests for future ground-based and space instruments to be built.

5.1. Structuring of the Atmosphere from the Photosphere to the Corona

There is, of course, no unique atmosphere in the Sun but different structures with the same basic feature, namely a physical continuity between a deep and cool region where the β of the plasma is high and an outer diluted and hot region where the

Table 1: Examples of diagnostics with CDS and SUMER taken in the "Blue" and "Red" Books.

Lines of Li-, Be-, B-, Na- and Mg-like ions: EM, flows	**CDS**
Lines from different ionization states of the same element: Fe VIII to XV (T)	**CDS**
Line pairs sensitive to density: SiX (356/347 Å), Fe XIII (318/320 Å).Mg VII (319/367 Å),	
Fe XIV (353/334 Å, Fe XI (308/353 Å), Fe XII (338/364 Å)	**CDS**
Line pairs sensitive to density: Si VIII (1440/1446 Å), S X (1196/1213 Å)	**SUMER**
Line pairs sensitive to density: O VI (1032-1037/173 Å)	**CDS and SUMER**
Line pairs sensitive to temperature: Fe XVI (335/263 Å), Ar VIII (700/180 Å), ..	**CDS**
Line pairs sensitive to temperature: O IV (790/554 Å), N III (991/686 Å)	**SUMER**

plasma β is very low. In the lower region, the magnetohydrodynamic is dominated by gas motions whereas in the upper region it is determined by magnetic field lines motions.

5.1.1. Coronal Loops

Detected with Skylab, they have been extensively observed with Yohkoh and NIXT. After the establishment of scaling laws derived from the simplest energy balance between conduction and radiation, it is important to investigate loops of different temperatures and the temperature variation along and across a loop in order to measure supplementary energy sources. It is also important to know the location of the feet of active region loops; are they located in spot umbra or penumbra? (see Figure 2 from Lites[24].

With SOHO, it is possible to have both density and temperature diagnostics along the loop and across the loop and to measure longitudinal flows.

With the addition of the IR, in continuum windows (1.17μ, 1.6μ, 2.2μ), one can connect the coronal to the photospheric plasma, the apex to the feet of the loop, and one can scan all temperatures from 5000K to 10^6K.

5.1.2. Cool (Internetwork) Atmosphere

The existence of CO bands[2] is clear evidence of very cool material above the photosphere and may indicate the existence of regions deprived of chromosphere.

The contradiction between the CO emission and the UV emission, (which cannot be explained without hot material) is explained, at least in non-magnetic regions[8,9], by the existence of a cool atmosphere which is periodically crossed by waves which transform into shocks and heat the plasma. This scenario seems to be confirmed by time-resolved Ca II profiles. Time-resolved line profiles measurements can be extended to a very wide range of temperatures with SUMER and CDS. In particular, the illumination by downward Lyman lines and continuum and taken into account in models. Simultaneous observations in 2.3μ should evidence the periodic *cooling* of the atmosphere.

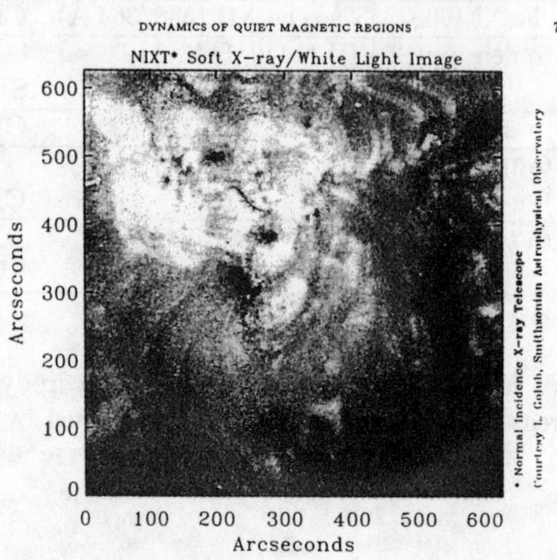

Figure 2: NIXT soft X-ray image showing coronal loops in active regions (courtesy L. Golub, Smithsonian Astrophysical Observatory). As mentioned by Lites[25], it is difficult to trace back loops feet to spots umbra or penumbra.

5.1.3. *The He Shell*

Its existence is now well established but its height is still discussed: disk and limb measurements in He I (10830 Å) give about 2400 km[28]. The formation of He I[1] is mainly determined by photoionization and then radiative recombination and is consequently very sensitive to UV (coronal) radiation. On SOHO, CDS, SUMER, and EIT can measure *all UV* lines and continua of He I (and He II) and the whole ionizing radiation. Disk and limb scans in "all" He lines, including 10830 Å, across quiet and active structures would be of prime importance.

5.2. *Oscillations (wave heating)*

The importance of simultaneous observations at different altitudes in magnetic and non-magnetic regions (network and Internetwork) is stressed in some presentations of this workshop.[5] The nature of the waves (standing? propagating?), the evidence of some power at high frequency waves and other properties such as the different behaviours of network and internetwork, should be studied in a large range of temperatures. The He I 10830 Å is an intermediate line between chromospheric and coronal temperatures. SUMER can observe cold (O I, C I, ...) to hot (O VI) material. Joint observations in IR (10830 Å) and UV (SUMER) could help to understand the waves themselves but high spatial resolution IR observations in lines such as Mg I (12μ) could also evidence the motions of loop feet which actually may be the source of waves excitation.

5.3. Sunspots

The understanding of sunspots has benefited from recent work in the IR: it concerns the properties of pores and penumbra and of the **magnetic field** measurements (complemented by extrapolations) must be made on the same spot at the same time. IR (1.6μ, 12μ) combined with CDS (Fe lines) and SUMER (C IV) offer unique possibilities to scan spots from the photosphere to the corona. Of course, EIT (SOHO) and TRACE would complement the spectral SOHO information with temporal and spatial information.

5.4. Diagnostic in the Corona

As mentioned previously, as the IR offers possibilities to diagnose the corona.

5.4.1. Hot Material

Contrary to EM determinations, line ratios offer a true density diagnostic:

Fe XIII (10747/10798 Å) can be measured from the ground[25]; note the PICO, a ground-based model of LASCO-CI[29] also measures this ratio. CDS (and SUMER) measure a lot of ratios representative of different temperatures (e.g. 318/320 Å of Fe XIII). The comparison of Fe XIII IR and SOHO measurements will allow a calibration of past ground-based observations in the IR along with an estimate of the effect of a varying spatial resolution.

Such density measurements will be used as inputs for the UVCS Doppler dimming technique.

Coupled with DEM (CDS, SUMER) and line-of-sight (LASCO) density measurements they will allow for the determination of the filling factor in coronal structures.

5.4.2. Cold Material

Whether in the form of cool loops, plasmoids, or prominences, cold material is known to exist in the corona. The determination of density in the 67000 to 10^5K temperature range is a difficult task because of the low emission of these structures, as compared to disk radiation. It can only be performed out of the limb, from the ground with a coronagraph.

The IR (1.6μ continuum, He I lines) may prove to be very useful to detect the lowest temperatures (4300 K[16]). Although a lot of CDS and SUMER optically thin lines cover a wide temperature range, we suggest the use of Lyman lines of H (SUMER, UVCS). They have the disadvantage of needing Non-LTE modeling for a clear diagnostic of the emitting structure. However, because of H abundance, any cool structure of measurable size (one arcsec for instance) and reasonable natural density (10^5 cm^{-3}) is optically thick in Lα, Lβ lines, .. and easily observed in contrast to the hot corona.

5.4.3. Coronal Holes

These open-field structures are the primary target of SOHO which will fly during a solar minimum. Basic questions concern the maximum temperature of coronal holes (value and altitude). CDS and SUMER are well suited for such studies which require a low scattered light for out-of-limb measurements. The He I 10830 Å is very useful to prepare and complement SOHO observations, especially at the borders and the base of coronal holes.

5.5. Explosive Events, Flares

Impulsive events trigger questions concerning their initiation: the different processes proposed (reconnection, currents, emerging fields) imply different regions, from the photosphere to the corona. Some processes imply twists and shears at very low levels; they also imply different fine structures (dipoles, kernels) only detected at a subarcsec resolution unreachable with SOHO.

The full temperature coverage obtained with the IR (continuum), the sub-mm, CDS and SUMER on SOHO, **TRACE** allows a simultaneous observation of the same phenomenon.

One can also add that *mm* measurements can provide the evidence of particle propagation in the very initial phase of flares[18]; such a measurement could be correlated with the detection of enhancement in the red wing of the Lα line.

Also the SUMER O V (1218 Å) line could be used as a proxy for hard X-rays.

5.6. Miscellaneous

1. **Abundances** measurements in magnetic and non-magnetic regions from low regions (*4f - 5g FeI in the IR*) to the corona (*Fe lines of CDS)* can be complemented by SOHO in-situ.
2. **Sun-Grazers**
   The potential of Si X (1.43 μ) is discussed by Kumar & Davila[22] and Penn

& Kuhn [26]. The IR observation could use LASCO (and its wide fov) as a forecaster.

3. **Helioseismology**
Global oscillations in the IR (1.6μ) have been reported[24]. They could complement the wealth of data coming from GBO networks and SOHO.

6. What's Next?

We distinguish between the immediate possibilities offered by present and scheduled instrumentation, the projects which need decision and funding and the distant future prospects.

6.1. Immediate Next

We stress again the importance of coordinated observations between ground-based IR instruments and SOHO (and TRACE). This is demonstrated in sec 5. We shall focus on two points:

6.1.1. The Importance of He I 10830 Å

It is basic for synoptic support, especially since it outlines coronal holes. It is also useful for special joint programs concerning limb, oscillations, bright points, sunspots (magnetic field) studies (see sec 5). The potential of the He I 10830 Å is evidenced by Figure 1 of Harvey[15]. Let us also mention here the possibility offered by this line for the detection of weak fields with the Hanlè effect.[6]

6.1.2. The Importance of Coronal Magnetic Field

A 40 G $B_{coronal}$ has been measured by Kuhn[20]. In radio, the measurements of gyroresonance frequencies has been made above sunspots[3] Any other magnetic measurements will be of paramount importance to complement SOHO diagnostics.

Moreover the spectral atlas obtained by CDS (and SUMER) should provide the intensities of EUV lines (e.g. Fe ions) and allow the prediction of intensities of IR lies which share the same upper levels.[10] It is a crucial step for the choice of IR lies for coronal polarimetry.

6.2. Future

In view of the many possibilities offered by a large polarization-free instrument for out-of-limb IR observations, a Large Mirror Coronagraph is presently needed and possible.[4] It should coincide with extended SOHO and TRACE operations.

The Large Mirror Coronagraph should include a coronal magnetograph (once the appropriate lines are selected). The ideal associated space instrumentation is not defined now, but it would probably combine spectroscopic and imaging (multi-layer) capabilities.

6.3. Distant Future

Although the space IR is made difficult by cryogenic constraints, much can be gained from atmosphere-free measurements: better solar IR atlas, polarimetry.

7. Acknowledgments

I express my thanks to J. Kuhn, M. Penn and local organizers for their invitation and my appreciation of the USAF support through WOS-94-2153 grant.

8. References

1. E.H. Avrett, J.M. Fontenla, & R. Loeser, in "Infrared Solar Physics", Syposium 154 of IAU, eds. Rabin, Jefferies and Lindsey, Kluwer Acad. Publ., p. 35, (1994).
2. T.R. Ayres, this workshop (1994).
3. T. Bastian, this workshop (1994).
4. J.M. Beckers, this workshop (1994).
5. K. Bocchialini & F. Baudin, this workshop (1994).
6. V. Bommier & S. Sahal-Brechot, in "Physics of Solar Prominences", IAU Colloq. 44, eds. E. Jensen, P. Maltby, and F.Q. Orrall, p. 87 (1979).
7. J.H.M.J. Bruls, S.K. Solanki, R.J. Rutten & M. Carlsson, *A & A* in press (1994).
8. M. Carlsson, M. & R.F. Stein, this workshop (1994).
9. M. Carlsson & Stein, R.F. in "Chromospheric Dynamics", ed. M. Carlsson, p. 47 (1994).
10. E.S. Chang, this workshop (1994).
11. V. Domingo & A.P. Poland, *the SOHO Mission*, ESA SP-1104, (1989).
12. C.B. Farmer & R. Norton (eds.): "A High Resolution Atlas of the Infrared Spectrum of the Sun and the Earth Atmosphere from Space", Vol. 1, *The Sun*, NASA RP-1224 (1989).
13. B. Fleck, this workshop, (1994).
14. R.A. Harrison "The CDS for SOHO" ('Blue Book'), SC-CDS-RAL-SN-93-0007 Report (1993).
15. K.L. Harvey, in "Infrared Solar Physics", Symposium 154 of IAU, eds. Rabin, Jefferies and Lindsey, Kluwer Acad. Publ. p. 71 (1994).
16. T. Hirayama, Y. Nakagomi & T. Okamoto, in "Physics of Solar Prominences", IAU Colloq. 44, eds. E. Jensen, P. Maltby and F.Q. Orrall, p. 48 (1979).
17. P. Kaufmann, this workshop (1994).

18. S. Koutchmy in "Infrared Solar Physics", Symposium 154 of IAU, eds. Rabin, Jefferies and Lindsey, Kluwer Acad. Publ. p. 239 (1994).
19. J.R. Kuhn, this workshop (1994).
20. J.R. Kuhn, H. Lin, P. Lamy, S. Koutchmy, & R. Smartt, in "Infrared Solar Physics", Symposium 154 of IAU, eds. Rabin, Jefferies and Lindsey, Kluwer Acad. Publ. p. 185 (1994).
21. C.K. Kumar & J. Davila in "Infrared Solar Physics", Symposium 154 of IAU, eds. Rabin, Jefferies and Lindsey, Kluwer Acad. Publ. p. 81 (1994).
22. Li Rufeng, Ye Binxum, Chen Hailin, Liu Shaohua, Deng Bailian, Ma Jagu, H.A. Hill & P.H. Oglesby, in "Infrared Solar Physics", Symposium 154 of IAU, eds. Rabin, Jefferies and Lindsey, Kluwer Acad. Publ. p. 283 (1994).
23. B. Lites, in "Chromospheric Dynamics" ed. M. Carlsson, p. 1 (1994).
24. M. Penn, this workshop (1994).
25. M. Penn & J. Kuhn, *Sol. Phy.* **151**, 51.
26. D. Rabin, in "Infrared Solar Physics", Symposium 154 of IAU, eds. Rabin, Jefferies and Lindsey, Kluwer Acad. Publ. p. 449 (1994).
27. W. Schmidt, M. Knöelker & Westendorp Plaza, C. *A & A* **287**, 229 (1994).
28. R. Schwenn, this workshop (1994).
29. J.-C. Vial "The Solar Corona from SOHO", *Adv. Space Res.* **14**, 4, p. 181 (1994).
30. K. Wilhelm, "SUMER", The 'Red Book', MPAE Report (1994).

PICO

A MIRROR CORONAGRAPH ON PIC DU MIDI

A. Epple, R. Schwenn
Max-Planck-Insitut für Aeronomie, Max-Planck-Straße 2,
D-37191 Katlenburg-Lindau, Germany

ABSTRACT

On Pic Du Midi we use a mirror coronagraph for observation of the emission line corona in the green, red and infrared coronal emission lines. The instrument is essentially similar to LASCO-C1, flown on SOHO from July 1995 on. The field of view ranges from 1.1 $R_\odot$ to 2.1 $R_\odot$ above the suncenter. Image acquisition is done with a 1024×1024 CCD-Array. We present an overview about the installations and first observational results.

1. INTRODUCTION

Since the invention of the coronagraph[6], ground based observatories have used internally occulted systems for a long time. These classical Lyot coronagraphs are telescopes with low scattering chromatic objective lenses to avoid artificial stray light in the instrument. However, residual instrumental stray light and the scattered light in the terrestrial atmosphere make it difficult to detect faint coronal features out to large distances from the solar limb and therefore classical systems are mostly limited to observations very close to the limb. To allow observations further out to study large coronal structures one has not only to limit the instrumental stray light level but possibilities of photometric corrections (subtraction) of the sky background must also be applicable.

Concerning the first point, Newkirk and Bohlin[9] already pointed out in the 1960's that the instrumental stray light can be considerably reduced by using reflective mirror optics instead of lenses. However only in the mid 80's optical workshops were able to polish and coat the requested mirrors to a sufficient quality. Today, mirrors with a stray light level down to a few 10^{-8} $B_\odot$ can be ground. The first coronagraphs designed with reflecting optics were LASCO-C1, the innermost of the three SOHO visual light coronagraphs, and a prior prototype of C1: PICO.* As SOHO will be flown not before July 1995 we have recently installed PICO on Pic Du Midi in the French Pyrenees to perform ground based observations prior to and accompanying LASCO.

PICO incorporates all the advantages of the mirror design such as compactness, low level of instrumental scattered light and optical achromacy, to name the most important. Complementary to SOHO we started observations of the two strong lines of [FE-XIII] in the near infrared. CCD imaging techniques allow us to perform a background correction for the sky.

*Pic Du Midi Coronagraph

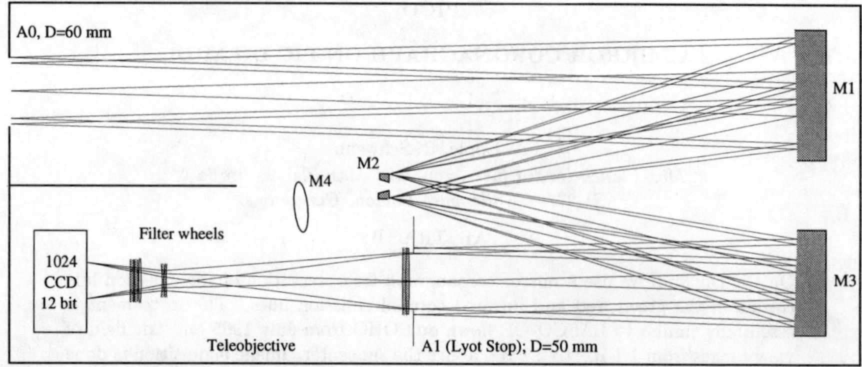

Figure 1: Optical system of PICO. The sketch is compressed in the horizontal direction by a factor two.

2. INSTRUMENTATION

The basic optical layout of PICO is outlined in Fig. 1. The light enters the instrument at the entrance aperture A_0 and is focused on the field mirror M_2 by M_1, a super polished off-axis parabolic mirror. M_2 is called an "inverse occulter" since the light of the solar disk is not stopped by a solid cone but falls through a hole drilled in the center of the mirror and leaves the coronagraph. Thus heat disturbances ("occulter seeing") next to the occulting cone used in Lyot coronagraphs are avoided. The apparent diameter of this hole (1.1 $R_\odot$) gives the inner border of the observational region. Changing M_2 may result in different sizes of the occultation diameter. The coronal light, reflected by M_2, is collimated by M_3, an identical mirror to M_1. In fact, M_1 and M_3 form one single parabola. By giving M_2 a slightly convex surface the entrance A_0 is imaged on the Lyot stop, the classical way to stop out diffracted light of the entrance aperture. The proper coronagraphic system is thus established only of three mirrors. A more detailed description of the LASCO-C1 optical system can be found in Brueckner et al., 1992[1].

Following the Lyot Stop a simple two element achromatic teleobjective (f/26) image the corona on the $1024 \times 1024 \times 12$ CCD-Array. The outer limits of the field of view is thus limited by the focal length to 2.1 $R_\odot$ at the edges of the camera. Since we use a $D = 50\ mm$ Lyot Stop the pixel size of 4 arcsec. limits the spatial resolution to 8 arcsec.

PICO is designed to produce filtergrams of the emission line corona. The narrow band interference filters are mounted in two filter wheels right in front of the image plane. The FWHM of the filters are not constant over the field of view but broaden from 2.5 Å at 1.1 $R_\odot$ to 3.5 Å at 1.5 $R_\odot$ for the visual light filters due to the beam geometry. In the IR the filters are less sensitive to the beam geometry with a FWHM of 6 Å over the whole field.

A big advantage of the inverse occulter design is that the sunlight itself can be used to monitor the solar disk. After having passed the inverse occulter M_2, it is rejected by a 45° auxiliary flat mirror (M_4), collimated and filtered by a 0.6 Å H_α filter. A video camera and video tape recorder display the image on a control monitor and record it on tape. The purpose of the image is to observe the location of active regions on the disk as well as for fine pointing control of the instrument as there is normally a considerable overocculation by M_2.

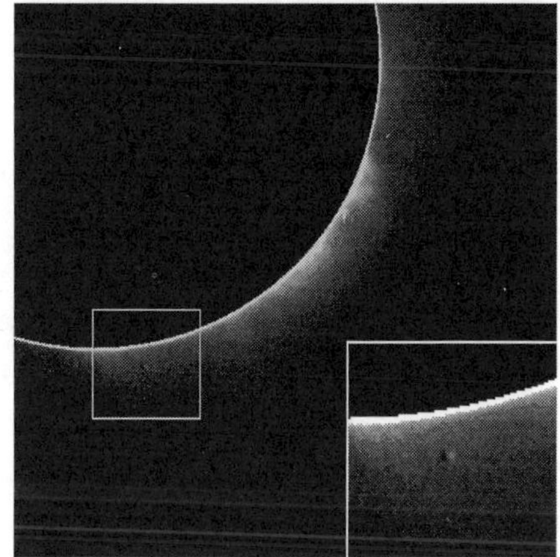

Figure 2: Mercury prior to superior conjunction on Apr 30, 1994. Green coronal line [Fe-XIV] λ5303 Å.

Figure 3: Annular solar eclipse on May 10, 1994. See text for explanations

3. OBSERVATIONS

In the visible spectral range we observe the green ([Fe-XIV] λ5303 Å) and the red ([Fe-X] λ6374 Å) coronal lines. As the two ions are formed at different temperatures the ratio of their intensities can be used to determine the local coronal temperature[3,4]. A first attempt has been undertaken to observe the infrared coronal lines of [Fe-XIII] (λ10747 Å and λ10798 Å). For two reasons these are very interesting for ground based observations: the scattered light in the "infrared" sky is essentially reduced, compared to the visible. Thus, the contrast between the coronal structures and the background is increased. Further the ratio of intensity of the two coronal lines of [Fe-XIII] is sensitive to the coronal electron density[2,7] which allows a determination of n_e. Previous observations have mostly been done by heliospectrographic techniques[10] rather than imaging filtergrams. Two auxiliary filters centered on λ5100 Å and λ10700 Å in the nearby continuum of the observed spectral lines allow an observation of the visual and

Figure 4: Green coronal line [FE-XIV] $\lambda 5303$ Å on May 10, 1994. The bright sharp structures in the lower left are instrumental defects.

Figure 5: Infrared coronal line [FE-XIII] $\lambda 10747$ Å on May 10, 1994

infrared background (sky, residual instrumental stray light and K-corona) which can be subtracted after appropriate scaling from the image comprising the emission line corona. Finally prominences are observed in H_α $\lambda 6563$ Å as well as HE-I $\lambda 10830$ Å.

For the observations we use a 1024×1024 CCD array coded on twelve bits. Corresponding to the poor f-ratio of f/26 exposure times of 30 seconds have to be chosen for the visual spectral range. As the infrared lines are at the very end of the sensitivity of the CCD, much longer exposure times of about 10 to 15 minutes are required for [FE-XIII].

First observational results of PICO on Pic Du Midi are presented in Fig. 2 to 5. They result from an observational campaign during April and May, 1994. The orientation of all frames show *equatorial* north at top, west at right. Fig. 2 shows an image of the green coronal line on April 30, 7:28 UT, only one hour prior to the superior conjunction of Mercury. The separation between planet and suncenter is 18 arcmin. at $P = 170°$, next to an active region behind the solar limb. The 'black shadow' on the left side of Mercury is due to background correction and thus an artifact of the image treatment. The planet's apparent brightness is estimated to -2^{m}. But not only inner or outer planets on their orbits regularly get close to the sun, as observed here, also other celestial bodies, especially comets, cross the field of view of a coronagraph. Since the first observation of a comet on collision course with the sun by the SOLWIND coronagraph on P78-1[8], several comparable events observed by space borne coronagraphs have never been confirmed by ground based observations. Although the cited comet was quite bright (-3.5^{m}) fainter comets should be observable by routine control of coronal observations.

In Fig. 3 the difference of two images taken with the green line filter of [Fe-XIV] on May 10, 1994, right before the first contact of an annular solar eclipse, is shown. On Pic Du Midi the maximum phase of 70% was reached only 10 min before sunset. The delay between the images is 5 min. As the moon moves through the field of view it partially hides coronal light. After subtracting one image from the other the white light and emission line corona at $\lambda5303$ Å become visible in a crescent shaped region. At the location of the occulter a white light image of the solar disk, as seen simultaneously by the auxiliary video camera described above, is mounted in the frame. The central disk brightness is normalized to $50 \cdot 10^{-8} B_{\odot}$ to be compared with the coronal brightnesses. These range between $50 \cdot 10^{-8} B_{\odot}$ and $2 \cdot 10^{-8} B_{\odot}$ at 2 $R_{\odot}$ before vanishing in the noise. In the right part of the image the corona can be seen out to almost 2 $R_{\odot}$ whereas above the solar equator (the position angle of the rotational axis is $P= -22°$, therefore south western limb is overlying the solar equator) it vanishes already at 1.6 $R_{\odot}$. Taking into account a sky brightness of $33 \cdot 10^{-6} B_{\odot}$ at 1.3 $R_{\odot}$ coronal features down to a few tenths of a percent below the sky brightness are shown.

The emission line corona above the eastern limb, taken at 9:00 UT in the morning of the same day, is shown in Fig. 4 (green line [Fe-XIV]) and Fig. 5 (IR-line [Fe-XIII] $\lambda10747$ Å). This region turns out to be essentially the same as the one shown in Fig. 2 having been observed at Apr 30, one and a half weeks earlier. A system of loops and radial structures with an underlying prominence (not displayed) extends over 40° above the western equator. Structures can be seen out to 1.5 $R_{\odot}$in both images. Despite a poor image correction of the infrared image, it is obvious that the structures in both images are similar but not identical. However detailed analysis or comparison of such images requires better image correction: instrumental improvements will be undertaken in the near future.

4. CONCLUSIONS

We present first observational results using a prototype of LASCO-C1, a small, low scattered light mirror coronagraph adapted to ground based observations on Pic Du Midi. We were able to detect the emission line corona out to 1.5 $R_{\odot}$ in the green and infrared coronal lines of [Fe-XIV] and [Fe-XIII] using tools of image treatment and background correction. Although mainly limited by the brightness and temporal variability of the sky, there are some major advantages compared to space borne observations which justify to drive forward ground based observations in the future: higher spatial resolution can be obtained using larger apertures, which will be available using coronagraphs with reflective optics. The temporal resolution is not affected by a limited telemetry. New spectral ranges as the infrared with a reduced sky brightness and therefore more favorable for ground based observations can be observed in complement to space borne systems using new detector techniques.

5. References

1. G. E. Brueckner et al., The Large Angle Spectroscopic Coronagraph (LASCO): Visible Light Coronal Imaging and Spectroscopy, *Proceedings of the First SOHO Workshop, Annapolis* (1992)
2. D. R. Flower, G. Pineau des Forêts, Excitation of the [Fe-xiv] Spectrum in the Solar Corona, *Astron. Astrophys.*, **24**, 181 (1973)
3. M. Guhathakurta, R. R. Fisher, R. C. Altrock, Large Scale Coronal Temperature and Desity Distributions 1984–1992, *Astrophys. J.*, **414**, L145 (1993)
4. R. C. Altrock et al, submitted to Advances in Space Res., Sept. 1994
5. S. Koutchmy, Space-borne coronagraphy, *Space Science reviews* **47**, 95 (1988)
6. B. Lyot, A Study of the solar corona and prominences without eclipses, *M.N.R.A.S* **99**, 580 (1939)
7. H. E. Mason, P. Young, [Fe-xiii]: UV and Infra-red Lines, *SOHO III workshop, Estes Park, Co.*
8. D. J. Michels, N. R. Sheeley jr, R. H. Howard, M. J. Koomen, Observations of a comet on collision course with the sun, *Science* **215**, 1097 (1982)
9. G. Newkirk and D. Bohlin, Reduction of Scattered Light in the Coronagraph, *Appl. Opt.* **2**, 131 (1963)
10. J.-E. Wiik, B. Schmieder, J.-C. Noëns, Coronal environment of quiescent prominences, *Solar Phys.* **149**, 51 (1994)

MIRROR CORONAGRAPHIC DEVICE AND ITS APPLICATION

IRAIDA S. KIM
Sternberg State Astronomical Institute, Moscow State University
Universitetsky pr. 13, 119899 Moscow, Russia

OLEG I. BOUGAENKO and VASIL V. BROUEVITCH
Sternberg State Astronomical Institute, Moscow State
Universitetsky pr. 13, 119899 Moscow, Russia

SERGE KOUTCHMY
Institute d'Astrophysique de Paris, CNRS

98 bis Bd Arago, 75014 Paris, France

DONALD F. NEIDIG and RAYMOND N. SMARTT
National Solar Observatory/Sacramento Peak, Sunspot, NM 88349 USA*

and

OLEG A. EVSEEV
Institute for Space Research, Russian Academy of Sciences
Profsoyuznaya ul. 84/32, 117810 Moscow, Russia

ABSTRACT

A mirror coronagraphic system is described. This device can be attached to any conventional telescope to operate as a coronagraph with an aperture given by the device and a focal length of that of the "parent" telescope. Two off-axis parabolic mirrors serve as primary mirror and collimating mirrors, while the parent telescope forms the final image in this compound system. An inverse occulting disk and Lyot stop are in the same plane but laterally displaced. The abberations are mainly astigmatism and field curvature. An angular resolution of 0.5-1.0 arc seconds is expected for a field-of-view of 40 arc minutes when the parent telescope is a Ritchey-Chretien design.

1. Introduction

To observe a faint astronomical object angularly close to a bright source requires removal of the direct and scattered/diffracted light due to the bright source. Usually imaging of planetary satellites, close binaries, novae and supernovae remnants, the solar corona, etc., has involved occulting the bright object and reducing the level of instrumental scattering by a coronagraphic technique,[5] but where a standard telescope is used, as in nighttime studies, this procedure is usually not very effective. However, a

*Operated by the Association of Universities for Research in Astronomy (AURA), under cooperative agreement with the National Science Foundation.

special technique has been suggested for this that can give true coronagraphic quality while using a normal (non-coronagraphic) quality telescope[?] (Kim et al., 1994).

Coronal observations mostly employ singlet-objective Lyot coronagraphs. The objective is specially polished to have an extremely small amount of scatter. An occulting disk at the primary image plane blocks the direct solar image, while a pupil-plane aperture (Lyot stop) serves to block the diffracted light that originates at the rim of the entrance aperture. But, for nighttime observations, it is usual to attach an optical system at the output of a normal telescope. Such a system serves to occult the bright source in the relayed image, while masking (at a pupil plane) the scattered light from the entrance aperture[9,?,1] However, the performance of such telescopic systems for low-scattered-light studies is limited fundamentally by the scattering contributions of the primary and secondary mirrors to the image of the faint object of interest. Hence, these systems will not be as effective as a true coronagraph, especially in cases where extremely faint emission (in comparison with a nearby dominant source) must be discriminated.

The current state of advanced technology in the superpolishing of optical surfaces has given an impetus to reviving the idea of reflecting coronagraphs proposed by, for example, Newkirk and Bohlin[6] and Zirin and Newkirk.[8] The effectiveness of this technology for reflecting coronagraphs has been successfully established by the development of ground-based instruments such as the Mirror Advanced Coronagraphs (MAC I, II and III),[7] the new Pic du Midi coronagraph (PICO),[2] as well as a space-borne version.[1] Spherical off-axis super-smooth zerodur and fused-silica primary mirrors are used for MAC I and II, and an off-axis parabolic silicon super-smooth primary mirror is planned for the space-born version. Here we describe a novel coronagraphic device that has the intrinsic quality of a conventional coronagraph, but that can also be used with conventional telescopes.

2. Optical Configuration of the Coronagraphic Device

The overall design consists of an aplanatic (absence of coma) two-mirror, afocal (F=) optical system with unit magnification and a field of view of 40 arcmin (Figure 1). The primary super-smooth mirror (M1) and a similar collimating mirror (M3) are identical off-axis, f/10 paraboloids. For solar applications, the inverse occulting disk is a flat annular mirror (M2) with a central hole of 1.01 of the solar mean diameter, located at the primary focal plane. The primary image light passes through this hole and is directed via a flat diagonal super-smooth mirror (M6), to a guiding system not shown in Figure 1. The diaphragm (D1, entrance pupil) is located before the primary mirror in the plane spatially coinciding with the plane of the M2 annular mirror. As a consequence, the diameter of the exit pupil equals D1 and occurs in the same plane. Since the diaphragm (D2) in this pupil plane functions as a Lyot stop, its aperture is slightly smaller than that of D1. The plane diagonal mirrors, M4, M5, direct the collimated coronal radiation to a parent telescope whose optics serve as a relay system. Replacement of the M4 and M5 mirrors by corresponding secondary optics converts this device to a separate reflecting coronagraph. The M1 and M2 mirrors are mounted in the same cell; alignment can thus be achieved by cell rotation. A system to provide an electrostatically-cleaned air flow across the M1 and M2 mirrors has been designed to protect them from contamination.

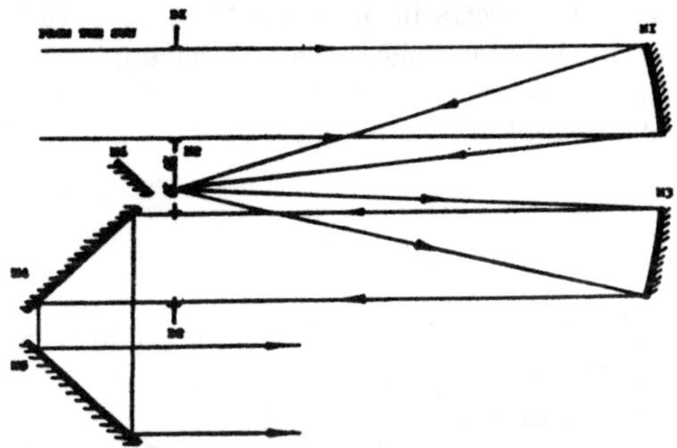

Figure 1: Optical sketch of the coronapraphic device.

A prototype of a post-focus coronagraph is currently under development at CSAT. Two halves of a 20-cm on-axis Si super-smooth (rms=5 A) parabolic mirror (D-200 mm, F=2000 mm) are used as a primary and a collimating mirror. Silicon has the advantage of high thermal conductivity and relatively high specific stiffness. The mirror cell features a radial thermal compensation system. The Serrurier's truss consists of 16 carbon-epoxy tubes. The fold mirrors mirrors are light-weight designs. The combined weight of these components is 25 kg with dimensions of (2250x310x310) mm. This portable device can be attached to any conventional telescope to operate as a coronagraph with an aperture of the device and a focal length of the parent telescope that is used in a normal operating mode for data recording.

3. Applications

Since this device is achromatic, it has the advantage, compared with a classical singlet-lens objective coronagraph, that multi-wavelength observatories, including the IR, can be carried out simultaneously. The collimated beam from M5 can feed a variety of analyzing instrumentation via relay optics, such as spectrographs, spectroheliographs, and tunable, multiple-line filters. Observations could be carried out in a patrol mode for disk or limb studies, as well as specific observational programs. For instance, simultaneous imaging of prominences, the chromosphere, and coronal regions in the light of "hot" and "cold" emission lines located in different spectral intervals could help to resolve the question of distribution of ions representing different stages of ionization. For this, color film and narrow-band filters that simultaneously transmit several chosen lines could be used, or multiple detectors to sample images corresponding to different spectral lines, the images isolated via dichroic beamsplitters.

Space debris surveys can be carried out using either ground-based or space-born coronagraphs. The coronagraphic device could be used for this purpose, either attached to a "parent" telescope or operating as an invididual reflecting coronagraph. Finally, this coronographic device could have wide applications in critical nighttime studies such as reflectance spectra of planetary rings and studies of satellites, lunar occultations of stars, binary stars where the pair have greatly differing luminosities, mass exchange in such systems, and studies of other faint emissions associated with stellar and other galactic and extragalactic objects.

4. References

1. Brueckner, A.E.: 1993, SOHO, The scientific payload, S/93/86, p. 47.
2. Epple, A. and Schwenn, R.: 1994, this issue.
3. Kim, I.S., Bougaenko, O.A., Brouevitch, V.V., Evseev, O.A.: 1995, Izvestiya of Russian Academy of Sciences, seriya fizicheskaya v.59, N 6 (in press).
4. Lyot, B.: 1930, C.R. Academy Sci. 99, 580.
5. Newkirk, G., and Bohlin, D.: 1963, Appl. Opt. 2, 131.
6. Smartt, R.N. and Koutchmy, S.: 1992, Proceedings of the Seventh Cambridge Workshop on Cool Stars, Stellar Systems, and the Sun, Tucson, Arizona, 9-12 October 1991, Giampapa, M.S. and Bookbinder, J.A., eds., San Fransicso, p.660.
7. Zirin, H., and Newkirk, G.: 1963, Appl. Opt. 2, 977.
   bibitem Breckenridge, J.B., Kuper, T.G. and Shack, R.V.: 1984, Opt. Eng. 23, 816.
8. Vilas, F. and Smith, B.A.: 1987, Appl. Opt. 26, 664.
9. Wang, S., Owensby, P.D., Toomey, D.W., Brown, R.H., Stahlberger, W.E., Cavedone, C.P., Hua, R., and Ftaclas, C.: 1994, SPIE, 2198, 578.

THE OPTICAL DESIGN OF AN INTERFERENCE SPECTROMETER WITH A HIGH RESOLVING POWER FOR SOLAR INFRARED.

E.S. Kulagin
Main astronomical observatory of RAS,
Pulkovo, ST-Peterburg 196140, Russia.

and

P.G. Papushev
Institute of Solar-Terrestial Physics,
P.O.Box 4026,Irkutsk 33, 664033, Russia.

ABSTRACT

For the IR-range, we believe that the use of the design of an interference spectrometer, previously suggested and tested by one of us [1], would be promising. The high resolving power and the ordinary form of the spectrum (without strip-channels) is achieved by a multiple phase-coherent reflection of the grating in the etalon. The brightness of the spectrum obtained is considerably increased at the expense of the free entrance of the beam from the grating into space between the plates of the etalon and it is nearly equal brightness obtained in a usual diffraction spectrograph of the same telescope with the same normal with the slit and the same dispersion.

1. Rationale

The IR spectral region offers scientific and technical advantages for the measurement of magnetic fields and line-of-sight velocity at the upper level of the solar atmosphere. To take full advantage of the merits of the IR-range and state-of-the-art matrix photoelectric detectors($2D$-array), when making observations, use must be made both of dedicated IR-telescopes and spectral equipment of a high resolving power. It is easy to demonstrate that, in order to measure the line-of-sight velocity with an error not more than $0.5km/s$ (r.m.s.) or weak magnetic fields ($H \sim 10G$), it is necessary to have spectrometers with resolving powers $R = \lambda/\Delta\lambda$ greater than $\sim 10^5$. The disadvantages of the currently used Fourier-spectrometers and spectrometers with Fabry-Perot scanning etalons are universally known. Modern large-format array detectors are quite suitable for recording spectra in the "usual" mode, i.e. we could use one detector dimension spectrally and the other spatially. The main difficulties to obtaining such spectra is the excessively large size of the grating. With currently available detection sensitivities, photon noise from the background is the dominant source of the noise. Also, with an increase of R, the relative contribution from the background of warm spectrometer elements would be dominant at ever shorter wavelengths. The instrumental background effect is eliminated by cooling down the optical elements of the spectrometer. As shown by estimates of the magnitude of the instrumental background from different optical elements of the spectrometer, when

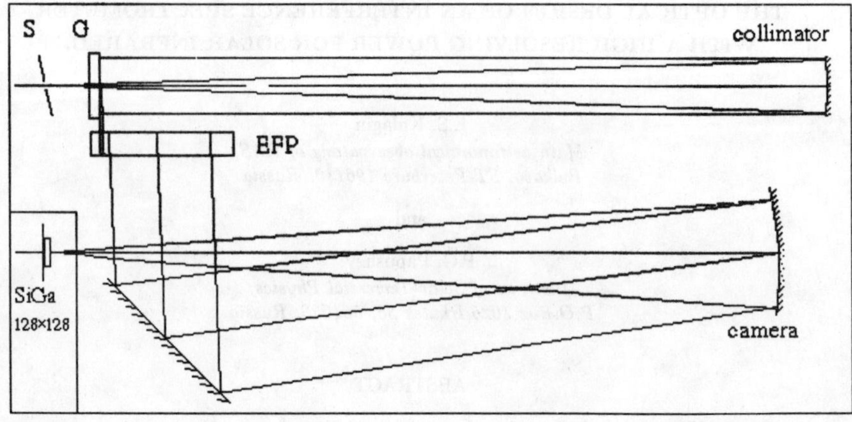

Figure 1: Optical layout of the interference spectrometer.

measuring such relatively weak sources as IR-emission lines of the chromosphere and corona at the solar limb, all spectrometer elements, beginning with the slit, must be cooled down to a temperature of about 100°K.

So that, as detector systems improve and spectrometer resolutions increase it becomes increasingly advantageous to use a cooled spectrometer.

In this paper, we discuss the optical system of a small-size solar spectrometer which has already been successfully used in the visible spectral region.

2. Optical design

In the spectrometer design under consideration (Fig. 1), the small diffraction grating is illuminated through the collimator with the light from the straight entrance slit of a normal width. The grating operates near normal incidence of the beams. The grating constant is chosen to be close to a mean wavelength of the spectral part under investigation. In this case, the diffraction grating well concentrated the light with the plane of polarization normal to the rules at large diffraction angles. With the diffraction angle of 85°, more than half of the incident light in the indicated polarization is concentrated. The light diffracted on the grating is incident on the Fabry-Perot etalon (FPE).

The high spectral resolution and the usual shape of the spectrum (without channel-bands) are achieved by multiple phased re-images of a small diffraction grating in the Fabry-Perot etalon. The etalon plates form a roughly right angle with the grating plane. Under these conditions, angular dispersions of the grating and the etalon are equalized for a considerably large part of the spectrum.

To increase the brightness of the spectrum obtained, the first (along to the beam path) etalon plate has, near the edge, a sharp rectilinear boundary between the operating surfaces covered and not covered with a reflecting layer. The bundle of light from the grating with the diffraction angle of about 85° easily enters the spacing between the etalon plates. The window for free entrance of the beams is formed by the rectilinear boundary of the reflective coat on the first etalon plate and by its image in the second plate. If the thickness of the gap of the etalon is designated as t, then the free entry to the etalon is accomplished from the part of the diffraction grating of the size $2t$. Therefore, the diameter of the illuminated part of the diffraction grating must be only somewhat larger, of about $3t$.

The significant increase of the region of free dispersion of the etalon and obtaining of the usual shape of the spectrum with high resolution may be explained in the following illustrative manner. If the slit is a normal, all rules of the diffraction grating which operates at normal incidence of the beams, emit diffraction waves in phase. Multiple grating images in the etalon perpendicular to the grating plane, lie in the same plane and form one large grating, whose effective size is $2tN_e$, where N_e is the finesse in the etalon. This large diffraction grating has the same resolving power as the etalon, while the etalon itself plays the role of the window, through which the spectrum from this large grating is observed.

It is possible to pass from one part of the spectrum to another through a small deviation of normal incidence of beams on the grating by suitably slightly rotating the etalon until a new equalization of the angular dispersions. Proceeding in such a manner it is possible to vary the wavelength by about ±10% without any channel-bands appearing in the observed part of the spectrum. A significant change of the observed wavelength is possible only by replacing the grating.

By way of example, we consider the main parameters of the above interference spectrometer for 10 micrometer spectral region with a spectral resolution $R = 10^5$. It will be assumed that the spectral resolution element, (FWHM), corresponds to two pixels of the total size of 100 micrometers. We first determine the desired gap of EFP which ensures a given R:

$$t = R\frac{\lambda}{2N_e} \tag{1}$$

when $Ne \approx 30$, the desired $t = 16.7mm$.

The diameter of the illuminated part of the diffraction grating must exceed $2t$ and is taken to be equal to $50mm$. With the relative aperture of the feed telescope 1/16, the collimator focal length is $800mm$.

The frequency of grating ruling is $100rules/mm$. The diffraction angle $\varphi = 85°$. The width of the wave front entering the etalon is $2.9mm$. After 30 effective reflections, the wave front broadens to $87mm$. In this way, the light size of FPE in the dispersion direction must be no less than $90mm$.

Angular dispersions of the grating and the etalon are equal and are determined by the diffraction angle on the grating. Hence the focal length of the spectrometer camera is:

$$F_k = \frac{\lambda}{\tan\varphi} \frac{dl}{d\lambda} \tag{2}$$

where the dispersion: $dl/d\lambda = 0.1R/\lambda = 0.1mm/\mathring{A}$. Consequently, $F_k = 875mm$.

3. Discussion

The small size of the interference spectrometer (IS) permits it to be housed in the optical dewar of a medium size. The real optical system and design of the cryostat will be described elsewhere. Here we only discuss the principles of design and the peculiarities of the IS optical system. Formerly, IS was not a fully wide-band instrument. It is, of course, important to pass from one part of the spectrum to another through a small deviation from normal incidence of the beams on the grating, with a suitable small rotation of FPE until the angular dispersions are equalized. In this way, it becomes possible to vary the wavelength by about ±10% without any channel-bands appearing in the observed part of the spectrum. If it is necessary to observe parts of the spectrum outside these limits, then the use of several small gratings mounted on a turntable should be provided.

4. References

1. E.S.Kulagin, *Soviet Astronomical Journal*, **57**, No 1, (1980) 200-210.

DETECTION OF A DUAL "DISPARITION BRUSQUE" AT PIC-DU-MIDI

J.M. NIOT
Observatoire du Pic-du-Midi, URA 1281 du CNRS
65200 Bagnères-de-Bigoree, France

J.C. Noëns

Observatoire du Pic-du-Midi, URA 1281 du CNRS
65200 Bagnères-de-Bigoree, France

D. Romeuf
Associated HACO observers team, Obs. Pic-du-Midi
65200 Bagnères-de-Bigoree, France

and

S. Koutchmy
Institut d'Astrophysique de Paris, Bd. Arago, 75014 Paris, France

ABSTRACT

An H-alpha survey above the solar limb is now performed with the H-Alpha (HACO) coronagraph at Pic-du-Midi. On June 29, 1994, well known events named "disparition brusque" were detected at the same time at the two opposite limbs. Presentation of this dual event is made in this paper.

1. Introduction

To participate to the general ground support during the SOHO mission a devoted H-Alpha coronagraph was installed at Pic-du-Midi. Now this instrument is equipped with a CCD 1024 x 1024 detector to produce H-Alpha images with FITS format for the entire limb at 1 R_o distance from the Sun center. With this field the spatial resolution is 3"/pixel. The time resolution may be adjusted according to the solar coronal events from 4 to 60 images/hour. An adapted numerical mass treatment permits to obtain calibrated intensity, altitude and position values for the Hydrogen emission regions.

A team of around 50 amateurs, named Associated Observers, is organized to obtain a daily survey of the solar events. The observations presented here were obtained by one of the members of this team.

2. Observations

The observations were obtained on June 29, 1994 from 6:00 to 15:30 UT. A spectacular event was seen at the same time on the two opposite solar limbs, East and West. Two limb prominences which appeared at PA 75 and PA 255 on June 28, presented a quite similar evolution on June 29, with an increasing phase followed by a "disparition".

In Figure 1, obtained at 1313, during the increasing phase we can note the general aspect of these two prominences at this step of their evolution. Figure 2 presents the sequence of the general evolution of the two opposite regions from 6:35 to 12:23 UT; the two prominences are in an increasing phase for both intensity and volume development. After 12:23, the West limb prominence begins to decrease and this evolution is accelerating at 13:13. The West limb prominence has totally disappeared at 13:53 and did not reappear after that time. The evolution of the East limb prominence is similar but the increasing phase is still evident 13:13. At that time, the disappearance occurred with a longer time lapse. The East limb prominence had totally disappeared at 15:00 UT.

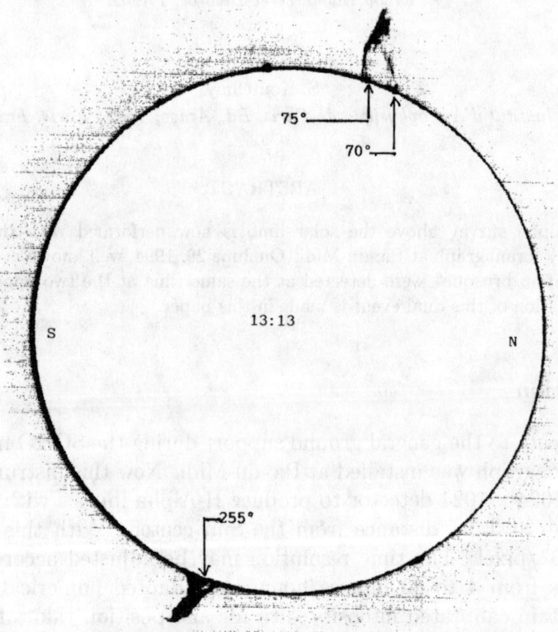

Figure 1: Full field-of-view image obtained on June 29, 1994 with the H-Alpha coronagraph. This image was registered during the increasing phase before the two "disparitions brusques".

With an adapted numerical procedure it is possible to determine the sky brightness and instrumental light scattered level on each image. The substraction of this diffused light level (which is a function of sky and instrumental conditions, distance to the limb, and position angle around the limb) is an important step to get quantitative information on the cool Hydrogen emissive regions. After this operation, we are able to obtain well calibrated intensities in the H-Alpha line for each image. Then it is possible to determine the total H-Alpha emissivity in selected regions. In Fig. 3, we compare the evolution of the total emissivity of the PA 60-85 and PA 240-265 regions from altitudes of 346000 and 238000 Km. These regions show the evolution of the two large prominences as in Fig. 1 and 2. The time scale is on the X-axis and the Y-axis is the total brightness of each region expressed in terms of Bo/pixel. Two different Y-axis scale were chosen for a more comprehensive view of the phenomena.

Fig. 3 shows at first that there are some similarities for the general evolution of the two opposite limb prominences. We can consider that on each side there are three different phases.

1. A pre-event phase in which there seems to be an instability in the total value of the emission level. In fig. 2, we can note that during this phase there is an increase of the volume of the prominence material. This phase was studied by Mouradian et al[1] on several filaments on the disk.
2. The phase I is followed by a rapid increasing of the total value of the emission level in the H-Alpha line. This phase II takes two hours for the two prominences and is starting approximately at the same moment. The end of the phase II is reached when the total emission value is at its maximum. We can note that the evolution of the West limb during the phase II is a half hour in advance of that on the East limb.
3. At the end of phase II there is a rapid decreasing of the total emissivity of the two regions. This is the phase of the "disparition brusque". For this phase III, we can note some differences between the evolution of the two prominences. This phase is shorter for the West limb prominence: in 70 minutes the total emissivity of these region reach zero level. The duration of the phase III for the East region is of the order 120 minutes. These differences of the evolution of the two opposite regions may be also seen in fig. 2. It appears that the "disparition" of the West limb prominence is very rapid between 13:33 and 13:53 UT. On the contrary, the East limb prominence slowly disappears, with an altitude extension phase till 15:05 UT. This seems to appear like a thermal disappearance as it was noticed by Mouradian and Soru-Escaut.[2]

3. Discussion

We observed almost simultaneously two "disparitions brusques" of large prominences situated on the opposite limbs of the Sun. There is no possibility to claim that these two phenomena belong in fact to the same magnetic disturbance produced in the inner parts of the Sun. But the numerous similarities in the different phases of these two phenomena suggest that the destabilization effect responsible for the double, diametrically opposite DB is a global phenomenon. The chronology of events

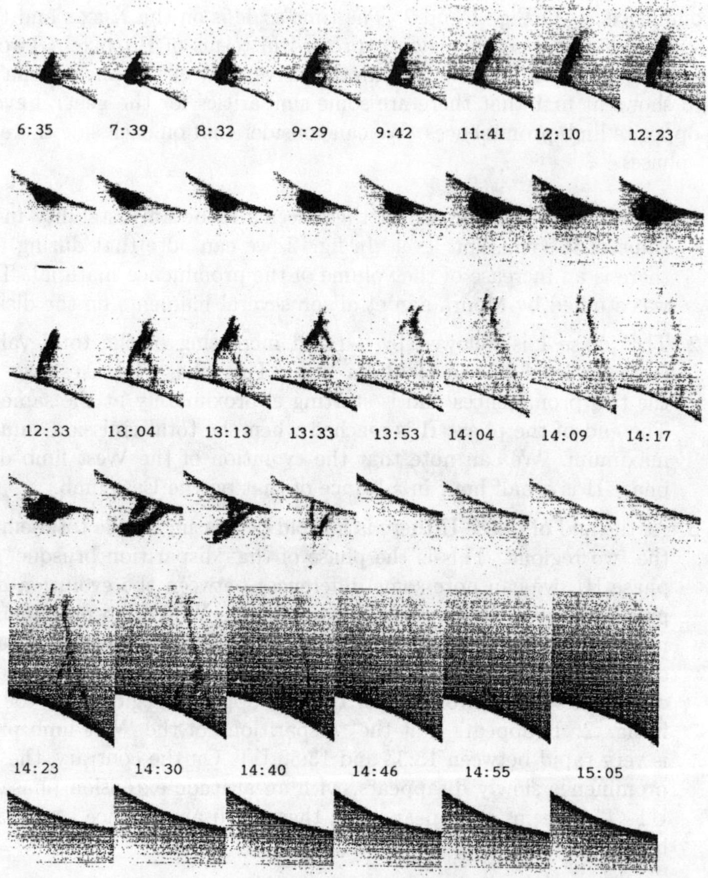

Figure 2: Mosaic obtained after the extraction of the two East and West studied regions. The East and West limb are shown at the top and the bottom of each raw.

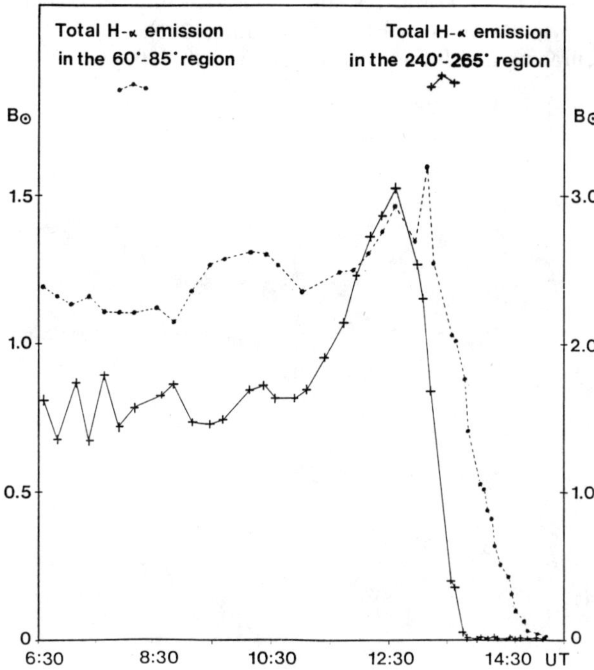

Figure 3: Evolution of the total emission level in the H-Alpha line for the two East and West limb regions. The three phases of the phenomena are clearly shown.

seen at the same time with SXT (H. Hudson, private comm.) suggests that several phenomena were simultaneously occurring around the Sun like flares, ejections, cusp appearance.

4. References

1. Z. Mouradian, M.J. Martres, and I. Soru-Escaut, "Proceedings of the Japan-France seminar on Solar Physics", p. 195 (1981).
2. Z. Mouradian and I. Soru-Escaut, *Astron. & Astrophys.*, **210** 410 (1989).

NEAR-INFRARED CORONAGRAPHIC DETECTION OF SPACE DEBRIS

DONALD F. NEIDIG
Phillips Laboratory (AFMC), Geophysics Directorate
National Solar Observatory/Sacramento Peak, Sunspot, NM 88349, USA*

RAYMOND N. SMARTT
National Solar Observatory /Sacramento Peak, Sunspot, NM 88349, USA*

IRAIDA S. KIM
Sternberg State Astronomical Institute, Moscow State University
13 Universitetsky Pr., 119899, Moscow, Russia

and

SERGE KOUTCHMY
Institut d'Astrophysique - CNRS
98 bis Bd. Arago F-75014, Paris, France

ABSTRACT

The use of a ground-based, large-aperture (4-meter class) reflecting coronagraph for detecting orbiting space debris is proposed. It is shown that, by taking advantage of the Fraunhofer diffraction from particles passing angularly close to the solar limb, the theoretical limiting detectable diameter is less than 1 mm for particles in low orbit. The latter limit assumes an effective angular resolution at the detector of 1 arcsec and near-infrared observations, where advantage can be taken of the reduced sky background. For particles of size ≥ 1 mm, the detection rate is expected to be about 1200 per hour, considering the current debris population. Coronagraphic detection of debris offers advantages in its independence of debris albedo, and its ability, at least in principle, to measure directly particle diameters based on the diffraction profile, independent of flux-distance relationship. The high detection rate for small debris may be useful in statistical studies of debris population, as well as changes in orbital or population statistics on short-to-long time scales; such changes would be associated with breakup events, atmospheric drag episodes, and collision cascade effects. The limitations and technical difficulties encountered in coronagraphic observations of debris, as well as the background and motivation for measuring orbiting debris, are discussed briefly.

1. Introduction

The unique benefits and capabilities offered by large-aperture, wide-band, visible light and near-IR coronagraphic instrumentation may find application in surveying

*Operated by the Association of Universities for Research in Astronomy, Inc. (AURA) under co-operative agreement with the National Science Foundation. Partial support for the National Solar Observatory is provided by the USAF under a Memorandum of Understanding with the NSF.

orbital debris. It is shown here that the debris-detection capability of a large coronagraph may exceed that of existing conventional methods relying on radar measurements or optical backscattering measurements made during twilight and nighttime.

Several space-based coronagraphs, including the Skylab SO52[1] and Solar Maximum Mission C/P experiments, have demonstrated the capability to observe debris at close range using very small telescope apertures. These detections were entirely serendipitous, since the aim of these experiments were coronal observations, and the instruments and detectors were not optimized for debris detection. The detections were made possible in large measure by the wide bandpasses and large fields of view afforded by the externally-occulted coronagraphic systems that are often chosen for space experiments, and by the relatively small angular velocity of local debris floating in the vicinity of the spacecraft.

Debris detection using ground-based coronagraphs, however, has not yet been achieved, owing to two principal constraints. First, conventional coronagraphs are internally-occulted, which generally forces the use of singlet lens objectives to reduce scattered light; the resulting chromatic aberration restricts the bandpass and therefore limits the photon flux. Second, the high angular velocity of orbiting debris as seen from the ground requires high rates of photon collection in order for successful detection to occur. With the advent of superpolished mirror technology, both of these limitations can be overcome by the significantly larger apertures and achromaticity that are now possible with all-mirror systems. Because coronagraphic observation of debris is background-limited by the daytime sky, it is advantageous to achieve high spatial resolution in addition to large photon statistics.

This paper considers only the forward scattering (Fraunhofer diffraction) from particles passing angularly close to the sun. The following sections discuss the motivation for space debris observations, the basis for the coronagraph's sensitivity in observing debris, and a calculation of the expected detection limits and rates. Further discussion on a proposed 4-m coronagraph can be found elsewhere in these proceedings.[2]

2. Background and Motivation for Space Debris Observations

The presence of numerous objects in low earth orbit has become a concern to military and civilian space programs, relating primarily to collisions between these debris objects and operating spacecraft or astronauts. With a typical collision velocity of about 10 km s^{-1}, the kinetic energy dissipated upon impact is impressively large. Dust-sized particles, for example, produce significant surface erosion which can degrade external sensors; mm-sized particles can damage or destroy external spacecraft systems (antennas, solar panels), and would be lethal to astronauts during EVA; collisions with cm-sized particles would be catastrophic, resulting in spacecraft breakup and the subsequent production of numerous new debris particles to be added to the already growing population. To date, several breakup events may have been caused by collisions with debris. The majority of orbiting particles of size greater than 0.1 cm are anthropogenic, having their origin in booster separations, accidental and deliberate detonations in space, deterioration of paints and coatings, propellant residue, and a variety of cast-off materials. Below 0.1 cm the population is dominated by meteoroids.

The possibility of a collision cascade phenomenon, or debris runaway effect, has been discussed widely in the literature. The phenomenon would be characterized by an exponential increase with time in the population of small debris, produced by a self-sustaining process of debris-spacecraft and debris-debris collisions. Modeling of this effect, which is highly dependent on the assumed rate of increase in space traffic as well as the effectiveness of atmospheric drag in removing the smaller particles (the latter, in turn, is dependent on the level of solar activity), suggests an onset sometime in the next century, with number densities of small debris possibly increasing by several orders of magnitude or more over current values. Thus there is a clear and direct need to monitor the debris population and to collect statistics. Small debris, due to its relatively larger number density, would be particularly useful in this regard.

Space debris exceeding several centimeters in size, in low orbit, has been demonstrated to be observable using medium-aperture optical telescopes during twilight; debris exceeding several tenths of a centimeter can be observed at any time of day using large radar telescopes (for debris that is radar reflective). In addition to remote sensing, in-situ measurements are obtained from micro-cratering on recovered spacecraft. Due to the relatively small collisional flux for large particles, the latter technique is practical only for debris smaller than 0.1 cm where the number of particles, and hence the collision flux, is relatively large. The size range between approximately one-tenth and several centimeters remains difficult to sample accurately.

In low earth orbit, which we define here to be 500-1500 km, the majority of debris is in nearly circular orbit, and nearly all orbital inclinations are represented. Based on recent estimates[3] the cross sectional area flux, in low earth orbit, is about 2 x 10^{-2} yr^{-1} m^{-2} for particles with diameter ≥ 0.1 cm, and drops to about 10^{-4} yr^{-1} m^{-2} for diameter ≥ 1 cm. The flux $\Phi(\geq a)$ and the density $N(\geq a)$ for particles with diameter $\geq a$, over the range $0.1 \leq a \leq 1$ cm, can be adequately represented by power laws:

$$\Phi_{1990}(\geq a) \approx 2 \times 10^{-16}\ \mathrm{a}^{-2.5}\ \mathrm{s}^{-1}\ \mathrm{cm}^{-2}, \tag{1}$$

$$N_{1990}(\geq a) = \frac{\Phi_{1990}(\geq a)}{\nu_c} \approx 2 \times 10^{-22}\ \mathrm{a}^{-2.5}\ \mathrm{cm}^{-3}, \tag{2}$$

where $v_c \approx 10$ km s^{-1} is the mean collision velocity in low earth orbit. Thus the total number of objects of size ≥ 0.1 cm in low orbit exceeds 10^7. For diameters > 1 cm the flux and number density exceed the values given by Eqs. (1) and (2). The number of monitored objects (generally having diameter ≥ 10 cm currently tracked and cataloged by the U.S. Air Force Space Command is about 7000.

The foregoing discussion constitutes a brief overview of space debris as it relates to environmental concern and to remote sensing by ground-based instruments. The topic of space debris has engaged the interest of a wide body of researchers; for further discussion and references the reader is directed to other works.[4,5]

3. Detection of Space Debris Via Fraunhofer Diffraction

The diffracted flux due to a particle located angularly close to the sun is calculated here, and compared with the sky background at the same point. Particles with diameters ≤ 10 cm in low earth orbit will be spatially unresolved; thus the signal-to-background ratio will be determined by the ratio of the diffracted flux to the flux contributed by the sky background per resolution element.

In order to calculate the debris signal we consider a particle with circular cross section, diameter a, distance r from the observer, and located at position angle R from the center of the solar disk and at angular distance x from an arbitrary point on the solar disk (Figure 1). The solar flux dF_λ incident on the particle from an element of solar surface subtending solid angle $d\omega = \rho\ d\rho\ d\phi$ at the particle will be

$$df_\lambda = I_\lambda(\rho)\ \rho\ d\rho\ d\phi, \tag{3}$$

where $I_\lambda(\rho)$ is the solar intensity at angular distance ρ from disk center. An accurate analytic approximation for the intensity is given by[6]

$$I_\lambda(\theta) = I_\lambda(0)[1 - u_2(\lambda) - \nu_2(\lambda) + u_2 \cos\theta + \nu_2 \cos^2\theta], \tag{4}$$

where θ is the heliocentric angular distance ($\approx \sin^{-1}\rho/R_o$, to high accuracy), $I_\lambda(0)$ is the intensity at disk center, and u_2 and ν_2 are tabulated constants. The flux element $df_\lambda(x,a)$ in the diffraction pattern, contributed by dF_λ is given by [7]:

$$df_\lambda(x,a) = \frac{dF_\lambda \pi^2 (a/2)^4}{\lambda^2 r^2}\left[\frac{2J_1(\beta(a,x))}{\beta(a,x)}\right]^2. \tag{5}$$

Eq. (5) invokes Babinet's principle, which states that the diffraction pattern produced by an obstructing disk is identical to that produced by a circular aperture of the same diameter. $J_1(\beta(a,x))$ is the first-order Bessel function with argument

$$\beta(a,x) = \frac{\pi\ a \sin x}{x}. \tag{6}$$

For the small angles encountered here we can use $\sin x \approx x$ with no significant loss of accuracy. The total flux $f_\lambda(R,a)$ in the diffraction pattern produced by a particle at a fixed position angle R is then obtained by integrating over the solar disk:

$$f_\lambda(R, a) = \int df_\lambda(x(\rho, \phi), a), \tag{7}$$

where the variable x is related to the position angle, R, through the coordinates ρ, ϕ on the solar disk:

$$x = [\rho^2 + R^2 - 2\rho R \cos\phi]^{1/2}. \tag{8}$$

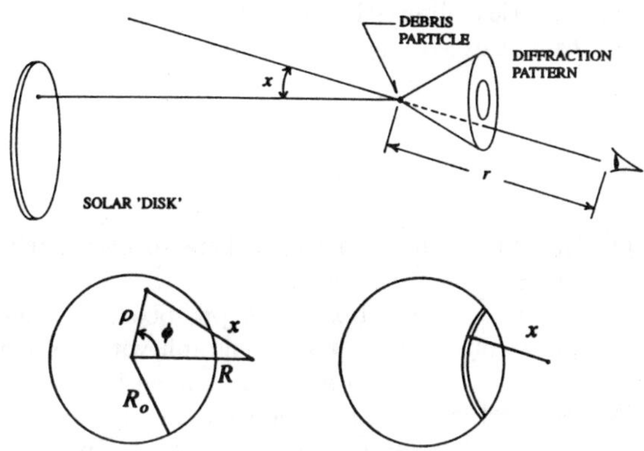

Figure 1: Top: sketch showing the relationship between the observer and the diffraction pattern produced by a particle illuminated by an arbitrary element of the solar surface. Bottom: depiction of the quantities used in calculating the disk-integrated diffraction pattern for a particle at angular distance R from disk center.

Duc to the circular symmetry of the diffraction pattern for a particle with circular cross section, the flux contribution from each x originates from an arc of radius x on the solar disk (Figure 1). Eq. (7) is solved numerically with x varying over the range (R - R_o) to (R + R_o) in which the individual flux contributions dF_λ are sorted according to their x-values and binned into an array $x, \Delta F_\lambda(x(\rho, \phi))$ and used in a final summation:

$$f_\lambda(R, a) = \frac{\pi^2 (a/2)^4}{\lambda^2 r^2} \sum_x \Delta F_\lambda(x(\rho, \phi))[\frac{2J_1(\beta(a, x))}{\beta(a, x)}]^2. \tag{9}$$

In the numerical integration, increments $\Delta\rho$ =10 arcsec and $\Delta\phi$ = 0.5 deg are used. The integration procedure is tested by demonstrating, as a by-product, that the correct value of the integrated solar flux at wavelength λ is obtained.

For spatially unresolved debris the flux $f_\lambda(R,a)$ in the diffraction pattern can, with an appropriate plate scale, be focused onto a single pixel in the detector. In this case the debris signal, expressed in electrons per pixel, will be

$$S_\lambda(R,a) = f_\lambda(R,a)\Delta\lambda \, A \, t_e \, \tau_\lambda \, Q_\lambda, \tag{10}$$

where $\Delta\lambda$ is the bandpass, A is the collecting area, τ_λ is the atmosphere-instrument transmittance, Q_λ is the detector quantum efficiency and t_e is the exposure time per pixel. For an orbiting particle observed at zenith, the exposure time per pixel, in the context of Eq. (10), cannot exceed the pixel crossing time, t_c:

$$t_c = \frac{\Omega \, r}{\nu_O}, \tag{11}$$

where Ω is the pixel angular dimension (taken here to be equal to the telescope's spatial resolution) and v_O is the orbital velocity.

In Figure 2 we plot the photon flux per Å per pixel, i.e. the signal given in Eq.(10) divided by Q_λ, as observed by a coronagraph with a 4-meter aperture, an atmosphere-telescope transmittance 0.5 and pixel dimension (spatial resolution element) 1 arcsec. We assume the debris particle to be in circular orbit at altitude 500 km; the corresponding pixel crossing time is 3.2 x 10^{-4} s, and the exposure time is taken to be identical to this. Figure 2 considers three wavelengths and three particle diameters for comparison. Note that for angles exceeding 20 arcsec above the limb, a small advantage in debris signal is gained by observing near 1.5 μm. An even greater advantage would be gained in practice, generally, by the reduced background in the near-IR (see below).

In assessing the detectability of the debris signal we compare it with the noise associated with the signal, the detector, and the background contributions of the corona, sky, and instrumentally-scattered light. The sky is by far the brightest of the contributors to the background at shorter wavelengths, with scattering of sunlight by atmospheric aerosols (the solar aureole) being dominant. For aerosol scattering we adopt a relative value 20 x 10^{-6} of the mean solar intensity at a distance 60 arcsec above the limb, at 0.55 μm. The latter value refers to the aureole brightness observed with high quality coronagraphs[8]; thus it already includes the contributions due to instrumental scattering by optical imperfections, dust on the optical surfaces, and residual diffraction effects. Because laboratory measurements suggest that the combined instrumental contribution would be considerably less than the total 20 x 10^{-6} quoted here, we will assume the latter contribution to be due entirely to aerosols. Then, to be conservative, we assume an additional scattering contribution of 3 x 10^{-6} due the instrument itself. Finally, another 3 x 10^{-6} is added for Rayleigh scattering. For comparison, the corona at the sun's poles (where it is least bright) has a brightness of about 1 x 10^{-6} of the solar spectral intensity, independent of wavelength.

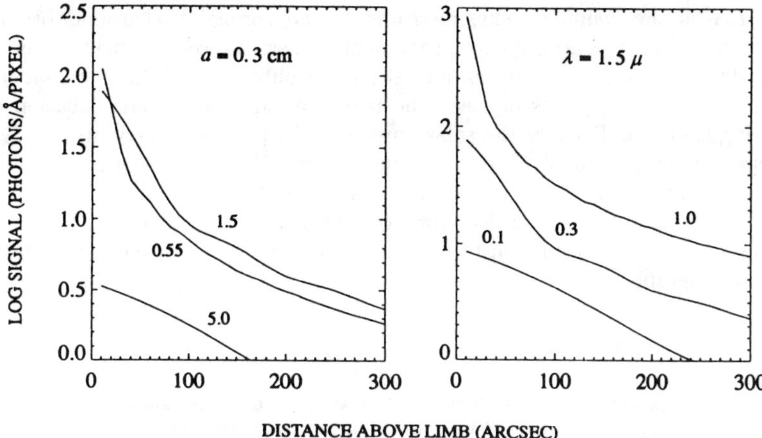

Figure 2: Log signal from debris in a 500-km orbit, observed at zenith with a 4-m coronagraph and 1-arcsec pixels. Left: result for particle diameter 0.3 cm and wavelengths of 0.55, 1.5, and 5.0 μm. Right: result for wavelength 1.5 μm and particle diameters 0.1, 0.3, and 1.0 cm.

Given the above values at 0.55 μm as references, we apply the following wavelength dependences to estimate the background in the near IR. Rayleigh scattering varies as λ^{-4} and thus becomes negligible. The instrumental scattering, for a Gaussian roughness distribution of the surface of the coronagraph objective, varies as $(4\pi\sigma/\lambda)^2$, where σ is the rms surface roughness.[9] Aerosol scattering varies as $\lambda^{-\alpha}$, where α usually falls within the range 0.5 to 1.5, depending on the aerosol size distribution (smaller particles lead to larger α). Measurements made in Sendai, Japan[10], indicate $\alpha \approx 0.9$, at most, over the wavelength range 0.56 - 1.04 μm, while those made near Boston[11], indicate $\alpha \approx 1.1$ over the range 0.5 - 1.6 μm. Neither of these measurements was made under coronal observing conditions. At continental, high-altitude mountain sites, the density of large aerosols is expected to be low, so that larger α values would be expected at these locations. Thus we assume $\alpha = 1.5$ at a good coronal site. Measurements of the solar aureole extending beyond 1.6 μm are not available to us. However, it is expected that the power law dependence will hold at least out to 3 μm[12]; in the computations leading to Table 1 we assume it will hold out to 5 μm. At 5 μm the sky brightness would then be comparable to the polar K and F corona.

The background signal, as measured by a coronagraph, will be

$$B_\lambda(R) = I_{\lambda,B}(R)\ \Omega_\lambda\ \Delta\lambda\ A\ t_e\ \tau_\lambda\ Q_\lambda, \tag{12}$$

where $I_{\lambda,B}$ is the combined sky, instrument, and coronal background brightness. Strictly speaking, the atmosphere-instrument transmittance, τ_λ, includes an atmospheric factor only for the corona and sky contributors (with the sky contributors requiring only the air mass between the telescope and the primary location of the scattering sources). Because the solar aureole is the primary contributor to $I_{\lambda,B}$ the atmospheric component of τ_λ, as it applies to aerosols, is assumed in Eq. (12). It is especially important to note that the exposure time per pixel, t_e, in Eq. (12), is always identical to the actual detector exposure time, whereas in Eq. (10), $t_e \leq t_c$. This distinction is fundamental in considering a strategy for space debris detection using coronagraphs.

TABLE 1.

CALCULATED SIGNAL-TO-BACKGROUND AND SIGNAL-TO-NOISE RATIOS FOR SEVERAL WAVELENGTHS AND PARTICLE DIAMETERS

Observation at zenith, particle altitude 500 km, circular orbit, 4-m aperture, 1-arcsec resolution, 1 arcmin above the solar limb, 1000-electron read-out noise*

$\lambda/\Delta\lambda$ (μm)	Particle dia. (cm)	S/B[a,b]	S/N[a,b]
.55/.2	.1	.00051	1.5
	.3	.0013	3.9
	1.0	.0041	12
1.5/.5	.1	.0064	8.1
	.3	.024	30
	1.0	.067	83
5.0/1.5[c]	.1	.025	2.0
	.3	.22	17
	1.0	1.0	79
5.0/1.5[c]	.1	.025	7.1*
	.3	.22	57*
	1.0	1.0	210*

*Last three lines of the table assume a detector with 30-electron read-out noise.

[a]Higher spatial resolution would lead to increased S/B and S/N in cases where the observation is background limited, as would apply to the first six lines of the table; in these cases S/B would increase as Ω^{-2} and S/N would increase as $\Omega^{-1/2}$, as long as $t_e \approx t_c$ and the pixel size is identical to the obtained spatial resolution.

[b]Both debris signal and background are calculated by multiplying the respective fluxes at wavelength λ by the given bandwidth, rather than by employing an actual convolution over the bandwidth.

[c]Bandpasses of this width may be unrealistic at this wavelength, due to atmospheric absorption bands (see Section 5).

One strategy, which we consider here, is to avoid the cumulative effect of the background by using very short exposures, with duration equal to the pixel crossing time, t_c. Because the background signal is very strong at the shorter wavelengths, we select a deep-well array detector with capacity 3 x 10^7 electrons and read-out noise 10^3 electrons. We then select a relatively wide bandpass, $\Delta\lambda$, in order to ensure that the total noise is dominated by photon statistics (we assume digitization noise will be less than photon noise). For wavelengths near 5 μ, where the sky background is low, we additionally consider a lower-noise detector with well-depth $\approx$ 3 x 10^5 electrons and read-out noise 30 electrons. Table 1 shows the resulting signal-to-background (S/B) and signal-to-noise (S/N) ratios for an assumed detector quantum efficiency of 50% and particle altitude and distance of 500 km (observation at zenith). It is emphasized that a spatial resolution of 1 arcsec is assumed in these calculations, and that the detection criteria apply to single pixels. Improved spatial resolution and/or multiple-pixel extraction algorithms would of course reduce the calculated detection limits.

4. Methodology, Measurements, and Detection Rate

We briefly describe here the application of the foregoing results in acquiring measurements of space debris. For simplicity we assume observations with the sun at zenith, in which case the particle distance r and orbital altitude are identical.

LIMITING DIAMETER. Table 1 shows that for 1-arcsec spatial resolution and debris distance r = 500 km, the theoretical detection limit (using the criterion S/N $\geq$ 3) corresponds to a diameter less than 0.1 cm, for observations made in the near IR. However, as the debris signal may be spread over several pixels we adopt 0.1 cm as a practical limit. For larger r the limiting diameter will increase approximately as r^2, for wavelengths out to 1.5 μm and somewhat beyond. The latter dependence derives from (1) the approximate linear relationship between debris signal and particle diameter (cf. S/B values in Table 1, for $\lambda \leq 1.5$ μm), and (2) the r^{-2} -dependence of the debris signal in cases where $t_e \leq t_c$ for all r considered, as applies to the strategy adopted herein. It is worth noting that for $t_e > t_c$, the r-dependence of t_c is manifested in Eq. (10), in which case the debris signal then varies as r^{-1}.

DETECTOR AREA AND READ-OUT. In order to capture the transits of individual particles we sample only a narrow arc just above the solar limb, at the edge of the occulting disk. Adopting a nominal arc width of 60 arcsec (60 pixels), with radius $R_o + 60 = 1020$ arcsec covering 90 deg of the solar limb near the sun's pole (total arc length 1602 arcsec), the active area of the detector therefore encompasses only 9.6 x 10^4 pixels and could be read out in 1 ms. (This currently non-existent detector could be a custom-made, arc-shaped device or a series of small rectangular arrays, each with individual read-outs, and arranged in the focal plane along the solar limb.) With an exposure time of approximately 0.3 ms (corresponding to the pixel crossing time for 500-km orbit), each debris detection would ideally consist of a dotted track of single exposed pixels each separated by two unexposed pixels. In practice, of course, the debris signal corresponding to each 0.3 ms exposure would ordinarily be spread over two or three pixels, owing to image size and motion relative to the pixel dimension.

ORBITAL INCLINATION. The orientation of the debris track readily provides a measurement of the orbital inclination.

ALTITUDE AND DIAMETER MEASUREMENTS. In the detection scheme above, the minimum angular velocity resolution (for a particle moving perpendicular to the limb and observed on a track spanning 58 pixels) would be 1 part in 58, or 0.017, which would correspond to an orbital altitude resolution of 9 km, in a circular orbit approximation. For particles with trajectories tangent to the occulting disk, a maximum velocity resolution of 0.0014, with altitude resolution 0.7 km, would be obtained. With the particle's distance r thus obtained, the diameter can then be calculated from the measured flux. Unlike measurements based on backscattering, this method of acquiring the particle diameter is independent of albedo. An additional method to be considered, made possible by Fraunhofer diffraction, might allow the particle diameter to be measured independently of absolute flux and distance; this would be accomplished simply by recording brightness as a function of angular distance above the limb, and matching the result with the calculated profile that is uniquely related to the diameter (cf. Figure 2). This method must assume, of course, a particle with circular cross section. For a non-spherical particle a characteristic dimension could be obtained only if the particle undergoes a negligibly small rotation during its transit across the detector.

ORBITAL ECCENTRICITY. Using the measured absolute flux and the particle diameter obtained from the shape of its diffraction profile, the particle distance and orbital altitude can be calculated independent of its angular velocity. The observed angular velocity can then be used to calculate the linear velocity without making use of the circular orbit approximation. The orbital eccentricity can then be calculated. However, considering the errors in the distance and velocity measurements, the eccentricity so derived will be correspondingly limited in accuracy.

DETECTION RATE. The expected detection rate of debris particles in low earth orbit is calculated from the product of the particle flux and the area of a plane extending between altitude 500 and 1500 km with angular width $\psi = 1442$ arcsec corresponding to the separation of the extreme points of the detector arc. The differential detection rate dG corresponding to the limiting observable diameter a_{lim} at height r and area element $dA(= r\ \psi\ dr)$ will be:

$$dG = \Phi(\geq a_{lim}(r))r\ \psi\ dr, \tag{13}$$

where $a_{lim}(r) \approx a_o(r/r_o)^2$ (see above), with $a_o \equiv a_{lim}(r_o)$ and $r_o = 500$ km. Substituting the latter approximation for $a_{lim}(r)$ in Eq. (1) and integrating Eq. (13), we obtain

$$G \approx 2\text{x}10^{-16}\text{a}_\text{o}^{-2.5}\ \text{r}_\text{o}^5\ \psi \int_{\text{r}_\text{o}}^{\text{r}} \text{r}^{-4}\ \text{dr s}^{-1}.$$

Taking $r = 1500$ km as an upper bound we find $G \approx 0.35$ s^{-1} (1270 hr^{-1}) for $a_o =$ 0.1 cm. Most of these detections will be due to particles near altitude r_o.

5. Technical Difficulties and Limitations

Several issues relevant to actual observing situations have not been included in the foregoing discussions, and these deserve mention here. First, there remains a question of background noise other than detector and photon statistical noise. The sky background itself may be intrinsically variable as a result of a non-homogeneous aerosol spatial distribution, or even discrete background events arising from nearby particles passing through the field of view. If the resulting variability occurs on spatial or temporal scales small enough to admit inclusion among the fundamental sources of noise, i.e., a background component which cannot be subtracted, then the S/N ratios in Table 1 must be reduced accordingly. On the other hand, individual particulates sufficiently large to produce signals that compete with those produced by debris can perhaps be discriminated on the basis of track nonlinearity or prevailing wind direction. Also, nearby particles (in fact *any* atmospheric particle) will produce a significantly out-of-focus image in a large telescope. Indeed, depending on the focal ratio of the coronagraph, the orbiting debris itself may not be in focus simultaneously at all orbital altitudes.

Second, the computations of debris signal presented here assume particles with circular cross section, i.e. spherical particles. Actual debris will have irregular shapes, producing diffraction patterns significantly different from those assumed here. It is not clear to what extent the characteristic particle dimension could be extracted using deconvolution methods.

Third, the coronagraph must remain pointed at the sun; therefore, its application in space debris detection is limited to a stare-mode operation. Thus the objects cannot be tracked sufficiently long to acquire orbital elements for inclusion in debris catalogs. Considering the small sizes of the debris considered here, however, this may prove no disadvantage, as the number of objects above 0.1 cm, or even above 1.0 cm, may be unmanageably large; also, small particles in low orbit are highly susceptible to orbital changes induced by atmospheric drag, so that cataloging their orbital elements may be a futile exercise in any case.

Fourth, the extremely rapid cadences required for acquiring debris tracks, using the strategy presented here, will result in huge pixel rates on the order of 10^8 s^{-1}. This would quickly overwhelm any storage system without a real-time flagging routine to capture and save frames containing debris images. It may not be trivial to employ such routines under the best of circumstances; and with the S/B and S/N ratios encountered here (< 0.01 and < 10, respectively, for debris near the detection limit) this may prove to be a technical challenge. Also, if detection rates on the order of 10^3 hr^{-1} are actually achieved, the data processing and analysis may be a formidable task.

Finally, the estimates presented here for the background sky brightness in the near IR have not included any complicating effects due to atmospheric absorption bands. Fortunately, absorption bands are not sufficiently numerous to preclude the use of wide bandpasses for wavelengths less than 2 μm, where the coronagraph's debris-detecting performance, in terms of S/N ratio, has already maximized. At 5 μm, however, strong absorption bands may disallow wide bandpasses such as assumed in Table 1. Nevertheless, observations in the region near 5 μm, although providing no advantage in S/N ratio for particles near 0.1 cm, remain desirable because of the improved S/B ratio at longer wavelengths.

6. Summary of Advantages Offered by Coronagraphic Detection

Coronagraphic detection of debris offers several advantages not found in other methods. In principle it promises to provide albedo-independent detection of particles down to 0.1 cm in diameter. In addition, the measurement of the diffraction profile may provide information on particle diameter independent of absolute flux and particle distance.

Although the coronagraphic method would not be generally suitable for acquiring orbital elements sufficiently precise for cataloging, the detection rate for small debris would be quite high. This capability lends itself, particularly, to acquiring statistics on the evolution of debris population, size, and approximate orbital characteristics. These are precisely the quantities that would be useful in detecting changes due to major breakup events, variations in atmospheric drag owing to solar EUV flux or geomagnetic substorms, and secular trends relating to collision cascade phenomena.

The calculations presented here assume 60 arcsec above the solar limb as a nominal location for debris detection. Improved S/B and S/N ratios could be obtained closer to the limb. Performance would be further enhanced at spatial resolutions ≤ 1 arcsec, as might be routinely obtained if the coronagraph were equipped with an adaptive optics system. Operation in the near IR offers better atmospheric seeing than at visible wavelengths, and adaptive optics systems are generally simpler to employ at long wavelengths.

The prospects for using a ground-based coronagraph in detecting space debris are owned primarily to advances in superpolished mirror and detector technology. The successful detection of debris will require a large collecting area, wide bandpass, high spatial resolution, and low scattered light; an additional advantage is offered by operation in the near IR. These requirements point directly to a large-aperture, IR-capable, reflecting coronagraph.

7. Acknowledgments

The authors are indebted to Frederic Volz (Phillips lab), Rangasayi Halthore (Hughes-STX), and Jeff Kuhn and Craig Gullixson (National Solar Observatory) for useful discussions. Primary support for this work was provided by the U.S. Air Force Office of Scientific Research, Task 2311G3.

8. References

1. D.W. Schuerman, D.E. Beeson, and F. Giovane, *Appl. Opt.*, **16** (1977) 1591.
2. J.M. Beckers, "CLEAR: A Concept for a Coronagraph and Low Emissivity Astronomical Reflector," (these proceedings).
3. D.J. Kessler, in *Preservation of Near-Earth Space for Future Generations*, ed. J.A. Simpson (Cambridge Univ. Press, 1994), p. 19.
4. N.L. Johnson and D.S. McKnight, *Artificial Space Debris*, (Orbit, Malabar, 1987).
5. J.A. Simpson (ed.) *Preservation of Near-Earth Space for Future Generations* (Cambridge Univ. Press, 1994).
6. C.W. Allen, *Astrophysical Quantities*, 3rd Ed. (Athlone, London, 1973).
7. M. Born and E. Wolff, *Principles of Optics*, 4th Ed. (Pergamon, New York, 1970).
8. S. Koutchmy, S. Colley, R. Smartt, C. Nitschelm and J.P. Zimmerman, *SPIE Symposium on Astronomical Techniques and Instrumentation for the 21st Century*, **1235-1282** (1990), 849.
9. R.N. Smartt, S. Koutchmy, and J.-C. Noëns, in *Infrared Solar Physics*, IAU Symp. 154, eds. D.M. Rabin, J.T. Jefferies, and C. Lindsey (Kluwer, 1994), p. 603.
10. T. Nakajima, M. Tanaka, and T. Yamauchi, *Appl. Opt.*, **22** (1983) 2951.
11. F.E. Volz, *Appl. Opt.* 32 (1993) 2773.
12. R. Halthore, Private Communication (1994).

Coronagraphic Observations of Near-Stellar Environments

David Jewitt
Institute for Astronomy, 2680 Woodlawn Drive, Honolulu, HI. 96822

ABSTRACT

This talk will provide an overview of the application of coronagraphs to the study of astrophysical problems outside the sun. To date, the primary night-time application has been to study dust in the vicinity of young stars. The single best known result of stellar coronagraphy is the detection of a disk around the nearby star Beta Pic. Observations of this star will be discussed, as well as results of a survey for similar structures around other stars. According to a leading hypothesis, the Beta Pic dust disk is released by comets travelling on highly elliptical, and nearly star-grazing, orbits. A family of comets on plunge orbits has been coronagraphically detected about the sun, providing a close connection between our star and others. I will discuss future exploitation of coronagraphic imaging in astronomy.

CORONAGRAPHIC OBSERVATIONS OF GRAVITATIONAL LENSES

B. FORT
DEMIRM, Observatoire de Paris, 61 avenue de l'Observatoire
75014 Paris, France

and

Y. MELLIER and R. PELLO
LAT, Observatoire Midi-Pyrénées, 14 avenue E. Belin
31400 Toulouse, France

ABSTRACT

During the last decade, gravitational lenses appeared to be an important tool for observational cosmology. They allow to study the distribution of dark matter and magnified distant objects as well, in a way out of the scope of any classical instrumentation. Beside, the natural optical benches defined by the deflectors and the sources can be used for testing the actual geometry of the Universe and the value of cosmological parameters. Unfortunately, because lenses are often massive and far more brighter than the distant magnified sources, the detection of gravitational image in the vicinity of bright sources is an observational challenge. In this presentation we try to identify if a large coronagraph could be relevant for the observations of gravitational lenses where the angular scale of the lensing effect is in the range from 2 to 100 arcsec. Our conclusion is that although gravitational lenses cannot be a strong additional justification for a large solar coronagraph proposal, as stellar coronagraphy probably would, it could give a good opportunity to respond to a few key points risen by recent observations of gravitational lenses effects by large scale structures in the Universe.

1. Introduction

Gravitational lenses are highly non linear optical systems where the gradient of the potential (Φ) plays a similar role as the gradient of index (n) in classical optics with $n = 1 - 2\Phi/c^2$. They can deflect, multiply and (de)magnify images of background sources. The potential acts on *all scales* and it is possible to observe gravitational lens effects for a large range of deflector mass, from the lightest ($10^{-2}\ M_\odot$) to the heaviest ($10^{16}\ M_\odot$). An elementary description of the basic concepts of image formation of extended sources (distant galaxies) may be found in a recent review[1] from which we took our observational examples to illustrate the presentation. The simulation displayed on Fig.1 shows the various regimes of images that can be considered, depending on the efficiency of the lens and the location of the sources with respect to the optical axis of the lens: strong lensing with highly magnified multiple images having an arc geometry and weak lensing distortion with images slightly stretched in a direction perpendicular to the gradient of gravity field. Indeed the possibility to observe such images pattern depends on the spatial resolution of the telescope with respect to the angular scale of the phenomena.

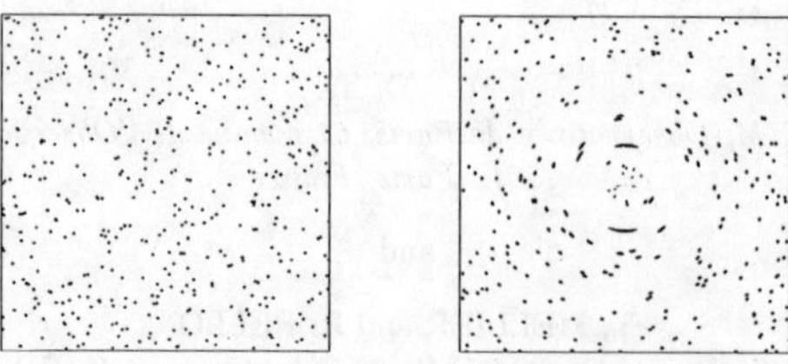

Fig. 1. Gravitational distortion induced by a rich cluster of galaxies. The left panel shows an unlensed random distribution of galaxies and the right panel the same fields lensed by a rich cluster. We can see the decrease of distortion with radial distant to the lens.

A second important property of gravitational images is that the value of the local surface brightness of an image point is always the value of the antecedent point of the source viewed without deflector. The contrast of a resolved source like a faint distant galaxy will not be enhanced on the sky. For a point like object like a QSO the global amplification of light actually comes from the *magnification* of the unresolved images. Therefore the observation of gravitational images of distant galaxies at a redshift larger than 1 requests the same exposure time than the observation of background galaxies in empty fields; that is typically more than 4 hours integration time with a 4 meter class telescope.

In the following, we consider several examples of gravitational lens observation with a large coronagraph to see if it helps to solve some questions which have been pointed out recently on the mass distribution in the Universe. The list is not exhaustive because the concept of coronagraphy is still new for a non solar astronomer despite the spectacular observation of the first circumstellar disk around the star Beta Pictoris[2].

2. The distribution of masses in clusters and galaxy halos

Franz Zwicky[3] first guessed that the velocity dispersion of galaxies in the nearby Coma cluster could only be explained if a huge invisible mass was sitting there. Fifty years later, it it widely demonstrated that masses of clusters estimated from virial analysis or from intra-cluster X-ray gas dynamics is larger than the stellar mass of the galaxies. A rich cluster of galaxy has a stellar and a X-gas masses that do not exceed respectively 5% and 20% of the total central dynamical mass ($r < 200$ kpc). Along with the discovery of flat rotation curve of spiral galaxies, this result contributed to rise one of the most amazing question of modern cosmology, the possible existence of large Dark Matter Halos (DMH) associated to galaxies and clusters of galaxies whose nature and distribution are still elusive.

The recent discovery of gravitational arcs in clusters of galaxies[4,5] is the most direct and spectacular confirmation that a huge concentration of invisible mass is located within the center of cluster of galaxies. In fact, the study of gravitational images of distant galaxies observed through rich clusters is the best way to recover their *total mass distribution*[1]. In the weak distortion regime, the images of faint distant background galaxies map the distribution of mass and a 2-D inversion can be performed even at large radius[6,7,8]. From recent results it seems that the invisible mass at large distance from clusters center increases tremendously up to M/L ratio larger than 1000! A large fraction of the mass of the Universe may be in clusters and group of galaxies. Extrapolated to the largest structures of the Universe like great walls and filaments around large voids, the result seems consistent with the recent dynamical analysis of large scale galaxy flows[9].

Despite these consistent results, we do not have yet a clear understanding of the formation and the dynamical evolution of clusters of galaxies. One important question is to understand how the total mass is distributed with respect to the radiating baryonic matter. Their respective behavior during the formation of the largest structures in the universe is a clue to peer at the actual possible nature of Dark Matter. We already know from gravitational arcs studies that at the center of cD-clusters the distribution of stellar light almost traces the distribution of the DM up to a distance of about 200 kpc. The total mass is also strongly peaked in the cluster center[1]. However, the ellipticity of the faint star cluster halos associated with the cDs is increasing with the radial distance while the gravitational potential derived from the lensing effect seem to have a constant ellipticity up to the arc radius. This could be due to a velocity anisotropy of stars and the existence of two mass components (baryons and invisible DM). With the observation of the distribution of the light of the very extended faint star halo of the cluster it might be possible to model the lens with three distinct mass components: the visible star mass, the gas mass and the invisible mass.

The collision and interaction between galaxies with merging and tidal effects are responsible for the building of a faint stellar component at the center of rich clusters. Theoretical simulations of the evolution of clusters[10] confirm the observation and predict that up to 30% of the total stellar cluster mass may be found in this component. The geometry and the eventual cut-off of the faint stellar light profile will be also a useful help to put constraints on the potential. But so far the observations have failed to detect accurately this halo because it can extend at a distance larger than 500 kpcs with a surface brightness below a few percent of the sky background level[11,12].

In order to get a good evaluation of the light distribution it would be necessary to detect the halo between the bright galaxies of the clusters up to 0.1% of the sky background. In principle the very high linearity of CCD allows the detection at such a low level of contrast against the sky background, but strong limitations come from the instrumental scattered light from a crowding of bright galaxies at the center of rich clusters. Canada-France-Hawaii Telescope observations with an excellent coating of the primary mirror shows that the diffuse light is at a level of R=26.5 at about 30 arcsec from a star with a magnitude R = 16.5. Although observers take a particular care to deconvolve the image from the spread function of the diffuse light, the result is very uncertain because at such a level of accuracy the slope of the spread function measured on a star has a spatial dependence on the field. It is is far better to block the light coming from every bright objects in the field at the first focus.

The most severe limitation arises from the light scattered by the CCD window of the camera and eventually from blooming effect of bright objects on the whole chip wafer, including external area. To lower the scattering light from the mirror itself it is indeed better to work in the far red or near IR. If a well designed coronagraph with a CCD housing at the end of a large vacuum chamber is built for a regular large telescope it would be possible to improve the measurement of faint extended halos. However up to now focci of large telescopes are equipped with fixed instrument that can hardly be adapted to stellar coronagraphy. A large coronagraph would be the most natural and efficient solution for this kind of observations.

Of a second major importance is the largest field of view of a coronagraph compare to regular telescopes. Coronagraphs allow to set a reference point for the sky background at a large distance from a bright central object.

Similar coronagraphic observations could be done on normal spiral galaxies. Recently the discovery of a very faint luminous halo around the spiral galaxy NGC 5907 has been reported[14]. The radial profile of the halo seems compatible with what is expected for the Dark Matter halo in order to recover a flat rotation curve. If confirmed this result is a major contribution to the long lasting problem of galaxy masses. It was said above that the faint star cD halo was almost tracing the component of Dark Matter at the center. Is this component related to the baryonic component apparently detected around NGC 5907? When the problems of instrumental halos will be properly handled we should be able to answer to the question. In conclusion it is quite certain that the possibility to use multimask plates on a very large coronagraph will open a new view of very faint extended halos in the Universe.

In this topic we have not considered observational searches of unknown galaxy deflectors responsible for multiply lensed QSOs because such a program only need an IR telescope with a good seeing or an adaptive optic system. The expected background galaxy is generally not so faint and the difficulty rather comes from its location within the seeing disks of the QSO images.

3. Lensing by large scale structures

A strong correlation has been found between QSOs at a redshift about 1 and the distribution of galaxies in the Lick or IRAS catalogs[14,15] on a scale of 20 arcminutes. Similar correlations have been found between the distribution of 1-Jansky radio sources and clusters[16,17]. Although the current observational status is not definitive it is now believed that some kind of magnification bias is working within Large Scale Structures (LSS) due to some gravitational lensing effect. Thomas et al (1994) said that the effect may be due to microlensing by small compact objects or stars in the galaxy halos[19], but Bonnet et al (1993) and Fort et al (1994) investigate the possibility of macrolensing effects by large condensation of Dark Matter with large (cluster) masses located within the LSS themselves[19,20]. Such a macrolensing effect is now detected in the field of a few very bright radio sources at a redshift about 1[20]. In these observations the bright radio sources have a V magnitude around 16. The scattered halo within about 20-30 arcsecond almost prevents the detection of very faint galaxies which are often associated with the QSO itself, and it is not possible to measure the local shear right on the line of sight of the QSO. The shear is rather determined on these location from a continuity argument based on the shear map detected on the whole CCD field. If the shear is strong we should see a very coherent pattern of aligned background galaxies which would provide directly the amplitude of the magnification bias[1] when the redshift of the lens can be guessed from the absorption lines in the QSO spectrum. These observation are crucial for the study of the mass function of Dark Matter Halos in the Universe and the comparison with the distribution of light.

Moreover the QSOs may be associated to a large condensation of mass. In that case it can produce an extra multiple arc system of a far more distant galaxy within less than 10 arcseconds from its center. It is therefore important to observe very faint extended object with a few arcseconds angular scale in the immediate vicinity of a bright star-like object (Fig. 2). The issues are similar to those encountered in stellar coronagraphy (see other communications during the workshop). The availability of a large 4 meter coronagraph with a possible use of adaptive optic in the far red will be a definitive asset for such programs. With such a coronagraph it is also quite possible that one try to use systematically the bright (massive) radio sources as large natural telescope to search very young (primeval) magnified galaxies in the near IR.

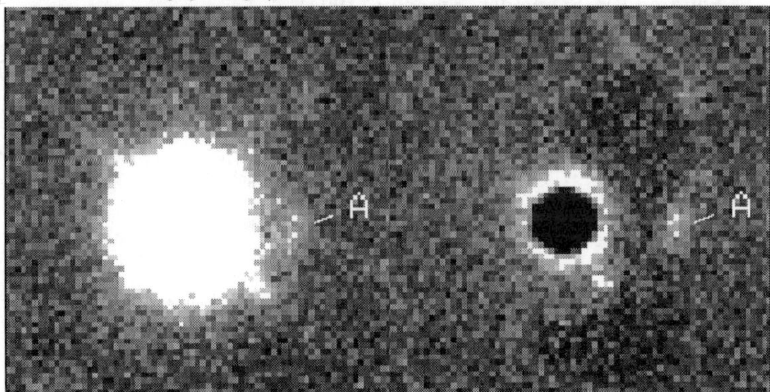

Fig. 2. Simulation of the detection of a gravitational arc by using a coronagraphic technique onto a bright radio source. The left image shows a real CFHT CCD image of a bright radio source co-added to the diffuse halo with a typical gravitational arclet (A) lying underneath. On the right panel we have simulated a coronagraphic observation of the same field with a halo ten time lower. The faint arc becomes clearly visible.

4. Acknowledgments

We thank I. Kovner for his encouragements and the numerous discussions on lensing. This work was supported by the EC HCM network CHRX-CT92-0044.

5. References

1. B. Fort and Y. Mellier, *A&AR* **5** (1994) 239.
2. B. A. Smith and R. J. Terrile,*Sciences* **226** (1984) 1420.
3. F. Zwicky, *Helvetica Physica Acta* **6** (1933) 100.
4. G. Soucail, B. Fort, Y. Mellier and J.-P. Picat *A&A* **172** (1987) L14.
5. R. Lynds and V. Petrosian, *BAAS* bf 18 (1986) 1014.
6. J. A. Tyson, F. Valdes, F. and R. Wenk, *ApJ* **349** (1990) L1.
7. N. Kaiser and G. Squires, *ApJ* **404** (1993) 441.
8. H. Bonnet, Y. Mellier and B. Fort, *ApJ* **427** (1994) L83.
9. A. Dekel, *ARAA* **32** (1994) 371.
10. E. M. Malumuth, *ApJ* **386** (1992) 420.
11. J. M. Uson, S. P. Boughn and J. R. Kuhn, *ApJ* **369** (1991) 46.
12. R. Vilchez-Gomez, R. Pello and B. Sanahuja, *A&A* **283** (1994) 37.
13. P. D. Sackett, H. L. Morison, P. Hardling and T. A. Boroson *Nature* **370** (1994) 441.
14. M. Bartelman and P. Schneider, *A&A* **268** (1993a) 1.
15. M. Bartelman and P. Schneider, *A&A* **271** (1993b) 271.
16. Rodrigues-William and Hogan, *preprint* (1994)
17. S. Seitz and P. Schneider, *preprint* (1994).
18. P. A. Thomas, R. L. Webster and M. J. Drinkwater, *preprint* (1994).
19. H. Bonnet, B. Fort, J.-P. Kneib, Y. Mellier and G. Soucail, *A&A* **280** (1993) L5.
20. B. Fort, Y. Mellier, H. Bonnet and M. Dantel-Fort, *in preparation.*

OBSERVATIONS OF SOLAR SYSTEM OBJECTS WITH A 4M ALL-REFLECTING CORONAGRAPH

THOMAS H. MORGAN
Southwest Research Institute
6220 Culebra Road, P.O. Drawer 28510, San Antonio, TX 78228, USA

ABSTRACT

Coronagraphic techniques have proven useful for at least two kinds of observational problems which are quite common in planetary astronomy: 1) Detection and study of faint objects in the presence of a larger, brighter object; and 2) Detection and study of faint extended structures about a bright central object. Examples of the first are searches for satellites or rings about the outer planets. Examples of the second are studies of the neutral cloud about Io, the ionized torus about Jupiter, and the Na corona about the Moon. With very few exceptions, these observations have been made with existing astronomical telescopes retrofitted for coronagraphic observations. Performance is typically not equivalent to what would be expected from an optimized coronagraphic facility. Planetary astronomers have proposed space-based coronagraphs; however, to our knowledge, none have flown.

A variety of interesting problems in planetary astronomy could be addressed with a next generation large-aperture coronagraph. These include studies of Io, including the surface, inner atmosphere, and the innermost portions of the neutral cloud; studies of asteroids (satellites and comas); observations of comets to obtain inner coma distribution of important species; observations of the outer planets to study the inner rings; observations of the exosphere and the magnetosphere of Mercury, and many others.

In order to observe planetary system objects, the coronagraph must not be limited in declination, must be able to acquire and track objects whose motion in both RA and Dec differs significantly from that of the Sun, must include great flexibility in terms of occulting masks, and should have adaptive optics capabilities. Planetary observations will require a broad range of focal plane instrumentation including imaging spectrographs for all the important atmospheric transmission windows from 0.4 to 5μm.

1. Introduction

The fundamental problem which coronagraphic observations address is the detection of a faint object in the presence of a very much brighter one. The problem may be very similar to the classic problem presented by the study of the solar corona, or it may be very different. Thus, studies of the exosphere/corona of the Moon and the zenocorona about Jupiter are studies of structures with the same angular size as the solar corona, and an extended central source. On the other hand, searches for faint companions about asteroids or for asteroidal coma, present an observational problem not dissimilar to stellar applications of coronagraphic techniques. A third application of coronagraphic techniques is, to planetary studies, the reconstruction of the vertical distribution of major components of an atmosphere from observations made through a sequence of anodizing masks. In what follows, we hope to give an illustration of the class of problems in planetary studies which might be addressed by a modern, large, all-reflecting coronagraph.

2. Current Work

Almost all planetary "coronagraphic" observations are taken with existing astronomical telescopes, for which the primary mirror is the entrance pupil of the coronagraph, and its image is the exit pupil. An occulting mask is placed in the telescope focal plane to block the light from the planet or satellite and another is placed at the exit pupil of the system to remove the scattered light due to diffraction at the mirror edges and by the secondary supports. The requirement that we retrofit the design to existing primary optics necessarily limits the performance of the system, and none of these planetary coronagraphs will reach the performance standard associated with solar coronagraphs; namely, off-axis rejection approaching 10^{-7}. The limitations imposed by the design which principally limit instrument performance are the surface roughness of the primary and the use of achromatic lenses before the exit pupil. Another factor which frequently increases the scattered light in the observations of planetary astronomers, who are typically visitors to astronomical facilities operated by others, is that the primaries are often cleaned on an infrequent schedule. The scattering from dust on the mirror provides an additional source of scattered light. Nevertheless, impressive results have been achieved with these instruments, including the discovery of the β Pictoris disk[14], obtained with a portable coronagraphic system developed by Vilas.[17]

Planetary coronagraphic studies have been hampered by the lack of funding to construct new planetary instruments of any type. However, there are coronagraphic facilities available at several installations including the 2.3 m at the University of Hawaii and an IR coronagraphic camera is being constructed for the IRTF. A representative list of persons using coronagraphic techniques for planetary studies (which makes no claim for completeness) is included in Table 1. Representative planetary observations now being made which use coronagraphic techniques, are observational study of the neutral cloud about Io and the ionized torus[11], the Jovian zenocorona[4], and the lunar atmosphere/corona. [5,8] At present, much of the work done with coronagraphic instruments as part of the world-wide Jupiter/SL9 campaign is available only in abstract form. The Jupiter/SL9 effort has resulted in instrumenting a number of facilities with coronagraphic instruments.[3] Thus, in the near future, there may be an upsurge in coronagraphic observations. Although all of these are labeled coronagraphic instruments, only the last two groups in Table I are using instruments which would be considered coronagraphs by the Solar community.

Table 1: Representative list of current work

Principal Observers	Principal Instrument	Scientific Interest
N. Schneider (Univ. of Co) and J.T. Trauger (JPL)	1.5m Catalina telescope with Lyot mask	The Io neutral plasma torus
M. Mendillo (Boston Univ.)	0.1m f/1.25 telescope with focal plane mask	Jovian magneto-torus and Lunar Corona
S. Larson (UA) and many others	Coronagraphic designs for a number of telescopes	SL9 events, Saturnian rings, satellites
T.H. Morgan (SwRI) and A.E. Potter (JSC)	0.4m Evans Coronagraph	Lunar Corona
D.M. Hunten (UA) and A.L. Sprague (UA)	0.12m Coronagraph	Lunar Corona
S.A. Stern	2.2m with prime focus camera	Asteroid Satellites

3. Characteristics of Typical Planetary Targets

The apparent diameter of the Sun is 0.5°, and at some epochs, the portions of the corona may extend 1 or 2° away from the surface of the Sun. There are several planetary structures which have similar apparent size, the lunar atmosphere/corona and the Jovian zenocorona. On the other hand, the search for asteroidal satellites and for detailed studies of the atmosphere of Io, the angular separations involved, approach those required for detection of stellar companions. Most of the work done to date has been with instruments for which the instrumentally scattered light at R/R_o 1.2 is 10^{-5} F_o, where R_o and F_o are the radius of the occulted object and its central intensity.

3.1. Representative Problems

In what follows, we shall consider several representative problems of particular significance which illustrate the application of a high-performance, all-reflecting 4-m coronagraph. As we shall note below, to realize the full potential of the large all-reflecting coronagraph, most of these problems will require a telescope with adaptive optics capabilities and with the advanced imaging spectrographs now going in service at some observatories. A sequence of anodizing occulting masks with graduated half-power diameters will also be required.

3.1.1. Io

The apparent diameter of Io is 1.2", and the visual magnitude of Io is 6.7. The satellite is very bright ($m_v = 6.7$). These conditions have made study of the atmosphere of Io as well as the inner reaches of the neutral cloud difficult, as the scattered light due to Io in the vicinity of the satellites dominates all but the emission in the strongest emission lines (see Figure 1). Combined coronagraphic/spectroscopic observations of the atmosphere could determine the base density of the Ioian atmosphere and the composition of the lower atmosphere. It may be possible to detect surface-feature related atmospheric events using the coronagraphic techniques since Ioian volcanos are quite bright in the 3-5μm region, given the apparent wealth of molecular species now thought to be present on the surface.[10]

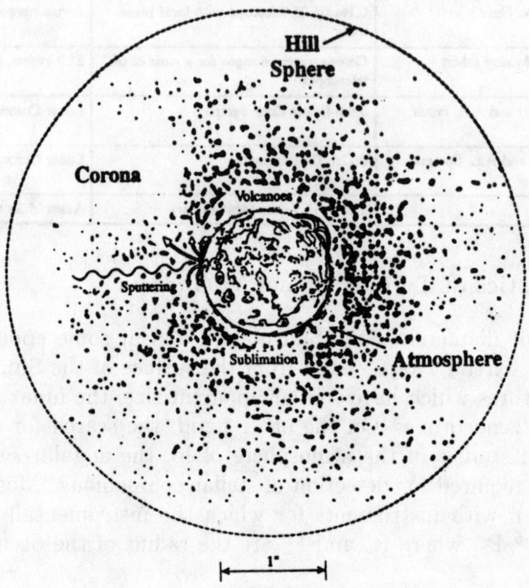

Figure 1: Cartoon of Io including the immediate environment of the planet illustrating some of the important structures and processes. See Schneider et al.[12] for a clear and extensive discussion of the nature and extent of the neutral and ionized species in the neighborhood of Io.

3.1.2. Saturn and Titan

There is less *in situ* data available for the Saturnian, Uranian, and Neptunian systems than there is for Jupiter and the Jovian system. Similarly, there is much more earth-based data (from traditional observatories and from orbiting observatories like the Hubble Space Telescope (HST), the International Ultraviolet Explorer (IUE), or the Extreme Ultraviolet Explorer (EUVE)) available for the trans-Jovian planets, than there is of Jupiter. It is precisely in these less studied systems where new and totally unexpected results may be obtained. However, even more so than in the case of Io discussed above, significant new work, even with a 4-m class instrument, will require adaptive optics capabilities, an extensive collection of occulting masks, and efficient imaging spectrographs for both visible and IR (out to 5μm) wavelengths.

An immediate observing task for the new 4-m all-reflecting coronagraph, would be Titan (0.8" at opposition). With the appropriate sequence of occulting masks, it would be possible to take a series of observations which could be used to "sound" the atmosphere. Since many of the known and likely species present in the atmosphere of Titan have strong features in the IR, an all-reflecting coronagraph could be used to determine the approximate mixing ratio of these species as a function of depth. While such measurements might give only crude height resolution and that indirectly, the ability to perform these observations over many years would provide an important body of new data for understanding Titan and the Saturnian neutral torus.

In a similar vein, large-scale studies of the Saturnian environment with a 4-m, all-reflecting coronagraph could lead to exciting new discoveries. Only recently has the OH torus about Saturn been detected.[13] Other simple molecules and radicals with weak transitions in the visible and IR out to 5μm, might also be present. Not only is the planet proper bright, but the Saturnian ring system leads to an even greater scattered light problem at most epochs. A 4-m all-reflecting coronagraph could be used, first to search for other faint emissions, and second to map them. While one might think that observations with HST, which can reach such strong features as the OH transitions near 3200 Å, would always be preferable to ground-based observations, one should note that the scattered light in most multi-purpose, multi-element optical systems will greatly reduce the capabilities of multi-user instruments like HST for these difficult observations – and do so for the OH measurements discussed earlier.

3.1.3. Asteroid Satellites

There is considerable indirect evidence for asteroidal satellites: anomalously slow rotation rates; shape differences in the photometric light curves of some asteroids; elongated asteroid images of some asteroids; the statistical frequency of double craters on the Moon and the Earth; and radar evidence for bifurcated satellites. However, the first direct evidence for one was the discovery of a satellite about 243 Ida. Preliminary calculations indicate that the 4-m coronagraph would be able to detect satellites about NEAs (at close approach to Earth) as small as 10-m in diameter, and satellites as small as 200-500-m in diameter in the main belt.

The number, frequency and nature of asteroid satellites is quite important. These data will tell us a great deal about asteroidal origins, break-up histories, and about the bulk physical properties of asteroids. This is particularly true if coronagraphic and spectrographic techniques can be combined in order to identify the surface compositions of both the main asteroid and the companion. The break-up history of asteroids may provide important indirect data for the models of the early solar system as well.

3.1.4. Mercury: Corona, Tail, and Magnetosphere

The Na exosphere almost certainly includes an extended component with varying tailward extension.[15] Additional support for the presence for such a structure can be found in the remarkable extension of the lunar Na corona.[5,6] The magnetic field of Mercury provides a modest magnetosphere[9] and best picture of the magnetosphere in relation to the planet is shown in figure 2. Although there has been considerable progress imaging the Mercurian exosphere,[7] there has been no successful effort to detect the extension. Images of the extended Na corona about Mercury could be used to determine the loss the mechanisms and the gas-surface interaction.

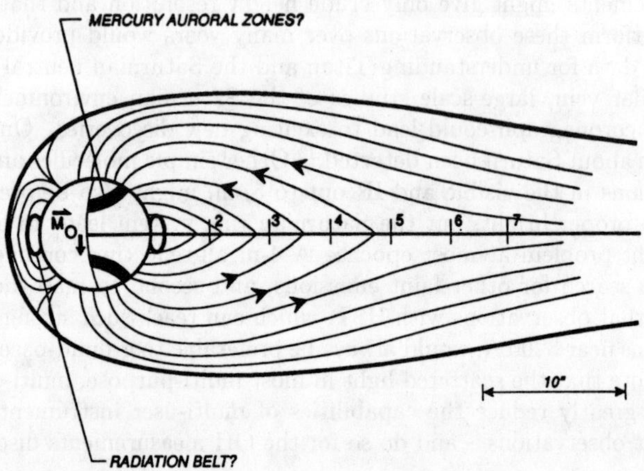

Figure 2: Representation of Mercury to illustrate the approximate appearance of the planet and the magnetosphere near greatest elongation. The magnetosphere is based on Russell et al.[9] The atmosphere is not shown here because there is presently no available data to show the vertical extent of the atmosphre around the planet. However, a large extended structure can be expected.[1]5

The most likely source mechanisms for the exosphere of Mercury are sputtering and impact vaporization, which supply both neutrals and ions from the surface.[1] While the column density of neutral Ca is small,[16] this may indicate that much of the Ca comes off the surface as CaO.[13] In any event, the ultimate sink for any Ca bearing species entering the atmosphere is via photo-ionization followed by outward drift along the field lines. It may be possible to observe emission lines due to Ca^+ and CO^+ (derived from the incoming meteoroid population in the magnetosphere). Discovery of any ionic line would give us the capacity to map the magnetosphere of Mercury.

Mercury is a bright object, thus a coronagraph is needed to reduce the scattered light near the planet sufficiently to allow for the detection of faint extended structures. The requirement for a large coronagraph to detect a faint, extended structure about a bright object like Mercury, derives from the short exposure times allowed. Mercury is never more than 29° from the Sun; thus dark sky observations will be limited to 20 to 30 minutes under the best conditions (many of the emission features that we are looking for are also present in the twilight airglow). Since we are in the case of Na looking for the fainter portion of the extended Na structure (< 50kR), and since the lines for Ca^+ and CO^+ are near 0.4μm, it is impossible to use the techniques which have produced images of the deep Na exosphere).[7] Thus, we must observe with a small occulting disk for short periods when Mercury is near maximum elongation. Smaller coronagraphs simply will not give us the sensitivity (< 1kR) in short exposures (10 - 15 minutes).

4. Spaceborne Coronagraphic Observations

Although there is a "coronagraphic finger" in the focal plane of the faint object camera on HST, there is currently no spaceborne coronagraphic capability available to planetary astronomers. There have been several proposals to fly coronagraphic instruments designed to observe planetary targets as earth orbiters, but to the knowledge of the writer, none have been flown. There are many potential advantages to such an attempt. First, is access to the ultraviolet. The resonance lines from many important elements and simple compounds fall in the UV (< 3200 Å - the atmospheric cutoff). These include resonance lines due to neutral atomic species (Mg, S, O, H, and C) as well as important transitions from simple molecules (OH, CO, and SO). In addition, important ions such as those of Mg, S, and O have strong transitions in this spectral region. The advent of all-reflecting coronagraphs opens a large part of this spectral regime for coronagraphic studies. Although the immediate priority for such an instrument would be the Jovian system, other important projects ranging from mapping OH about Saturn[13], to mapping O about Mercury would also be of immediate interest.

As the interpretation of these data require detailed measurement of the UV spectrum of the Sun at the time of the measurement as well as an understanding of the Solar wind in the vicinity of the planet for their interpretation, these problems form a natural area for collaboration between planetary astronomers and solar astronomers. In addition, it is this area of planetary astronomy which provides the most immediate area for comparative planetary global change studies.

5. Conclusion

Planetary astronomers have never had access to a large instrument with the performance characteristics of the proposed large all-reflecting coronagraph being discussed today within the solar community. Coronagraphic techniques have proven useful in the past in planetary astronomy for two types of observational problems: 1) detection and study of faint objects in the presence of a larger, brighter object; and 2) detection and study of faint extended structures about a bright central object.

A variety of interesting problems are available in planetary astronomy for a next-generation large-aperture coronagraph. These include: studies of Io's surface, inner atmosphere, and the innermost portions of the neutral cloud; observation of Titan's atmosphere and neutral cloud as well as Saturn's torus; studies of asteroids (satellites, comas); observations of comets to obtain inner coma distribution of important species; observations of the outer planets to study the inner rings; and observations of the exosphere and the magnetosphere of Mercury. In addition, there are a wide range of interesting problems available for space-based coronagraphs, and the interpretation of the data taken with these instruments requires the involvement of the solar astronomy community.

6. Acknowledgments

The author would like to thank a number of people for helpful discussions and useful suggestions. These include S.M. Larson (University of Arizona), A.E. Potter (Johnson Space Center), N.M. Schneider (University of Colorado), S.A. Stern (Southwest Research Institute), E.S. Barker (University of Texas), T.A. Livengood (Goddard Space Flight Center), D. Jewitt (University of Hawaii), and M. Mendillo (Boston University).

7. References

1. R.E. Johnson & R.Baragiola, *Geophys. Res. Lett.*, **18**, 2169 (1991).
2. R.M. Killen & T.H. Morgan, *Icarus*, **104**, 33 (1993).
3. S.M. Larson, private communication.
4. M.J. Mendillo, M.J. Baumgardner, B. Flynn & W.J. Hughes *Nature*, **348**, 312 (1990).
5. M.J. Mendillo, M.J. Baumgardner & B. Flynn, *Geophys. Res. Lett.*, **18**, 2097 (1991).
6. A.E. Potter & T.H. Morgan, *Geophys. Res. Lett.*, **18**, 1515 (1988).
7. A.E. Potter & T.H. Morgan, *Science*, **248**, 835 (1990).
8. A.E. Potter & T.H. Morgan, *Geophys. Res. Lett.*, submitted (1994).
9. C.T. Russell, D.N. Baker & J.A. Slavin in "Mercury" (University of Arizona Press, Tucson: F. Vilas, C.R. Chapman, and M.S. Matthews, eds.) 514 (1988).
10. F. Salama, L.J. Allamandola, S.A. Sandford, J.D. Bregman, F.C. Witteborn & D.P. Cruidshank, *Icarus*, **107**, 413 (1994).
11. N.M Schneider & J.R. Trauger, *Ap. J.*, in press (1994).
12. N.M. Schneider, W.H. Smyth & McGrath, in "Time-Variable Phenomena in the Jovian System", NASA-SP-494 (M.J.S. Belton, R.A. West and J. Rahe, eds.), 75 (1987).
13. D.E. Shemansky, D. Matheson, D.T. Hall, H.-Y. Hu & T.M. Tripp, *Nature*, **363**, 329 (1993).
14. B.A. Smith & R.J. Terrile, *Science*, **226** 1421 (1984).
15. W.H. Smyth, *Nature*, **323**, 697 (1986).
16. A.L. Sprague, R.W.H. Kozlowski, & D.M. Hunten, *Icarus* **101**, 33 (1993).
17. F. Vilas & B.A. Smith, *Ap. Opt.*, **26**, 664 (1987).

NEEDS FOR A LARGE CORONAGRAPHIC SOLAR TELESCOPE VIEWED FROM THE TREND OF STUDIES PUBLISHED DURING THE LAST 20 YEARS

TOKIO TSUBAKI
Department of Earth, Science, Shiga University
2-5-1 Hiratsu, Otsu 520, Japan

ABSTRACT

With the purpose of emphasizing needs for a large Coronagraphic solar telescope, the trend of studies published in "Solar Physics" in 1973, 1983 and 1993 has been analyzed. The most remarkable points or changes during these twenty years are: (1) observational studies have decreased in share from 66% to 56%, (2) observations in IR region have always been less than 5%, and (3) the share of coronal studies has continuously increased up to 15%, nearly the same as that of flares which have always been the biggest target.

1. Introduction

Since the first issue of "Solar Physics" was brought out in 1967, considerable number of studies have been published in this journal. Although not quite few papers, in the field of solar physics, have also been published in other general journals, especially in the recent decade, a brief statistical analysis was made for the papers published in Solar Physics in 1973, 1983, and 1993 to take a glance at the trend of solar physics studies for recent 20 years.

The purpose of this paper is to derive positive needs for a large coronagraphic solar telescope from this analysis.

2. Observational or Theoretical?

By dividing the published papers into only two categories, that is, the observational and the theoretical, the numbers and the share of studies have been counted as tabulated in Table I. There have sometimes been such studies that include both observational and theoretical contents, such as review papers. In this case, the stronger impression was adopted for classification by reading through the papers.

As can be seen in Table I, there are two big changes found for these 20 years: (1) the share of observational studies has decreased from 66% to 56%, or the ratio of the observational to the theoretical has changed from about 2:1 to 1.3:1, and (2) the total number of papers published has decreased from around 250 in 1970's and 1980's to 150 in 1990's, that is, by about 40%, at least a part of which might be due to the reason that the publications of solar physics studies have probably increased in other general journals, such as Astrophysical Journal, Astronomy and Astrophysics, Publications of Astronomical Society of Japan and so on.

Table 1: Number and the share of observational and theoretical studies published in the journal, Solar Physics, in 1973, 1983 and 1993.

Year	1973		1983		1993		Total	
Number/Share	N	%	N	%	N	%	N	%
Observational	155	66.0	159	61.6	87	56.5	401	62.0
Theoretical	80	34.0	99	38.4	67	43.5	246	38.0
Total	235	100.0	258	100.0	154	100.0	647	100.0

3. Wavelength Range?

Among the papers published as observational studies, wavelength ranges employed for each observation were picked up and tabulated in Table II. The study done by using simultaneously two or more wavelength ranges, the range which looked most important for the study was adopted. Here, each range is treated as its widest meaning, for instance, the term Ultra Violet range contains all the range of XUV, EUV, and near UV, and the X-ray region includes the gamma-ray region.

Table 2: Wavelength range employed for the observations whose results were published as observational studies.

Year	1973		1983		1993		Total	
Number/Share	N	%	N	%	N	%	N	%
Particle	10	6.5	8	5.0	5	5.8	23	5.7
X-ray	15	9.7	48	30.2	9	10.3	72	18.0
Ultra Violet	10	6.5	14	8.8	5	5.8	29	7.2
Optical	76	49.0	65	40.9	47	54.0	188	46.9
Infra Red	5	3.2	2	1.3	4	4.6	11	2.7
Radio	39	25.1	22	13.8	17	19.5	78	19.5
Total	155	100.0	159	100.0	87	100.0	401	100.0

As seen in this table, the optical range is always used most frequently followed by the radio, X-ray (including gamma-ray), and UV ranges. The Infrared range however has always been less than 5% (only 2.7% in total of 3 years), the reason of which might be due to the fact that we do not have a solar telescope large enough and a high quality detector for the IR observations.

4. Research Fields?

Number and share of papers in each research field have been counted and tabulated in Table III, where the top 13 fields in total number for the three years, 1973, 1983 and 1993, are given. For the study which treated two or more objects, themes, or fields, the one that looked most important was selected by reading through the paper.

Table 3: Number and the share of papers in each research field published in the same period as in Table I. Shaded numbers represent top 3 in share.

Year	1973		1983		1993		Total	
Number/Share	N	%	N	%	N	%	N	%
Phys. Proc.	16	6.8	5	1.9	1	0.6	22	3.4
Photosphere	9	3.8	8	3.1	2	1.3	19	2.9
Sunspot	20	8.5	12	4.7	13	8.4	45	7.0
Mag. Field	5	2.1	14	5.4	23	**15.0**	42	6.5
Oscil. & Wave	12	5.1	42	**16.3**	9	5.8	63	**9.7**
Chromosphere	19	8.1	12	4.7	1	0.6	32	4.9
Active Region	10	4.3	9	3.5	8	5.2	27	4.2
Flare	30	**12.8**	74	**28.7**	26	**16.9**	130	**20.1**
M. Burst	29	**12.3**	15	5.8	11	7.1	55	8.5
Prominence	20	8.5	12	4.7	8	5.2	40	6.2
Corona	24	**10.2**	21	**8.1**	23	**15.0**	68	**10.5**
Sol. W. & IPS	11	4.7	8	3.1	8	5.2	27	4.2
Instrument	6	2.6	9	3.5	4	2.6	19	2.9
Others	24	10.2	17	6.5	17	11.1	58	9.0
Total	235	100.0	258	100.0	154	100.0	647	100.0

It is to be noted here that two special issues were published in 1983: one on "Problems of Solar and Stellar Oscillations" and the other on "Recent Advances in the Understanding of Solar Flares". Although only full papers of solar studies have been picked up, these two issues might contribute a relatively large portion of numbers in the two fields, that is, "Oscillations and Waves" and "Flare".

Note also that the following abbreviations are adopted for expressing research fields in the table:

- Phys. Proc.= Physical Process,
- Mag. Field = Magnetic Field,
- Osc. & Wave = Oscillations and Waves,
- M. Burst = Microwave Burst and

• Sol. W. & IPS = Solar Wind and Inter-Planetary Space.

As seen in this table, the share of "Flare" research has always been the biggest (20% in total of 3 years) followed by "Corona" (11%) and "Oscillations and Waves" (10%). Instead of "Oscillations and Waves", however, "Microwave Burst" was in the top 3 in 1973, and so was "Magnetic Field" in 1993. The similar result has been obtained for solar physics activities in Japan.[1] The effect of two special issues in 1983 mentioned above might have reduced the share of "Corona" comparatively to the ordinary issues at the same age. If we consider this fact, it follows that the share of "Corona" has continuously increased up to 15%, nearly the same as that of "Flare", which can be taken as one of the most remarkable changes for these 20 years.

Both for the observational and the theoretical studies, the same analysis has been done to investigate the difference of research trend between them and the results are tabulated, respectively, in Table IV and V.

Table 4: The same as Table III but for observational studies.

Year	1973		1983		1993		Total	
Number/Share	N	%	N	%	N	%	N	%
Phys. Proc.	4	2.6	0	0.0	0	0.0	4	1.0
Photosphere	5	3.2	7	4.4	1	1.2	13	3.2
Sunspot	14	9.0	9	5.7	12	**13.8**	35	8.7
Mag. Field	2	1.3	7	4.4	11	12.6	20	5.0
Oscil. & Wave	8	5.2	15	**9.4**	1	1.2	24	6.0
Chromosphere	17	**11.0**	10	6.3	0	0.0	27	6.7
Active Region	9	5.8	4	2.5	6	6.9	19	4.7
Flare	18	**11.6**	54	**34.0**	18	**20.7**	90	**22.4**
M. Burst	21	**13.5**	10	6.3	6	6.9	37	9.2
Prominence	17	**11.0**	7	4.4	3	3.5	27	6.7
Corona	16	10.3	14	**8.8**	12	**13.8**	42	**10.5**
Sol. W. & IPS	4	2.6	5	3.1	5	5.8	14	3.5
Instrument	4	2.6	8	5.0	4	4.6	16	4.0
Others	16	10.3	9	5.7	8	9.2	33	8.2
Total	155	100.0	159	100.0	87	100.0	401	100.0

As seen in these tables, there are not great differences recognizable between the two. The share of "Flare" research in total is the biggest for both tables and that of "Corona" is also at the top 3 for the both. The following characteristics can, however, be pointed out: (1) although many objects, such as "Sunspot", "Chromosphere", "Flare", "Microwave Burst", "Prominence", and "Corona" were nearly equally studied observationally in 1970's, the objects have been reduced into several restricted ones, such as "Sunspot", "Magnetic Field", "Flare" and "Corona" in 1990's, and (2) theoretical studies of "Magnetic Field" has become to get the top share in 1990's.

Table 5: The same as Table III, but for theoretical studies.

Year	1973		1983		1993		Total	
Number/Share	N	%	N	%	N	%	N	%
Phys. Proc.	12	**15.0**	5	5.1	1	1.5	18	7.3
Photosphere	4	5.0	1	1.0	1	1.5	6	2.4
Sunspot	6	7.5	3	3.0	1	1.5	10	4.1
Mag. Field	3	3.8	7	**7.1**	12	**17.9**	22	8.9
Oscil. & Wave	4	5.0	27	**27.3**	8	**11.9**	39	**15.9**
Chromosphere	2	2.5	2	2.0	1	1.5	5	2.0
Active Region	1	1.3	5	5.1	2	3.0	8	3.3
Flare	12	**15.0**	20	**20.2**	8	**11.9**	40	**16.3**
M. Burst	8	**10.0**	5	5.1	5	7.5	18	7.3
Prominence	3	3.8	5	5.1	5	7.5	13	5.3
Corona	8	**10.0**	7	**7.1**	11	**16.4**	26	**10.6**
Sol. W. & IPS	7	8.8	3	3.0	3	4.5	13	5.3
Instrument	2	2.5	1	1.0	0	0.0	3	1.2
Others	8	10.0	8	8.1	9	13.4	25	10.2
Total	80	100.0	99	100.0	67	100.0	246	100.0

5. Needs for a Large Coronagraphic Solar Telescope?

With the purpose of emphasizing needs for a large coronagraphic solar telescope, a brief analysis has been done for the studies published in the journal, Solar Physics, in 1973, 1983 and 1993. The results can be summarized as follows.

(1) The share of observational studies has decreased from 66% to 56%, or the ratio of the observational to the theoretical has changed from about 2:1 to 1.3:1 during these 20 years. Since the Skylab experiment was successfully carried out in 1970's, several powerful space observations have been done, such as, SMM, Hinotori, and Yohkoh to our knowledge. If we give attention to the ground based observations, however, there have not been any outstanding solar observation systems appeared newly in the recent decade although several 8 m aperture of stellar telescopes are simultaneously being constructed. This seems a strong reason why the share of observational studies has been decreased. Considering also the fact that nearly 50% of observations have been done with the optical telescope, we strongly feel that a large solar telescope should appear as soon as possible.

(2) Observations in the range of infrared have scarcely been carried out: the share

has always been less than 5% or only 2.7% in total. Most IR observations have been done in the near IR region, such as HeI 10830 line. To extend the wavelength range to a longer side, a large enough telescope should be constructed by developing simultaneously a better quality detector.

(3) The share of coronal studies has continuously been increased up to 15%, which is nearly the same as that of flare research. This must be due to the fact that there are many fundamental problems remaining in coronal physics, such as heating mechanism or magnetic structures. To clarify these problems, it seems absolutely necessary to make detailed observations of the corona with a large coronagraphic telescope.

As the final conclusion, we would like to appeal that a large coronagraphic solar telescope should appear as soon as possible for making coronagraphic observations in a wide range of wavelength to recover the activity of observational studies.

6. References

1. T. Tsubaki, *Scientific Bulletin* **12** No. 4, 51 (1987).

THERMAL BIFURCATION REVISITED

THOMAS R. AYRES*
Center for Astrophysics & Space Astronomy, University of Colorado,
Boulder, CO 80309, USA

ABSTRACT

Stigmatic imaging of carbon monoxide bands in the thermal infrared – now possible at the NSO McMath-Pierce facility – has opened a new chapter in the study of solar surface structure. Time-resolved intensity and Doppler maps of strong CO absorptions show pervasive fluctuations connected with p-mode oscillations, sparsely-distributed persistent network bright points, and rare localized cooling events perhaps associated with overshooting granules. Overall, the thermal and velocity perturbations are relatively modest. Numerical simulations of the surface morphology point to an average sub-canopy zone which is considerably cooler than the low chromosphere of the VAL C′.

1. BACKGROUND

The strong $\Delta v = 1$ bands of carbon monoxide fall near 4.7 μm in the thermal infrared. They are key diagnostics of the temperature-pressure stratification of the solar outer photosphere. The vibration-rotation lines are thermalized by low-energy collisions with atomic hydrogen, and consequently form close to LTE.[1] Although the individual oscillator strengths are small,[2] the large abundance of CO in the cooler layers causes the absorption cores to arise at high altitudes, especially for viewing angles close to the limb. The multiplicity of CO transitions provides a high degree of observational redundancy, and a wide range of formation heights with which to probe the temperature profile.

In the early 1970's, Robert Noyes and Donald Hall used the horizontal infrared spectrograph (IRSC) of the McMath telescope for exploratory studies of the solar $\Delta v = 1$ bands. They discovered the surprising "cool cores" of the strongest CO absorptions at the extreme limb, as well as the exaggerated response at disk center to global p-mode oscillations.[3] Noyes and Hall speculated that the very low CO limb brightness temperatures ($T \approx 3700$ K) were connected with the narrow subduction lanes of the photospheric granulation pattern. In the mid-1970's, much attention was focused on the solar outer atmosphere, particularly the $T_{\rm min}$ zone at the photosphere-chromosphere interface.[4] The curious behavior of the CO fundamental bands was dismissed as a phenomenon of the deep photosphere, and neglected until the commissioning of the McMath 1-m Fourier transform spectrometer (FTS) in 1978.[5] At that time, Larry Testerman and I obtained scans of the CO $\Delta v = 1$ bands with very high

*Visiting Astronomer, National Solar Observatory, National Optical Astronomy Observatories, which is operated by the Association of Universities for Research in Astronomy, Inc. (AURA) under cooperative agreement with the National Science Foundation

spectral resolution ($\omega/\Delta\omega \approx 2 \times 10^5$)† and very high signal-to-noise (S/N$\approx$ 3000). Somewhat to my surprise – as an advocate of a hot $T_{\rm min}$ [6] – the new observations not only confirmed the earlier result of Noyes and Hall, but also suggested that the cold gas was confined to high altitudes, *above* the $T_{\rm min}$ of popular models.[7]

In response, I devised a simple mechanism to explain the presence of cool material in the supposedly hot chromosphere.[8] The *thermal bifurcation* hypothesis proposed that the outer solar atmosphere could exist in either of two distinct thermal states: one cold, the other hot. The cold phase obtains where the local mechanical heating is low. There, the plasma temperature is dictated by a balance between mild radiative heating in the H^- continuum and strong radiative cooling due mostly to collisional excitation of the CO fundamental bands, themselves. If the mechanical dissipation increases above a critical level, however, the gas warms and the CO molecules begin to dissociate. As the molecular concentration falls, so does the CO cooling, and the gas heats up even more. The thermal runaway is quenched at $T \approx 6000$ K – yielding the hot phase – when partial ionization of hydrogen liberates enough electrons to catalyze atomic line cooling.

Thermal bifurcation is one face of the more general phenomenon of *molecular cooling catastrophes*.[9] These are implicated, for example, in the fragmentation and collapse of interstellar clouds,[10] the formation of dust in red giant winds,[9] and possibly even the fueling of "starbursts" in distant galaxies.[11] Although challenged by P. Mauas and collaborators,[12] key aspects of the mechanism have been validated by NLTE-blanketed radiative equilibrium simulations,[13] model chromospheres with prescribed heating laws,[14] and recent thermal instability calculations.[15]

Unfortunately, the McMath FTS lacks sufficient spatial resolution to probe the morphological behavior of the CO lines on the solar disk. The FTS, by itself, cannot assess directly the degree to which the low chromosphere is thermally bifurcated. Thus, many of us eagerly awaited the installation of a large IR grating – salvaged from the defunct IRSG – in the Main spectrograph of the McMath.‡ The new setup offered the possibility of long-slit stigmatic imaging in selected CO $\Delta v = 1$ lines down to the diffraction limit of the 1.5 m telescope ($0\overset{''}{.}8$ at 4.7 μm).

During tests of the "new" grating in 1993 April, Bill Livingston discovered the remarkable off-limb emissions of the CO lines. It was the first direct detection of cool molecular gas within the chromosphere proper.[16] Subsequently, in 1993 October, Han Uitenbroek and collaborators conducted the first CO observations using the new IR grating and a 256 × 256 InSb camera in the Near Infrared Magnetograph (NIM).[17] Although the CO lines are not magnetically sensitive, the NIM provides 2-axis stepping control over the telescope and a data acquisition system for the Amber Engineering camera.

The Uitenbroek *et al.* study verified the important role played by the *p*-modes

$^\dagger\Delta\omega$ is the wavenumber displacement to the first zero of the sinc function that describes the instrumental profile of the large interferometer.

‡The McMath recently was rededicated the McMath-Pierce facility. Not surprisingly, Keith Pierce played a key role in the Main spectrograph upgrade program, and IR solar astronomers owe him a debt of thanks for his efforts.

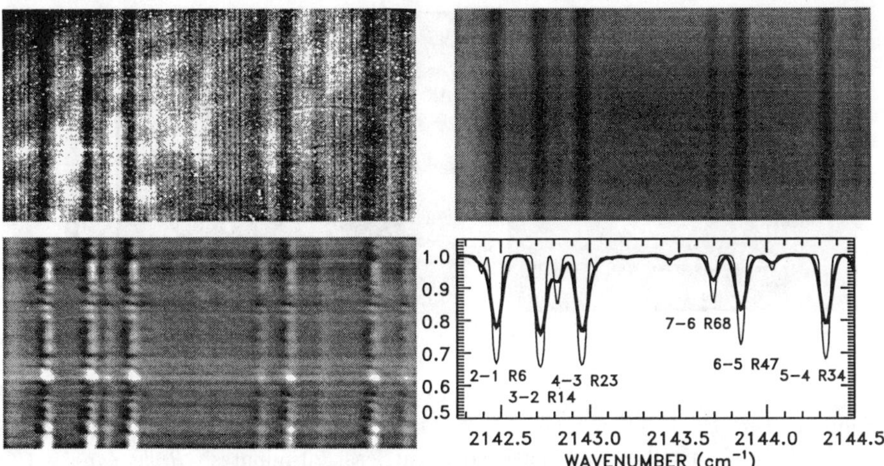

Figure 1: *Upper left* – Raw Amber frame (compressed vertically to emphasize systematics along slit). *Lower left* – background subtracted, flatfielded ratio map. *Upper right* – recovered spectrum. *Lower right* – Trace of central portion of recovered frame (thick curve) compared with synthetic spectrum (thin curve).

in controlling the thermal and dynamical character of the outer photosphere, hinted in earlier low spatial resolution studies with the FTS.[18] In more recent work, Uitenbroek and Noyes have pointed to examples of *exploding granules* reaching up to, and affecting, the CO-forming layers.[19] The authors suggested that the expansively cooling convective plumes might be responsible for the depressed core intensities of the CO lines at the extreme limb through *shadowing.*

In the present paper I describe results from my own CO work with the NIM. The emphasis of my observational experiments has been somewhat different from that of the Harvard group. I place my own slant on the interpretations, leveraged by numerical simulations with 2-D highly inhomogeneous models.

2. NEW OBSERVATIONS

The observations described below were obtained during the week of 1994 March 21–25, in collaboration with Doug Rabin of NSO. The NIM was operated in much the same way as during the first Uitenbroek *et al.* run, except that an auxiliary CCD camera was available to take narrow-band filtergrams in Ca K or Hα. The solar image was stepped across the 100″-long slit to build up large field-of-view maps of a 2.5 cm^{-1} interval of the spectrum (usually centered at 2143 cm^{-1}), or boustrophedonically scanned to build up a "movie" of a narrower spatial swath. The Amber frames were flatfielded by dividing the dark-subtracted image by a spatially-averaged

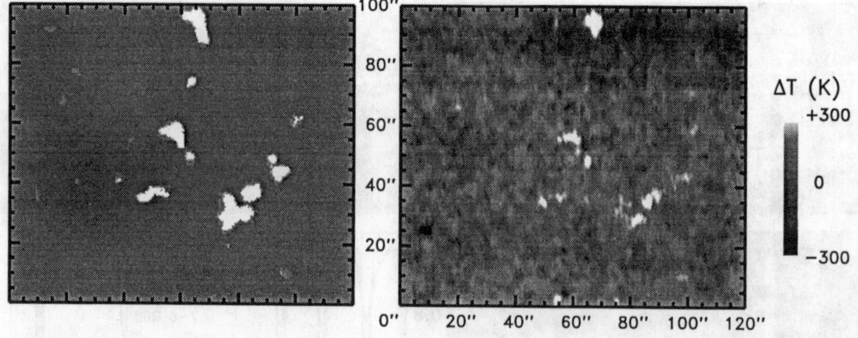

Figure 2: *Left panel* – Ca K map of 120″ × 100″ area near disk center. Light features are Ca K-bright patches that persisted for at least 20 minutes. *Right panel* – CO thermal map. Color bar at far right refers solely to CO.

spectrum. The latter was obtained by coadding 10 independent spatially-mixed exposures taken near disk center. The ratio frame was restored by multiplying it by a synthetic spectrum. Figure 1 depicts the reduction sequence. The strategy effectively suppresses telluric absorptions and sensitivity variations across the detector. Further details will be provided elsewhere.[20]

The empirical resolving power of the NIM is $\omega/\Delta\omega \approx 35{,}000$.[§] That is considerably lower than the 90,000 cited by Uitenbroek and collaborators, and is a factor of ≈ 5 less than routinely obtained with the FTS. The strong CO lines in the 4.7 μm region are only partially resolved, and one cannot use the NIM spectra to directly measure the absolute core depths (related to the thermal conditions in the $\tau_{\rm CO} \approx 1$ layers). Nevertheless, the resolution is sufficient to measure Doppler shifts and *relative* intensity fluctuations (recognizing that the apparent ΔT's will be diluted somewhat owing to the degradation of the line profiles).

Figure 2 illustrates a thermal map of the surface obtained on 1994 March 24. It was derived from a sequence of 240 consecutive slit positions, with steps of 0″.5 . The slit was oriented N/S in a quiet region near disk center. The field contained Ca K-bright network, identified with the CCD camera. The infrared seeing, dominated by image shake, was about 2″ (at best). The entire observation required 12 minutes of real time. The 256 individual spectra in each corrected frame were measured by a semi-autonomous Gaussian fitting algorithm.[18] The resulting parameters were averaged for CO lines of similar excitation. The companion Ca K image was based

[§] $\Delta\omega$ is the FWHM of the Gaussian instrumental profile.

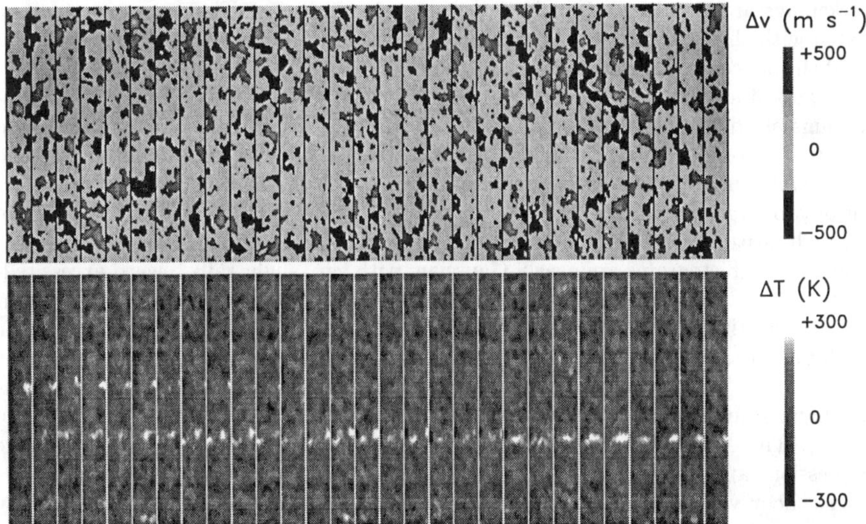

Figure 3: "Movie" of $10'' \times 100''$ swath near the middle of the previous field. *Upper panel* – Doppler velocities with extremes highlighted. *Lower panel* – Thermal map with highs and lows emphasized. Note persistent network bright points.

on CCD frames taken prior to and after the NIM sequence. It is depicted in a way that highlights persistent bright features. The CO map is coded to emphasize the highs and lows of the temperature range.

In the strong-line CO thermal map most of the fluctuations are within a few hundred degrees of the mean. Only a small percentage of the field is at the extremes. Most of the hot patches in the CO map coincide with long-lived Ca K bright points, presumably network fragments. In the matching weak-line CO map (not shown here), the point-to-point thermal fluctuations are reduced by a factor of two, but the network bright points still are visible.

Figure 3 depicts a movie sequence of Doppler shifts and thermal fluctuations obtained by repetitively scanning a $10''$-wide swath with $0''\!.5$ steps. The temporal resolution was about 2 minutes. The swath was located near the middle of the previous field, intersecting the network fragments there. Solar rotation was not compensated, so the scene drifted by about one swath width over the 1 hour observation. The Doppler time series clearly displays the $5''$–$10''$ *wavepackets* of the p-modes, although the cadence is such that only a few of the peaks are seen in consecutive swaths before the phase coherence is lost. The thermal movie again shows the persistent network bright points. A few transient warm and cool patches also are present. These probably are associated with extrema of the p-mode interference pattern. A curious dark feature arises near the southern network fragment just past the middle of the sequence. It appears in several adjacent swaths, indicating a lifetime of order 10 minutes. Similar darkenings are found occasionally in other movie sequences. They perhaps are associated with the exploding granule events identified by Uitenbroek and Noyes in their fixed-slit space-time diagrams.[19]

Figure 4 summarizes the thermal and velocity behavior of the time series. Histograms are depicted for a weak CO line and the average of three strong lines. Also shown is a contour plot pitting temperature against velocity. The weak and strong lines exhibit nearly the same velocity amplitude. The thermal fluctuations are a factor of two larger for the strong features, however, reflecting the enhanced response of the high-altitude gas to the compressions and rarefactions of the global oscillations.[18] The velocity histograms are nearly Gaussian, with very slight tails toward excess redshifts. The strong-line temperature distribution also is quite Gaussian, with slight tails at positive and negative excesses. The warm tail is more pronounced in the weak-line distribution. It is due to the network elements deliberately included in the field-of-view. The contour diagrams are elliptically symmetric, as expected for a near-adiabatic standing wave (where temperature and velocity fluctuations are 90° out of phase). There are no pronounced extensions, for example, toward large temperature depressions at large blueshifts (a potential signature of cool overshooting plumes).

I concur with Uitenbroek and colleagues that most of the thermal and velocity structure seen in the outer photosphere is connected with dynamical phenomena, principally the *p*-mode oscillations. Nevertheless, I challenge the authors' inferences concerning the importance of the velocity fields in controlling the physical state of those layers. The rms Δv is very subsonic, and the rms ΔT is only a few percent of the local temperature. The correct view – in my opinion – is that the high photosphere seen in strong CO cores is relatively placid and unstructured, aside from the sprinkling of network bright points and the rare small-scale darkenings of unknown origin.

Figure 5 illustrates an example of the off-limb emissions of the CO bands. The specific frame was selected from a long series taken near the East limb with an inclined slit. The choice was based on an edge sharpness criterion that only a handful of the $\approx$ 1000 candidate limb images taken during the run could satisfy. Even so, the apparent blurring during the 140 ms exposure was about $1''.5$, significantly larger than the diffraction limit.

The background of the diagram depicts the stigmatic spectrum, clipped at low intensities to graphically show the off-limb extensions of the CO lines (the "peaks"), relative to adjacent continuum intervals (the "valleys"). The dots illustrate the bin-by-bin intensity roll-off points measured at different levels (20%, 40%, and 60%) relative to the average near-limb intensity (evaluated at $\mu = \cos\theta = 0.10{-}0.15$). Prior to normalizing the fluxes, a scattered light level was determined several arcseconds outside the limb and subtracted. In this example, which is typical, the strongest CO lines extend about $0''.6$ beyond the continuum edge, while the weaker lines (e.g., 7–6 R68) reach about half as far. The extensions are similar to those measured by Alan Clark and collaborators during the 1994 May eclipse.[21] They used the NIM in a rapid readout mode to record the residual CO spectrum as the sharp lunar edge progressively covered the solar limb. The tactic circumvents the diffraction limit of the McMath-Pierce. Unfortunately the observation cannot be performed routinely!

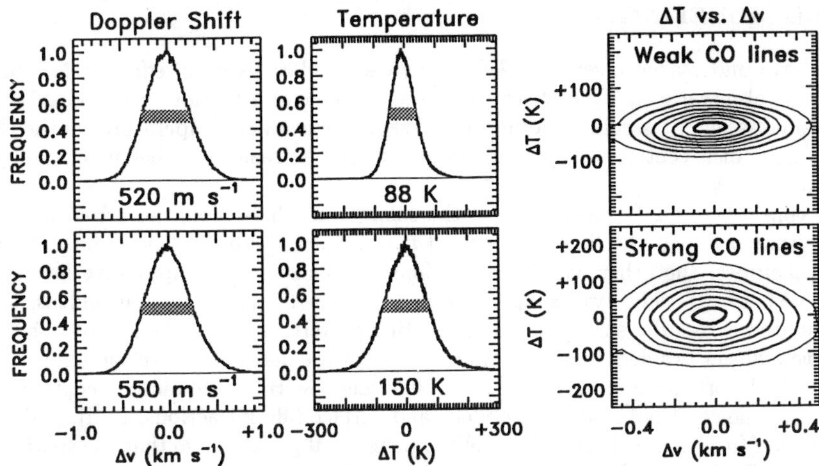

Figure 4: Histograms of Doppler shifts and temperature fluctuations, and joint contour plot, based on preceding movie sequence. *Upper traces* – weak 7–6 R68 line. *Lower traces* – average of strong 2–1 R6, 3–2 R14 and 4–3 R23 lines.

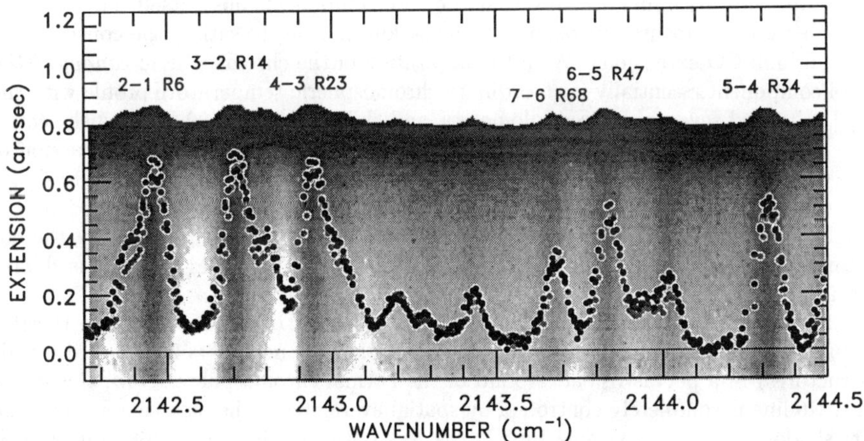

Figure 5: Off-limb emissions of CO in 2143 cm^{-1} region. Background is intensity-clipped frame (not to scale in spatial direction). Dots are monochromatic roll-off points of normalized spectrogram at several reference intensity levels.

3. 2-D MODELING

I have undertaken a series of 2-D numerical simulations in an effort to interpret the new observations. One effect that deserves special attention is the possibility that small-scale ultra-cold structures could skew the average temperatures of the CO cores near disk center, and perhaps dominate the extreme limb behavior through shadowing.[22]

I considered a 2-D radial slice of the solar atmosphere along an arc of a great circle that included the limb. I specified two distinct thermal models: one cool at high altitudes, the other with a conventional chromospheric temperature inversion (the VAL C′ [23]). The two components were assigned specific angular widths, and distributed across the surface in an alternating pattern. I solved the line-of-sight radiative transfer problem in spherical geometry, on the disk and above the limb. I assumed LTE formation for the CO $\Delta v = 1$ lines and the H^--*ff*–dominated continuum. For each choice of hot and cold models, and surface filling fractions, I adjusted the $T(h)$ and $\xi(h)$ distributions of the cool component – if possible – so that the spatially-averaged CO line depths and Doppler widths matched the accurately-measured FTS center-to-limb values.[18] I also endeavored to reproduce the observed $0''.6$ off-limb extension of the strong 2–1 R6 line at "eclipse" resolution (i.e., blurring $\lesssim 0''.2$), taking into account the low spectral resolving power of the NIM.¶ Figure 6 illustrates two different realizations of the broad spectrum of models I considered.

Scenario Zero – preferred by the aficionados of thermal bifurcation – has a pervasive cool component with a small filling fraction of normal chromosphere ($f = 20\%$ in 1 Mm-wide structures). Note that the cool component must, itself, have a chromospheric temperature inversion at $h \approx 900$ km in order to satisfy the constraint of the off-limb CO extensions. With the imposition of the chromospheric *canopy*,[16] the cool component essentially is an ordinary chromospheric temperature profile with the $T_{\rm min}$ elevated several hundred kilometers, and chilled about 1,500 K. At disk center, the calculated scene is relatively uniform, dominated by the large filling fraction of the cool component (which nevertheless is identical to the VAL C′ inward of its $T_{\rm min}$). At the limb, the scene also is relatively uniform, and even more dominated by the cool component (which effectively shadows the hotter one). In this scenario, the hot component plays only a minor – largely cosmetic – role. It could be dispensed with entirely, as far as the CO spectrum is concerned.

Scenario One – favored by those who live in a state of denial concerning thermal bifurcation – has a small filling fraction of the cool component ($f = 20\%$ in 1 Mm-wide structures) in a pervasive background of the normal chromosphere. Now, the strong 2–1 R6 line is completely controlled by spatial averaging in the disk-center scene and by shadowing at the extreme limb. The less optically thick 7–6 R68 transition is barely affected, however.

¶Low spectral resolution has a doubly insidious influence on the off-limb extensions of the CO features. It *raises* the on-disk absorption core intensities, and *suppresses* the off-limb emission peaks, thereby forcing the apparent half-power point closer to the limb.

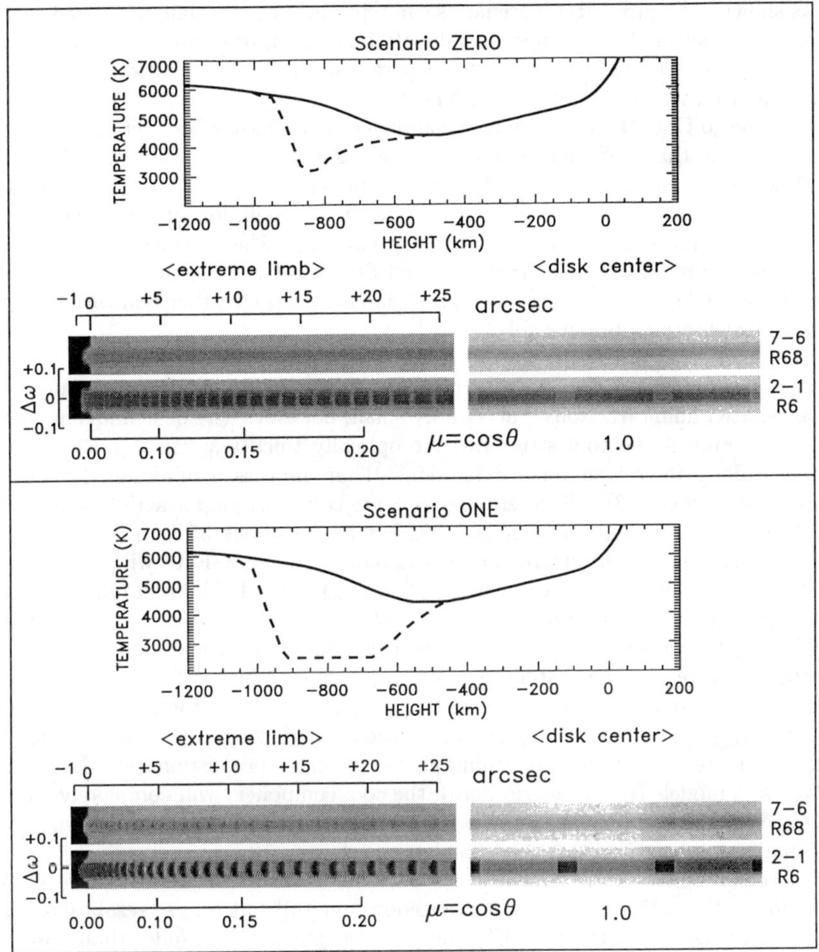

Figure 6: Thermal profiles and predicted stigmatic spectra. *Upper panel* – Scenario Zero: 80% cool component + 20% normal chromosphere (arrayed in 1 Mm wide structures on 5 Mm centers). *Lower panel* – Scenario One: dominant hot component, minority cool component. f's and Δx's are reverse of Scenario Zero.

As shown in Figure 7, both scenarios can reproduce the spatially-averaged center-to-limb behavior of the CO lines, and the off-limb extensions. How then can they be distinguished observationally? Somewhat paradoxically, the main discriminant is the spatial distribution of intensity at *disk center*.

In Scenario One, the minority cool component must have a broader cooler "$T_{\rm min}$" compared with that of Scenario Zero. The depressed thermal profile strongly affects the disk-center core intensity of 2–1 R6, dropping it essentially to the boundary value. Consequently, the scenario predicts a 20% population of ultra-dark points ($\Delta T \approx -1000$ K) in a disk-center CO strong-line map. The absence of such features in the empirical thermal scans indicates that Scenario One – as simulated here with granule-size (1 Mm) cold elements – is not viable. Are there other realizations of the scenario which are more palatable?

The most straightforward way to rescue Model One is to shrink the cold points below the resolution of the NIM observations. One cannot make the structures – or the associated filling fractions – arbitrarily small, however. The near-limb shadowing is effective only if the cool structures are optically thick (e.g., in 2–1 R6) at high altitudes along their transverse dimension. That imposes a minimum lateral size of about 250 km ($0''.3$). Furthermore, the *projected* covering fraction of the cold structures near the limb (say at $\mu = 0.2$) must be about 35% in order to yield the observed –700 K core depression in spatially averaged 2–1 R6 with a maximum thermal contrast of –2000 K (i.e., with $T_{\rm min}^{\rm cool} \approx 2500$ K, the RE boundary temperature with CO cooling[13]). The projected length of the cool "shadows" is $h \tan\theta$, where $h \approx 350$ km is the vertical thickness of the cold zone. The spacing between the cold structures is given by their lateral extent $\Delta x_{\rm cool}$ divided by the filling fraction $f_{\rm cool}$. If the lateral extent is its minimum value, $\Delta x_{\rm cool} \approx 250$ km, then the filling factor must be $f_{\rm cool} \gtrsim 5\%$ in order to provide effective shadowing at $\mu = 0.2$. Note that as $\Delta x_{\rm cool}$ increases, the required filling factor increases proportionately. In the case of $f_{\rm cool} \approx 1$ models (i.e., Scenario Zero), the cool component will completely shadow the minority hot component in strong CO lines at $\mu < 0.2$ if $\Delta x_{\rm hot} \lesssim 2$ Mm.

Simulations based on $\Delta x_{\rm cool} = 250$ km and $f_{\rm cool} = 10\%$ *can* reproduce the disk-center core intensity of 2–1 R6 and its behavior close to the limb. However, the scenario predicts a 2–1 R6 off-limb extension of only $0''.4$ at eclipse resolution. That is similar to the $0''.3$ of the VAL C′ alone, and is significantly smaller than observed. Ironically, the small predicted limb extension devolves primarily from the relatively low resolving power of the NIM. The $f_{\rm cool} \ll 1$ scenarios give rise to emission profiles off-limb that are barely optically thick (owing to the minimal satisfaction of one of the shadowing conditions and to the small filling factor). Such profiles are narrower than their opacity-broadened counterparts in the $f_{\rm cool} \approx 1$ scenarios, and thus are more adversely affected by the NIM's low resolution.

Furthermore, even a 5% population of ultra-cold structures with lateral diameters $\gtrsim 0''.3$ should be readily visible in high-resolution cyanogen images (0.388 μm) taken under conditions of excellent seeing. To my knowledge, no such dark points are seen in existing CN pictures, although like the CO maps there is a scattering of bright points associated mainly with network elements.[24]

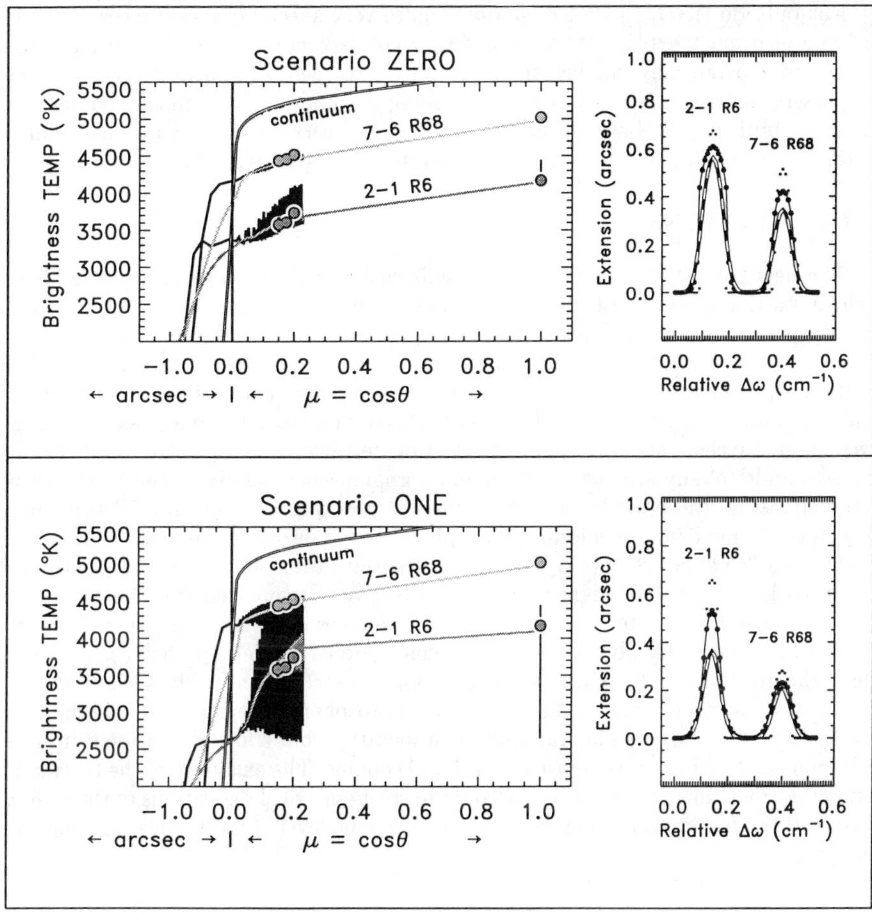

Figure 7: Center-to-limb behavior and off-limb extensions. *Upper panel* – Scenario Zero. *Lower panel* – Scenario One. Dark bars show point-to-point thermal contrasts at 0″.1 resolution; lighter shading, at 1″ resolution. Large dots represent FTS observations. Off-limb extensions were calculated as for Fig. 5, taking into account NIM spectral resolution and spatial smearing of 0″.2 ("eclipse" – large dots) and 1″ ("best NIM" – dashed). Small dots indicate pristine simulation.

Not only do the $f_{\rm cool} \ll 1$ scenarios require very special circumstances to work, but one also must explain the origin of very cold structures at high altitudes with diameters substantially smaller than the convective cells of the deep photosphere. A sparsely distributed population of exploding granules[19] is an unlikely source, for example, both on the basis of size ($\gtrsim$ 1 Mm) and very small time-averaged filling factor (cf., the "darkening" events identified in the disk-center movies).

4. CONCLUSIONS

The new studies of the CO $\Delta v = 1$ bands with the NIM, by several groups, have yielded a striking picture of the thermal morphology and dynamics of the solar outer atmosphere immediately beneath the chromospheric canopy. The velocity fields of those layers are mainly the few hundred m s^{-1} global *p*-mode oscillations, in contrast to the deeper layers where km s^{-1} convective flows dominate. The major thermal contrasts appear to be between the unmagnetized ambient atmosphere and the long-lived magnetic elements of the supergranulation network (whose positive temperature excesses undoubtedly are due to enhanced internal mechanical dissipation). The thermal response to the *p*-modes seen in strong CO absorptions is substantial compared with lines formed in the middle photosphere, yet it is rather modest in absolute terms. The "CO" layers appear to be comparatively stable and only infrequently are disturbed by events that might represent convective overshooting from below.

The model-based interpretation of the CO images suggests that the stable cool layers can extend well above the $T_{\rm min}$ of conventional models, perhaps pervasively filling the underside of the chromospheric canopy ($h \lesssim 1000$ km). The *COmosphere*[25] simply is a natural outward extension of the photosphere; a region where mechanical energy deposition becomes an important consideration, but a low-T solution continues to be competitive because of optically-thin CO cooling. The existence of the thermally bifurcated zone must be taken into account in semi-empirical dissections of the energy budget of the low chromosphere,[26] otherwise the true level of mechanical heating will remain elusive.

5. FOR THE FUTURE

Just a few runs with the NIM have produced a flood of new CO $\Delta v = 1$ data. The stigmatic spectrograms have provided key insights into the *p*-mode excitation of the outer photosphere,[17,19] and the off-limb extensions of the CO lines.[21] Unfortunately, however, the spectral resolution of the NIM is insufficient to study *absolute* CO temperatures, or to measure the line Doppler widths, particularly off-limb. The low resolving power also encourages blending in the 2143 cm^{-1} region (which is crowded with both solar and telluric absorptions), and obscures the true continuum level.

Outfitting the FTS with, say, a 60×60 camera would solve the spectral resolution problem, and provide reasonable spatial coverage. The downside would be poor time resolution ($\gtrsim$ 1 minute: unacceptable in terms of image blurring) and a horrendous data stream (several Gbytes per "image"). Increasing the resolving power of the NIM would require the use of double-pass and narrower entrance slits. The cost would be severely reduced light levels at the detector. One would obtain low S/N with short-enough exposures to freeze the seeing, or unacceptable image smearing with deeper exposures of adequate S/N. The present Amber camera is a high-background device. Some S/N could be recovered by replacing it with a low-background astronomical-quality (i.e., high-price!) sensor.

Of course, the S/N problem could be mitigated by installing a larger primary in the McMath-Pierce (or in a special-purpose instrument like the Large Reflecting Coronagraph discussed at the present workshop). Equally important, the smaller diffraction limit would bolster studies of fine surface details and the off-limb extensions of the CO bands (and those of other molecular and/or atomic species). A 2.5 m main mirror in the McMath-Pierce, for example, would have a diffraction limit of $0''.5$ at 4.7 μm (and $0''.1$ at 1 μm), and would collect nearly 3 times as much light as the present telescope. Such an upgrade of the McMath-Pierce – accompanied by modernization of the telescope control system and image motion compensation – would carry the present revolution in solar IR astronomy to new heights, particularly in the exploration of solar surface phenomena by CO "thermography."

6. Acknowledgments

I thank my collaborator Doug Rabin for his help in this effort. Key technical assistance was provided by Dave Jaksha and Claude Plymate. This work was supported by the National Science Foundation through grant AST-9218063 to the University of Colorado.

7. References

1. T. R. Ayres and G. R. Wiedemann, *Astrophys. J.* **338** (1989) 1033.
2. K. Kirby-Docken and B. Liu, *Astrophys. J. Suppl.* **36** (1978) 359.
3. R. W. Noyes and D. N. B. Hall, *Bull. Am. Astron. Soc.* **4** (1972) 389; ______. *Astrophys. J.* **176** (1972) L89.
4. J. E. Vernazza, E. H. Avrett and R. Loeser, *Astrophys. J.* **184** (1973) 605; ______, *Astrophys. J. Suppl.* **30** (1976) 1.
5. J. W. Brault, *Osservazioni e memorie dell' Osservatorio Astrofisico di Arcetri* **106** (1979) 33.
6. T. R. Ayres and J. L. Linsky, *Astrophys. J.* **205** (1976) 874.
7. T. R. Ayres and L. Testerman, *Astrophys. J.* **245** (1981) 1124.
8. T. R. Ayres, *Astrophys. J.* **244** (1981) 1064.
9. D. O. Muchmore, J. A. Nuth III and R. E. Stencel, *Astrophys. J.* **315** (1987) L141.
10. A. E. Glassgold and W. D. Langer, *Astrophys. J.* **204** (1976) 403.
11. N. Scoville and B. T. Soifer, in *Massive Stars and Starbursts*, (STScI, Baltimore, 1990).
12. P. J. Mauas, E. H. Avrett and R. Loeser, *Astrophys. J.* **357** (1990) 279.
13. L. S. Anderson, *Astrophys. J.* **339** (1989) 558.
14. L. S. Anderson and R. G. Athay, *Astrophys. J.* **346** (1989) 1010.
15. M. Cuntz and D. O. Muchmore, *Astrophys. J.* **433** (1994) 303.
16. S. K. Solanki, W. Livingston and T. Ayres, *Science* **263** (1994) 64.
17. H. Uitenbroek, R. W. Noyes and D. Rabin, *Astrophys. J.* **432** (1994) L67.
18. T. R. Ayres and J. W. Brault, *Astrophys. J.* **363** (1990) 705.
19. H. Uitenbroek and R. W. Noyes, in *Proceedings of the Oslo Mini-Workshop on Chromospheric Dynamics*, ed. M. Carlsson, (*in press*, 1994).
20. T. R. Ayres and D. M. Rabin, *in preparation.*
21. T. A. Clark, C. A. Lindsey, D. M. Rabin and W. C. Livingston, *these proceedings.*
22. T. R. Ayres, in *Mechanisms of Chromospheric and Coronal Heating*, eds. P. Ulmschneider, E. R. Priest and R. Rosner (Springer, New York, 1991), p. 228.
23. P. Maltby, E. H. Avrett, M. Carlsson, O. Kjeldseth-Moe, R. L. Kurucz and R. Loeser, *Astrophys. J.* **306** (1986) 284.
24. N. R. Sheeley, *Solar Phys.* **9** (1969) 347.
25. G. R. Wiedemann, T. R. Ayres, D. E. Jennings and S. H. Saar, *Astrophys. J.* **423** (1994) 806.
26. J. E. Vernazza, E. H. Avrett and R. Loeser, *Astrophys. J. Suppl.* **45** (1981) 635.

TWO-COMPONENT MODELING OF THE SOLAR IR CO LINES

EUGENE H. AVRETT
Harvard-Smithsonian Center for Astrophysics
60 Garden Street, Cambridge, MA 02138, USA

ABSTRACT

One-dimensional hydrostatic models of quiet and active solar regions can be constructed that generally account for the observed intensities of lines and continua throughout the spectrum, except for the infrared CO lines. There is an apparent conflict between a) observations of the strongest infrared CO lines formed in LTE at low-chromospheric heights but at temperatures much cooler than the average chromospheric values, and b) observations of Ca II, UV, and microwave intensities that originate from the same chromospheric heights but at the much higher temperatures characteristic of the average chromosphere. A model M_{CO} has been constructed which gives a good fit to the full range of mean CO line profiles (averaged over the central area of the solar disk and over time) but this model conflicts with other observations of average quiet regions. A model L_{CO} which is approximately 100 K cooler than M_{CO} combined with a very bright network model F in the proportions $0.6L_{CO}+0.4F$ is found to be generally consistent with the CO, Ca II, UV, and microwave observations. Ayres, Testerman, and Brault found that models COOLC and FLUXT in the proportions 0.925 and 0.075 account for the CO and Ca II lines, but these combined models give an average UV intensity at 140 nm about 20 times larger than observed. The $0.6L_{CO}+0.4F$ result may give a better description of the cool and hot components that produce the space- and time-averaged spectra. Recent observations carried out by Uitenbroek, Noyes, and Rabin with high spatial and temporal resolution indicate that the faintest intensities in the strong CO lines measured at given locations usually become much brighter within 1 to 3 minutes. The cool regions thus seem to be mostly the low-temperature portions of oscillatory waves rather than cool structures that are stationary.

1. Introduction

Because the solar surface can be studied at high spatial resolution, more is known about the atmosphere of the Sun than about any other stellar atmosphere. Even so, the basic structure of the solar atmosphere has not been determined well enough to explain all the available observations.

To a first approximation, a one-dimensional model in hydrostatic equilibrium can be constructed to account for the spectrum emitted from each type of feature seen at high spatial resolution on the solar surface, since the horizontal extent of most features is large compared with the vertical thickness of the emitting region. Thus, different cell and network components of quiet regions, various active regions, sunspot umbrae and penumbrae, and flaring regions can be modeled separately. Suppose the intensity spectrum computed from a one-dimensional model agrees with the observed intensity values from a given component region in lines and continua over a wide range of wavelengths, thus probing a wide range of atmospheric depths. Then we could conclude that the stratification of temperature and density in the model represents the actual conditions in the observed component region. Such models derived from observed spectra would give useful physical information about the atmosphere, such as the flow of energy in various forms and the heating mechanisms that are most important. However, only rough agreement has been achieved between such models and available observations even when the CO vibration-rotation lines are excluded.

The problem with the CO lines is that 1) they are formed in LTE (Ayres & Wiedemann[1]) so that the observed brightness temperature at each wavelength in a line should be close to the kinetic temperature at unit optical depth for that wavelength, 2) the strongest lines have central brightness temperatures around 4100 K at disk center, and values as small as 3700 K at the solar limb (Noyes & Hall[2]; Ayres & Testerman[3]), whereas 3) all other diagnostics of the solar temperature minimum (e.g., the Ca II and Mg II resonance line wings and the UV and microwave minimum brightness temperature ranges) indicate an average quiet-Sun minimum temperature of at least 4400 K. The CO lines show that part of the atmosphere is much cooler than indicated by these other diagnostics.

In this paper we exhibit a model that fits the average CO spectrum at disk center, and examine how various cool and hot models might be combined to agree with all diagnostics. In Section 4 we call attention to the recent observations of Uitenbroek, Noyes, and Rabin that show the temperature of the cool regions to vary with time.

The results reported here were obtained in collaboration with E. S. Chang, R. L. Kurucz, and R. Loeser. A more complete account is being prepared for publication in The Astrophysical Journal.

2. CO Observations and Modeling

In their Spacelab-3 ATMOS experiment on the Space Shuttle, Farmer & Norton[4] obtained disk-center solar spectra with high spectral resolution throughout the infrared wavelength range $\lambda = 2.1\text{-}16.5\ \mu\text{m}$. They give the intensity averaged over a circular region extending from disk center out to 0.28 solar radii for $\lambda < 4.97\ \mu$m, and out to 0.58 solar radii for $\lambda > 4.97\ \mu$m (2010.5 cm^{-1}). We have selected a set of 215 fundamental and first overtone lines of $^{12}C^{16}O$ and $^{13}C^{16}O$ between 2.3 and 6.9 μm (4350 to 1450 cm^{-1}) for the present study. The set includes very weak lines as well as the strongest ones.

Figure 1 shows the overall results of our CO line calculations based on an atmospheric model M_{CO} adjusted to fit these lines as well as possible. Brightness temperature is plotted as a function of wavenumber. Each of the 215 CO lines in the set is represented by a vertical line extending from the continuum value down to the calculated line-center brightness temperature. There are no CO lines in the gap between the fundamental lines on the left and the first overtone lines on the right. The first overtone lines are formed deep in the photosphere in the temperature range 5700-6300 K, while the fundamental lines have their continuua in the 5300-5700 K range and calculated line centers at brightness temperatures as low as 4200 K.

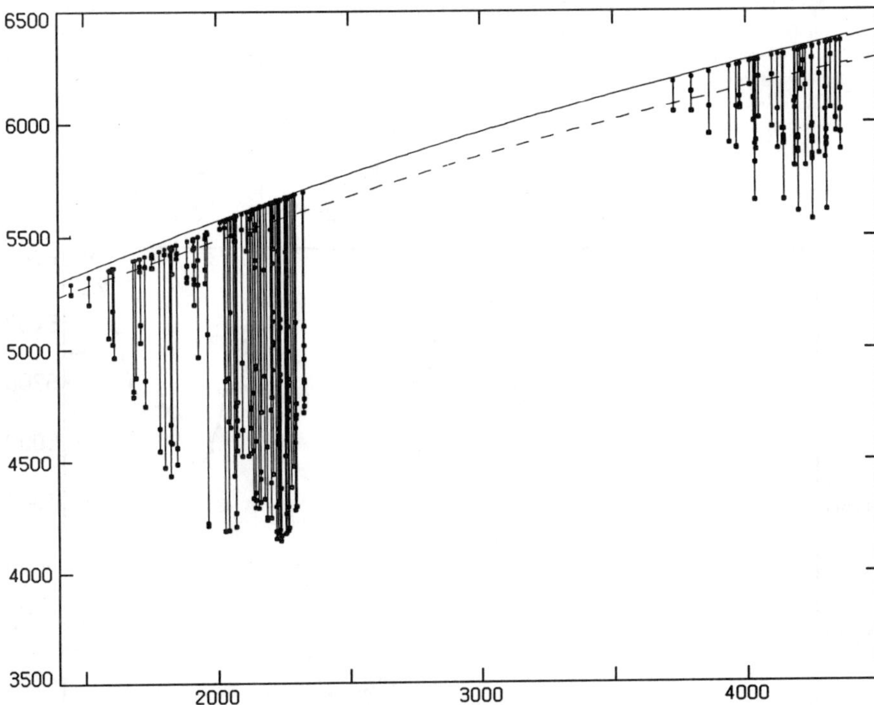

Fig. 1. Calculated brightness temperatures for the 215 CO lines in this study. The solid and dashed lines represent the calculated continuum at disk center and at 0.58 solar radii, respectively.

The solid curve in Figure 2 is the observed CO spectrum near 2239 cm^{-1} (4.466 μm) while the dots represent the spectrum calculated from model M_{CO}. The scale on the left is residual intensity. The corresponding brightness temperature, derived from the calculated continuum, appears on the right. The computed values represent the intensity integrated over the same area of the solar disk as observed. We included the effect of solar rotational broading and attempted to apply the same instrumental broading to the calculated spectrum as in the ATMOS experiment. We used what we considered to be the best available atomic and molecular data and solar abundance values (as will be discussed by Chang, Avrett, Kurucz, and Loeser, in preparation). The temperature distribution was the only function that was adjusted by trial and error in order to get a best fit to the set of CO lines. The corresponding density distribution is determined by assuming hydrostatic equilibrium. The CO lines are assumed to be formed in LTE but the calculated model takes account of non-LTE effects for H^- and the atoms that contribute to the opacity and to the electron density.

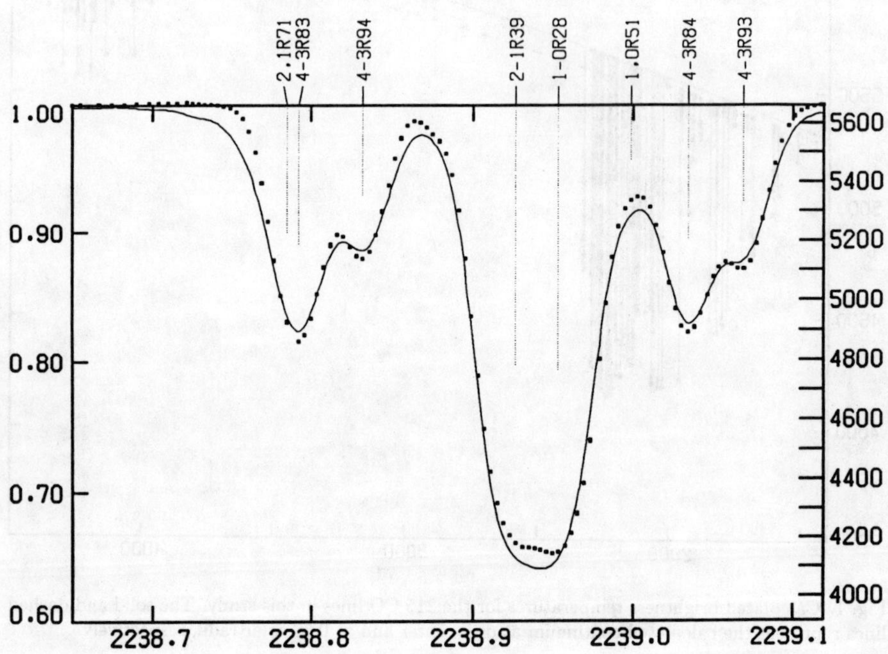

Fig. 2. CO spectrum near 2239 cm^{-1} (4.466 μm) from model M_{CO} (dots) compared with the ATMOS observations of Farmer and Norton (solid line). The scale on the left is residual intensity. The corresponding brightness temperatures appear on the right. Hyphens are used in the CO line designations at the top for $^{12}C^{16}O$, and commas for $^{13}C^{16}O$.

Figure 3 shows the temperature as a function of the continuum optical depth at 500 nm for model M_{CO} based on the CO line observations. Each vertical line segment in this figure extends from the central brightness temperature value calculated for a CO line to the observed value for that line, and is thus an approximate measure of the temperature change needed to match the observations. While some further adjustments could be made, the temperature values seem within about 100 K of the best distribution based on these observations. The disk-center CO lines do not give information about temperatures below 4100 K. We have extrapolated model M_{CO} to 3800 K to account for the lower values at the limb, and then introduced a rapid temperature rise to the much higher values in the upper chromosphere.

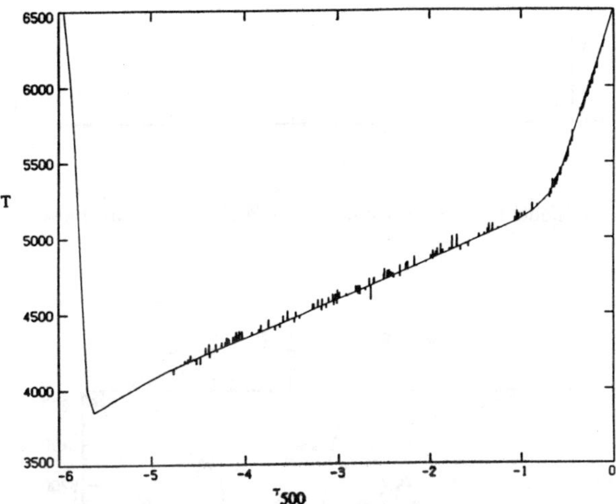

Fig. 3. T vs. τ_{500} for model M_{CO}. The vertical line segments indicate the differences between the calculated and observed brightness temperatures in the core of each CO line.

The temperature distribution in Figure 3 differs substantially from the one that roughly fits other diagnostics of the temperature minimum region and low chromosphere. Figure 4 shows the average quiet-Sun temperature distribution from Maltby et al.[5], and tabulated as model C by Fontenla, Avrett, & Loeser[6]. As before, the vertical line segments indicate the differences between the calculated and observed brightness temperatures in the core of each CO line.

Figure 5 compares the same observed spectrum as in Figure 2 with the one calculated with model C. Since the minimum temperature in the model is 4400 K the calculated values in the cores of the strongest lines cannot be any smaller. Note that there are also some systematic differences between the two models in the deeper layers.

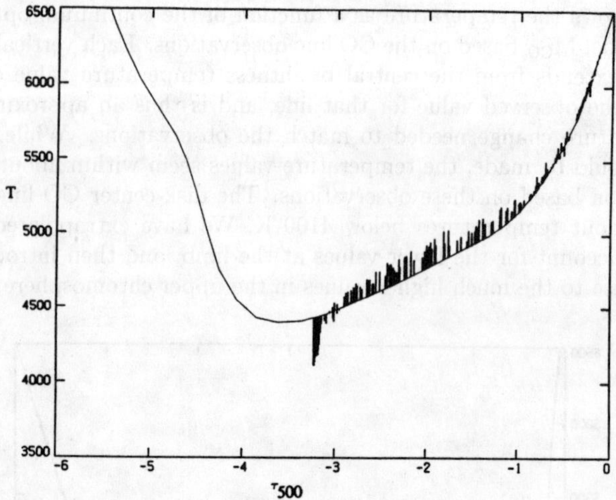

Fig. 4. T vs. τ_{500} for model C. The vertical line segments have the same meaning as in Figure 3.

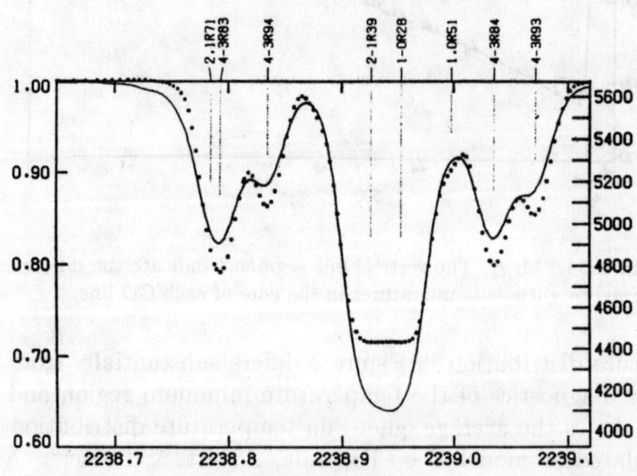

Fig. 5. Same as Figure 2 except that the calculated values are from model C.

3. Comparisons With UV Data

Now consider the observed and calculated intensities in the UV wavelength range 135-174 nm. The dots and crosses in Figure 6 are observations of the disk-center continuum intensity (between various emission and absorption lines) by Brekke[7] and Samain[8], respectively. The dashed lines indicate the brightness temperatures that correspond to the intensity at a given wavelength. Brekke & Kjeldseth-Moe[9] report that recent UV observations with a better absolute calibration than before imply that the minimum observed brightness temperature at disk center should be increased from 4450±50 K to 4520±25 K. (This upward adjustment has not been applied to the observations in Figure 6.)

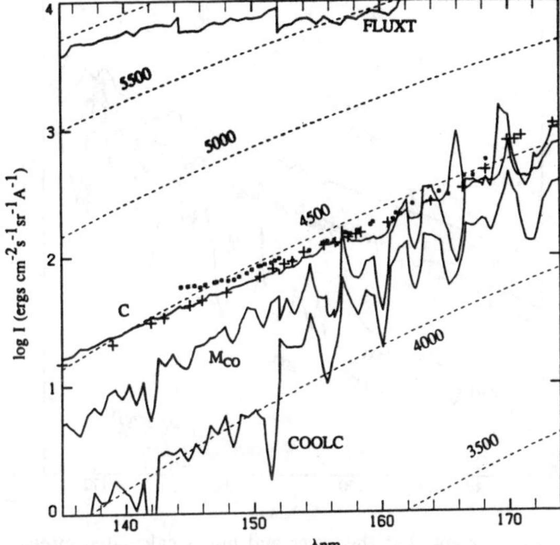

Fig. 6. Observed and calculated disk-center intensities in the UV wavelength range 135-174 nm. Brightness temperatures are indicated by dashed lines. The dots and crosses are continuum observations by Brekke and Samain, respectively. The calculated distributions include sampled opacities due to various lines. Calculated results are shown for models COOLC, M_{CO}, C, and FLUXT.

The curves labeled C and M_{CO} are the intensity distributions that are calculated from these two models. These are non-LTE calculations that include the overall effects of absorption and emission lines in the spectrum. Note that the model C intensities are in reasonable agreement with the observations while the model M_{CO} intensities are too low by a factor of 2 or more. We also plot the intensity distributions calculated from the COOLC and FLUXT models of Ayres, Testerman, & Brault[10]. They found that these separate cool and hot models when combined together with relative covering factors 0.925 and 0.075, respectively, approximately account for both the infrared CO line observations and the emission observed in the Ca II K line. However, the intensity at 140 nm from this combination of COOLC and FLUXT models is about

20 times larger than observed.

What other combination of cool and hot models might explain not only the CO and Ca II lines but also these UV observations? We have found that a model L_{CO}, roughly 100 K cooler than M_{CO}, combined with the very bright network model F given by Fontenla, Avrett, & Loeser[6], with relative covering factors 0.6 and 0.4 is consistent with the CO, Ca II, and UV observations. The UV results are shown in Figure 7. This combination is also consistent with observations in the far-infrared and sub-millimeter range where the emission also originates in the temperature minimum region and low chromosphere.

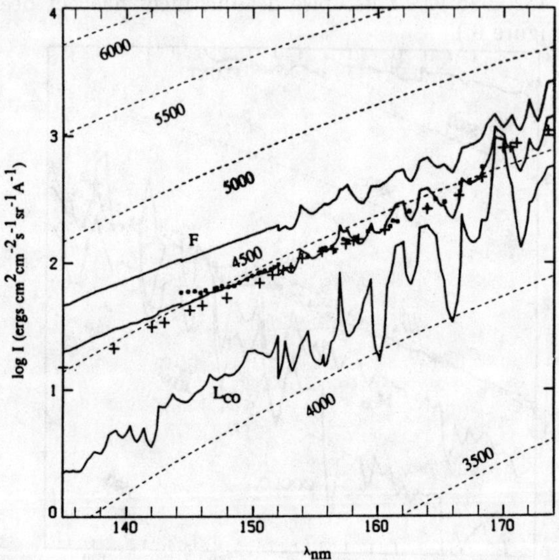

Fig. 7. Same as Figure 6 except that the lower and upper calculated intensities are from models L_{CO} and F, and the middle distribution represents $0.6L_{CO}+0.4F$.

4. Discussion

Observations of the infrared CO lines indicate the presence of a cool component in the atmosphere at low-chromospheric heights. The CO lines are highly sensitive to this cool component because of the rapid increase of the CO line opacity with decreasing temperature in the 3500-4500 K range. Other diagnostics are not as sensitive to these low temperatures, and do not require the two-component modeling discussed above. However, our $0.6L_{CO}+ 0.4F$ combined model seems to explain both the CO observations and the other diagnostics.

The contrast in Figure 7 that might be resolved at very high spatial resolution is greater than that observed by Foing & Bonnet[11]. Their spatial resolution was sufficient to detect any features of 1-2 arc sec size. They found the maximum brightness temperature between the bright network and the darkest internetwork features to be roughly 150 K at 160 nm. This may indicate that model F is too bright to serve as the hot component in this simple modeling procedure, or that we must consider the radiative interaction between the component regions in calculating the spectrum.

Another interpretation is that the two components are not spatially distinct and constant with time, but that local time variations are important. This view is strongly supported by recent observations by Uitenbroek, Noyes, & Rabin[12] and Uitenbroek & Noyes[13] who obtained spatially and temporally resolved CO spectra near 4.67 μm with a new infrared array detector at the McMath-Pierce solar telescope at Kitt Peak. Fig. 2 of Uitenbroek & Noyes shows a space-time map of the 3-2 R14 line-core brightness temperature along a 94 arcsec slit placed over a quiet region near disk center during a period of 23 minutes. They find peak-to-peak temperature fluctuations of approximately 400 K and substantial time variations at each position. These observations clearly show that dynamical effects play an important role in determining the low temperatures seen in the cores of the strong CO lines.

5. Acknowledgements

This research was supported by NASA Grant NAGW-3299.

6. References

1. T. R. Ayres and G. R. Wiedemann, *ApJ,* **338** (1989) 1033.
2. R. W. Noyes and D. N. B. Hall, *ApJ,* **176** (1972) L89.
3. T. R. Ayres and L. Testerman, *ApJ,* **245** (1981) 1124.
4. C. B. Farmer and R. H. Norton, *A High-Resolution Atlas of the Infrared Spectrum of the Sun and the Earth Atmosphere from Space*, NASA Pub. 1224, 1989, vol. 1.
5. P. Maltby, E. H. Avrett, M. Carlsson, O. Kjeldseth-Moe, R. L. Kurucz, and R. Loeser, *ApJ,* **306** (1986) 284.
6. J. M. Fontenla, E. H. Avrett, and R. Loeser, *ApJ,* **406** (1993) 319.
7. P. O. L. Brekke, *Observed Structure and Dynamics of the Solar Chromosphere and Transition Region Based on High Resolution Ultraviolet Spectrograms*, Ph.D. Thesis, University of Oslo, 1992.
8. D. Samain, *A&A,* **74** (1979) 225.
9. P. O. L. Brekke and O. Kjeldseth-Moe, *ApJ,* **431** (1994) L55.
10. T. R. Ayres, L. Testerman, and J. W. Brault, *ApJ,* **304** (1986) 542.
11. B. Foing and R. M. Bonnet, *A&A* **136** (1984) 133.
12. H. Uitenbroek, R. W. Noyes, and D. Rabin, *ApJ,* **432** (1994) L67.
13. H. Uitenbroek and R. W. Noyes, *Chromospheric Dynamics*, ed. M. Carlsson (Institute of Theoretical Astrophysics, University of Oslo, Norway, 1994), p. 129.

temperature between the bright network cells and the dark inter-network regions to be roughly 150 K at 435 km. This result means that model E is too bright to serve as the [illegible] component in these models, and when modeling we must consider the realistic interaction between the component regions in calculating the spectrum.

Another interpretation is that the two components are not spatially distinct and constant with time, but that local time variations are important. This view is strongly supported by recent observations by Uitenbroek, Noyes & Rabin[11] and Ayres & Rabin[12] who obtained spatially and temporally resolved CO spectra using a [illegible] array detector on the McMath Pierce solar telescope at Kitt Peak. [illegible] Noyes show [illegible] the strong CO lines [illegible] near disk center during a period of 25 minutes. They find peak-to-peak temperature fluctuations of approximately [illegible] K and substantial time variations at each position. These observations clearly show that dynamical effects play an important role in determining the low temperatures seen in the cores of the strong CO lines.

7. Acknowledgement

This research was supported by NASA Grant NAGW-[illegible].

8. References

1. T. R. Ayres and C. R. Wiedemann, ApJ 338 (1989) 1033.
2. R. W. Noyes and D. N. B. Hall, ApJ 176 (1972) L89.
3. T. R. Ayres and L. Testerman, ApJ 245 (1981) 1124.
4. [illegible] Farmer and R. H. Norton, A High Resolution Atlas of the Infrared Spectrum of the Sun and the Earth Atmosphere from Space, NASA Ref. Publ. [illegible] 1989, Vol. I.
5. [illegible] ApJ 306 (1986) [illegible].
6. [illegible]
7. [illegible]
8. [illegible] 464 [illegible]
9. [illegible] ApJ 433 (1994) L65.
10. [illegible] ApJ 204 (1975) [illegible].
11. [illegible]
12. H. Uitenbroek, R. W. Noyes and D. Rabin, ApJ 432 (1994) L67.
13. [illegible]

INFRARED APPLICATIONS FOR RADIATIVE TRANSPORT IN STOCHASTIC MEDIA

CHARLES LINDSEY
Solar Physics Research Corporation
National Solar Observatory, PO Box 26732, Tucson, AZ 85726-6732, USA

YEMING GU
National Solar Observatory, PO Box 26732, Tucson, AZ 85726-6732, USA

and

JOHN T. JEFFERIES
Solar Physics Research Corporation
National Solar Observatory, PO Box 26732, Tucson, AZ 85726-6732, USA

ABSTRACT

We describe a general statistical perspective for radiative transfer through stochastic media and consider applications in the infrared. The concepts we have developed are applicable to a broad variety of media, including those in which single stochastic structures by themselves can be optically thick. Such conditions particularly characterize the solar chromosphere and magnetic flux tube morphology extending downward into the photosphere over a broad range of the radiative spectrum. Because the applicability of LTE greatly simplifies computational tasks and modeling based on the theory, the infrared offers a particularly promising spectrum of applications.

1. Introduction

Much of astrophysics rests on understanding how radiation propagates through matter, a natural consequence of the fact that almost everything we know about astronomical objects depends on the radiation observed from them. As a result, we spend a lot of time creating models of extraterrestrial media and running computations to examine how radiation should behave in them. There is no question that the models we create are always simpler than the media themselves: The simplicity is not just a convenience; it is an essential element in our actual understanding of the problem. For example, when a turbulent medium made of discrete atoms and ions behaves in some important respect as a smooth, uniform model, this very quality becomes *part of* our understanding, even though the medium itself is neither smooth nor uniform. Pressure, viscosity and conductivity (thermal and electrical) are important examples of smooth-medium concepts at the foundation of our understanding of media that are blatantly non-smooth on an atomic scale. Similarly, when it is plain that an atmosphere is *not* smooth, either from direct observation or because plane-parallel models consistently fail in some important respect, then the prospect of models that characterize certain qualities of inhomogeneity becomes attractive. Before the Renaissance, the Sun was widely thought, for apparently religious and political reasons, to be perfectly smooth. However, since the advent of the telescope, direct observation has shown the solar atmosphere to be non-smooth at all levels to the finest scales resolvable. It is now clear that there exist aspects of stellar atmospheres in general that are best examined in the context of models that are stochastic in character. The purpose of this paper is to review recent efforts to characterize inhomogeneous models stochastically and to characterize their radiative properties similarly.

Radiative transfer for well-defined atmospheric models is usually reduced in some part to the mathematical context of solving the radiative transfer equation:

$$\frac{dI}{ds} = -\kappa\, I + \kappa\, S. \tag{1}$$

Here κ is the opacity of the medium and S its source function at a particular point on the optical path, and the equation describes the evolution of the specific intensity, I, along the direction of the optical path as one proceeds along it an incremental distance ds. When κ and S are known functions of position in the atmospheric medium, radiative transfer analysis reduces entirely to solving Eq. 1, which is straightforward given appropriate boundary conditions. Thus, atmospheric modeling using radiative transfer is easiest when radiative parameters—the source function and opacity—at a particular location in the atmosphere are completely determined by physical conditions—*e.g.*, density, temperature, degree of ionization—at the same location. This is usually the case for opacity, but for the source function it fails badly where scattering is important, as in strong visible lines. Terrestrial clouds viewed in daylight, for example, appear bright not because they themselves are locally hot but because the radiation we see from them is scattered light from a hot object somewhere else: the Sun. On the other hand media that scatter poorly, *e.g.*, soot, dark smoke, or smog, *i.e.* any thing black, appear appropriately dark when cool and light up only when they themselves are truly hot. This condition is called "Local Thermodynamic Equilibrium" (LTE), a term that has come to be applied to the source function of a particular species of radiation when it is in thermal equilibrium with the local kinetic motion of free electrons in the medium, even when there may be many other qualities of the medium, including other parts of the radiative spectrum, that are *not* in thermal equilibrium. LTE is one of the situations in which knowledge of the physical conditions in an atmosphere tells us the source function directly and thus reduces radiative transfer entirely to the straightforward proposition of integrating Eq. 1.

In practice, radiative-transfer analysts have concentrated most of their attention on the far more difficult non-LTE problems. In fact, one is generally doing well to handle non-LTE radiative transfer satisfactorily even for the simplest smooth, plane-parallel atmospheric geometries. It seems evident that the study of stochastic atmospheres will be greatly facilitated when LTE applies, and this important concept appears to be widely applicable in the infrared solar spectrum.

To be fair, it is mainly spectral lines, not the continuum, that depart from LTE in the visible. The primary mechanism of continuum opacity in the visible is absorption by the bound H^- ion, which destroys the ion, directly producing a kinetic free electron. Photons with wavelength, λ, exceeding 1.6 μm have insufficient energy to disassociate the H^- ion, and H^- bound-free opacity gives way to H^- free-free, arising from collisions between neutral hydrogen atoms and free electrons. H^- free-free opacity increases in proportion to λ^2 throughout the infrared spectrum. In both H^- free-free and H^- bound-free interactions, the photon is coupled directly to a free electron, and this connection is not pre-empted by the rapid radiative de-excitation that constitutes scattering. (Recombination is far slower than de-excitation.) Thus, solar continuum emission is in LTE and so serves as an excellent atmospheric thermometer in both the visible and infrared.

The free-bound opacity varies only moderately throughout the visible spectrum, so the visible continuum shows us only a fairly narrow layer at the base of the photosphere viewed at disk center. The infrared opacity, however, varies by a factor of several hundred as λ progresses from 2 to 40 μm. This strong variation in opacity allows the infrared continuum to sample a broad range of heights, making it central to photospheric and chromospheric modeling since the advent of the Harvard Smithsonian Reference Atmosphere (Gingerich *et al.* 1971[1]).

Visible lines in the solar spectrum are generally strongly influenced by resonant scattering, making the line radiation an unreliable local thermometer. With increasing wavelength, however, coupling of atomic states to radiation tends to decrease rapidly, leaving collisional excitation by free electrons dominant, so that the source function approaches the Planck function at the local free-electron temperature. Thus, the infrared offers us powerful radiative diagnostics in general and further offers us the simplicity of LTE, both in the continuum and (with some notable exceptions) in lines. These qualities make the infrared irresistibly desirable for studying complex atmospheres.

2. The Stochastic Perspective

Lindsey and Jefferies 1990[2] make a distinction between what they term the "macroscopic" and the "microscopic" perspectives that arise in radiative transfer in stochastic media. Eq. 1, for example, arises from a perspective that we characterize as microscopic: It is intended to be integrated by evolving I along the optical path in increments, ds, which individually manifest only a small, if not infinitesimal, optical depth. While this microscopic perspective will remain the focus of this paper, we note that a macroscopic formalism is often useful in computational work in a stochastic medium. The reader will probably note that when ordinary radiative transfer is applied to smog or smoke as a quasi-uniform medium, the medium is microscopically stochastic, and a simple macroscopic perspective has usually been applied to dispense with the fact that significant opacity is actually encountered only in individual smog particles, inside of which it is typically enormous. This macroscopic perspective allows the theorist to forgo the numerical work of microscopically evolving I through an optically thick structure such as a soot particle when the results of this labor are destined be obscured by just the surface layer of the particle. In media that contain stochastic structures with scale sizes similar to that over which an arbitrary optical path is *likely* to attain a significant opacity, the distinction between microscopic and macroscopic perspective becomes considerably more imposing than in quasi-uniform smog or smoke. Lindsey 1987[3], Braun and Lindsey 1987[4], and Jefferies and Lindsey 1988[5] have made use of a macroscopic concept termed the "equivalent smooth atmosphere."

The general context for studying stochastic atmospheres microscopically is different from that of Eq. 1, wherein κ and S are understood to be completely specified over the region of space to which it is to be applied. The proposition for the stochastic problem is that the atmosphere be characterized statistically. In principle, we can construct an ensemble of specific atmospheric models satisfying the stochastic qualities desired and do statistics on the properties of the emergent radiation. It is customary to express parameters that characterize an individual ensemble element in terms of random variables, $\boldsymbol{\xi}$. Random variables are most familiar as simple real values, sets of values, or perhaps vectors, such as may be used to to characterize Brownian motion of a single particle over a measured time interval. To characterize a stochastic atmosphere requires *functions*, in place of simple values, that vary randomly from one ensemble member to another. In full generality, $\boldsymbol{\xi}$ would be a pair of functions, $(\kappa, S)(\mathbf{x})$, of position, $\mathbf{x}$, in space representing the opacity and source functions for a single atmosphere in the ensemble. For simplicity, we may prefer to let $\boldsymbol{\xi}$ be the function pair $(\kappa, S)(s)$, specified only along a particular optical path rather than at all ponts in the atmosphere. In a broad variety of such problems, it is convenient to represent the behavior of κ and S in terms of a single parameter, $\boldsymbol{\zeta}$, that varies stochastically along the optical path. We could create an ensemble of such functions $\boldsymbol{\zeta}(s)$, solve the radiative transfer equation for each:

$$\frac{dI}{ds} = -\kappa(\boldsymbol{\zeta})I + \kappa(\boldsymbol{\zeta})S(\boldsymbol{\zeta}), \tag{2}$$

and do statistics on the emergent intensities that result. A considerable amount of insight has been gained from this general approach, which is included in the work of Frisch and Brissaud 1971*a*[6], Frisch and Brissaud 1971*b*[7], Auvergne, *et al.* 1973[8], Brissaud and Frisch 1974[9], Levermore *et al.* 1986[10] and Boisse 1993[11], among others.

The appropriate form of $\boldsymbol{\zeta}$ depends on the properties of the stochastic atmosphere represented. If, for example, κ and S were interdependent in such a way that the knowledge of one would determine the other, then $\boldsymbol{\zeta}$ could be a single real variable. Otherwise, $\boldsymbol{\zeta}$ must have added dimensions, such as in an array:

$$\boldsymbol{\zeta} \equiv (\zeta_1, \zeta_2, \ldots). \tag{3}$$

Because complete generality (for a given species of radiation) could be accomplished directly by letting ζ consist of the simple pair (κ, S), it should never be necessary for ζ to have more than two dimensions, but convenience may argue for more than two. On the other hand, when simplicity allows, a simple integer index, i, in a certain range might be preferred, as in the familiar case of an atmosphere composed of discrete species of media. A stochastic mixture of oil and vinegar, for example, might be represented by letting the domain of i be the simple set $\{1, 2\}$, where "1" characterizes points in space occupied by oil and "2" points occupied by vinegar.

In a stochastic model, the emergent intensity, I, along some optical path becomes a random variable, with a statistical distribution, $f(I)$. $f(I)dI$ expresses the fraction of occurrences in a statistical ensemble for which the intensity will lie in an interval dI centered at I. Figure 1 illustrates the determination of $f(I)$ by a Monte-Carlo computation for a simple two-component model atmosphere. The upper-left panel (a) of the Figure portrays a two-component medium composed of disks randomly superimposed into an ambient background. In this case, the disks are all the same size, but can overlap. Intensities are computed for horizontal optical paths through the medium with the path length indexed by the variable x and the vertical position of the path indexed by y. The evolution of the intensity looking into the medium from the left along a sample optical path, shown in Panel a, is plotted below in Panel b (solid curve), along with the average intensity along all paths (dashed curve). The horizontal dot-dashed curves show the sources functions, S_1 and S_2, of the ambient medium and the disk interiors respectively. In any particular component of the atmosphere, the intensity simply decays exponentially toward its respective source function. The emergent intensity is plotted as a function of y in the upper-right panel (c). The source functions of the two components are indicated here by vertical dot-dashed lines. The intensity profile has a noticeably billowy appearance, caused by the circular cross-sections of the embedded structures. The resulting intensity distribution is plotted in the lower-right panel (d). The statistics used to compute the actual distribution in Panel d are considerably more extensive than the segment shown in Panels a and c.

A more informative distribution is needed to characterize the *spatial* scale of the emergent radiation. An example would be a distribution of the form $f(I_1, I_2; \Delta y)$, determined by examining the intensity at two locations always separated by a distance Δy in the y-direction. $f(I_1, I_2; \Delta y)dI_1 dI_2$, then, expresses the fraction of occurrences in which the first intensity lies in the interval dI_1 centered at I_1 and the second intensity, emerging a distance Δy away from the first, lies in the interval dI_2 centered at I_2. It should be clear that for $\Delta y = 0$ this distribution will be collapsed into the infinitely thin sheet along $I_1 = I_2$ in $(I_1, I_2, \Delta y)$-space, since I_1 must be equal to I_2 when they are measured at the same place. As Δy is increased, the distribution will spread. An illustration is shown in Figure 2. How rapidly the distribution spreads is, of course, dependent on the spatial scale of the morphology that characterizes the stochastic quality of the medium being considered.

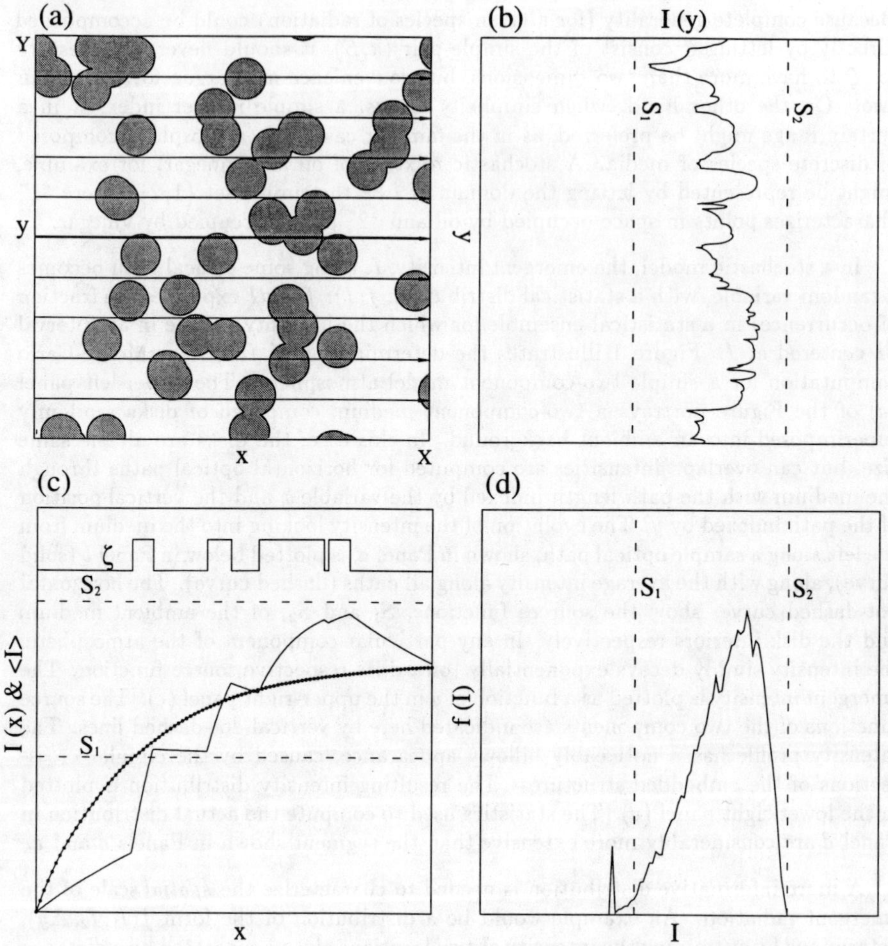

Fig. 1. Illustration of a Monte-Carlo computation for a simple two-component atmosphere. Panel a shows a stochastic model with a sample optical path (heavy solid line). Panel b shows the evolution of intensity along the sample path (solid curve) and the intensity averaged over y (dashed curve). Panel c shows the emergent intensity as a function of y. Panel d shows the distribution, $f(I)$, of the emergent intensity. Taken from Gu, Lindsey and Jefferies 1995[12].

Fig. 2. Surface relief representations of the distribution $f(I_1, I_2; \Delta y)$ for $\Delta y = 0.05D$ (left frame) and $\Delta y = 0.2D$ (right frame), where D is the diameter of the embedded structures. The shading assumes illumination of the surface from upper left. Horizontal axes represent I_1; vertical axes represent I_2.

Model calculations of the Monte-Carlo type are useful for examining statistics of the type illustrated above in a wide variety of models. However, it is important for a general understanding of how radiation behaves in stochastic media to examine the principles that govern the evolution of statistical qualities of radiation along the optical path. This has been the thrust of work by Jefferies and Lindsey 1988[5], Lindsey and Jefferies 1990[2], and Gu, Lindsey and Jefferies 1995[12]. They have considered simple multi-component media. Some of their work is reviewed in the following section.

3. Markov Inhomogeneity in Multi-Component Media

Lindsey and Jefferies 1990[2] consider how the intensity distribution, $f(I)$, evolves statistically in a simple multi-component medium represented by a field in which each point in space is characterized by one of a finite number of species of medium, indexed by an integer, i. The atmosphere is smooth and well defined within any region containing only a single species of medium. The stochastic nature of the atmosphere lies in the morphology of the network of boundaries separating different species.

The formalism of Lindsey and Jefferies (1990)[2] approaches the problem of how $f(I)$ evolves by examining the behavior of the distribution over both random variables, I and i. This is denoted by an array distribution, $\mathbf{f}$, literally an array of distribution functions whose individual components, $f_i(I)$, correspond to their respective species of medium. $f_i(I)dI$ is simply the fraction of instances in a statistical ensemble in which the intensity at a given point on the optical path falls in the interval of width dI located at I *and* the point considered lies in a region occupied by medium species i.

Note that

$$\sum_i f_i(I) = f(I), \tag{4}$$

and that

$$\int_0^\infty f(I)dI = 1. \tag{5}$$

If a particular point on the optical path is characterized by medium species i, then as we proceed an infinitesimal distance, ds, along the path, we suppose that there is a proportional probability, $P_{ij}ds$, of crossing a boundary to enter a region containing a medium of a different species, $j \neq i$. Lindsey and Jefferies 1990[2] assume that P_{ij}, while it may be a function of location and optical path direction, is independent of stochastic qualities, such as how far the point has progressed since the boundary crossing into the current medium, i. This is known as the Markov assumption. In microscopic detail, this assumption is likely to be quite inaccurate. Under the Markov assumption, for instance, the distribution in the lengths of contiguous path segments through any species of medium is a simple decaying exponential function of length, just as in Poisson statistics. For a medium such as that portrayed in Figure 1, for example, composed of disks embedded in an ambient medium, the distribution of path lengths through the interiors of single disks increases monotonically from a value of zero at length zero to infinity for lengths equal to the disk diameter, whence the distribution is truncated. This cannot be fit to any form of decaying exponential and is, therefore, clearly non-Markovian. In fact, it is straight-forward to extend the formalism of Lindsey and Jefferies 1990[2] to incorporate non-Markovian media. On the other hand, Gu, Lindsey and Jefferies 1995[12] explore a broad class of distinctly non-Markovian models in which the Markov assumption allows a highly informative first approximation of how radiation emerges from a stochastic atmosphere.

Under the Markov assumption, the possibility of boundary crossings causes $f_i(I)$ to evolve along the optical path simply by continuously *mixing* it with other components of $\mathbf{f}$ as one proceeds:

$$\left.\frac{\partial f_i}{\partial s}\right|_{\text{mix}} = \sum_j P_{ij} f_i. \tag{6}$$

We call the matrix $\mathbf{P}$ whose elements are P_{ij} the "mixing matrix." On the other hand, within a particular species of medium, the effect of absorption and emission along the optical path is simply to uniformly *smash* each component, $f_i(I)$—compressing it in the I-direction and extending it vertically so as to conserve the integral over I—about its respective source function, S_i:

$$f_i'(I) = Qf(I'), \tag{7}$$

where

$$Q = 1 + \kappa\ ds, \tag{8}$$

and

$$I' = S_i + Q(I - S_i). \tag{9}$$

The transport equation for this continual infinitesimal *mixing* and *smashing* of $\mathbf{f}$ as we procede along the optical path is simply

$$\frac{\partial f_i}{\partial s} = \kappa_i(I - S_i)\frac{\partial f_i}{\partial I} + \sum_j (\kappa_i \delta_{ij} + P_{ij}) f_i. \tag{10}$$

This equation comprises the analogy in a statistically defined Markovian medium to Eq. 1 in a specifically defined medium. For media with optically thick structures, the radiation transport equation above tends to give rise to sharp, quasi-discontinuous features in $\mathbf{f}$ that probably pose a challenge to some of the more powerful numerical techniques. However, the simple procedure of alternately smashing and mixing the components, f_i, of $\mathbf{f}$, yields accurate numerical solutions for $\mathbf{f}$ even when these solutions have sharp quasi-discontinuities. Lindsey and Jefferies (1990)[2] demonstrate this with a selection of simple, examples. Once the task of expressing the stochastic morphology of the atmospheric medium in terms of a mixing matrix is accomplished for the optical paths of interest, the integration of the radiative transport equation is accomplished simply and with a minimum of computation.

4. Characterizing Stochastic Morphology

By way of summary, we can say that a stochastic atmosphere is characterized, as fully as possible under the Markov assumption, by the mixing matrix, $\mathbf{P}$. To the extent that stochastic qualities can change with location, $\mathbf{P}$ is dependent on location, $\mathbf{x}$, in the atmosphere. Further, to the extent that the stochastic structure is non-isotropic, $\mathbf{P}$ at any given point, $\mathbf{x}$, can be dependent on the direction, $\hat{\mathbf{r}}$, of the optical path. The essential labor in preparing a stochastic model for radiative-transport analysis reduces to the determination of $\mathbf{P}(\mathbf{x}, \hat{\mathbf{r}})$. This is the basic task addressed by Gu, Lindsey and Jefferies (1995)[12] for multi-component atmospheres represented in terms of a limited number of species of medium. They consider atmospheres that consist of an ambient medium into which are introduced structures randomly located at various spatial densities, such as the non-Markovian model illustrated in Figure 1. They represent the mixing matrix elements for transitions from species i to j by assuming that each step along the optical path in medium i leaves one at a random place in the volume occupied by that medium.

If one chooses a random point in the interior volume, V, of a structure defined by a simple enclosing surface, ∂V, such as that represented in Figure 3 for example, then the probability that an infinitesimal step, ds, along $\hat{\mathbf{r}}$ will encounter the boundary is simply the fractional area represented in the Figure by the hatched region of the interior to the right of the structure. The result is that

$$P_{\text{out}\leftarrow\text{in}} = \frac{1}{V} \int_{\substack{\partial V \\ \hat{\mathbf{r}}\cdot\hat{\mathbf{n}} > 0}} da\, \hat{\mathbf{r}} \cdot \hat{\mathbf{n}}. \tag{11}$$

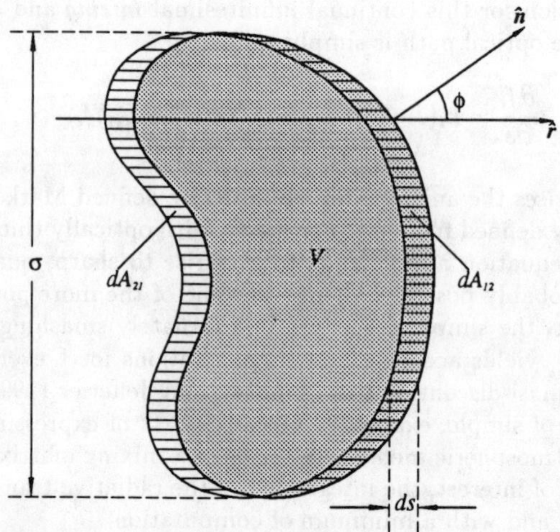

Fig. 3. Geometry for computing egressing and ingressing mixing matrix elements for simple closed structures embedded in an ambient medium.

The integral is taken over the surface, ∂V, but only over that part of the surface on which the optical path is *leaving* the structure ($\hat{\mathbf{r}} \cdot \hat{\mathbf{n}} > 0$). Here, da represents the differential element of surface area and $\hat{\mathbf{n}}$ is the unit outer normal to the surface, as illustrated in the Figure. For simple convex structures, this integral reduces simply to the projected cross-section, σ, presented by the structure to the plane perpendicular to the optical path, so in this instance, the mixing matrix element for transitions from inside to outside reduces to

$$P_{\mathrm{out} \leftarrow \mathrm{in}} = \frac{\sigma}{V}. \tag{12}$$

For the ingressing matrix element, $P_{\mathrm{in} \leftarrow \mathrm{out}}$, a similar expression results, where the integral is taken over the ingressing portion of ∂V ($\hat{\mathbf{r}} \cdot \hat{\mathbf{n}} < 0$), and the term $1/V$ is replaced by the density of structures in the ambient medium. The foregoing analysis assumes that stochastic properties such as morphology and structure density are uniform over the volume under consideration and that the density of structures is small enough that overlap is improbable. However, Gu, Lindsey and Jefferies (1995)[12] proceed to describe how to account for overlap for two component media with arbitrarily high structure densities. They point out that overlap results in interior path length distributions that are more nearly Markovian than when overlap does not occur. (For randomly placed structures that can overlap, the exterior (*i.e.* ambient) path length distributions are completely Markovian at any density.)

Gu, Lindsey and Jefferies (1995)[12] also consider non-isotropic media such as in atmospheres containing tubes and fluted structures, which may be used to represent magnetic flux-tubes or chromospheric spicules. In these atmospheres, it may be said that the stochastic properties of embedded structures are themselves a strong function of location in the atmosphere and can change on a physical scale similar to that of the structure itself. These concepts have powerful applications in atmospheric modeling of both the solar photosphere and chromosphere.

5. Future Directions

There is little question that radiative transfer in stochastic media is presently in its early infancy, an attribute that it seems to share with many other exciting topics such as turbulence, self-similarity and fractal qualities of complex physical systems, and chaos. The direction this area of research will take in the long term can probably be characterized as well as those of chaotic systems themselves: totally unpredictable, which makes it all the more interesting. In the near term, however, a number of promising avenues are already apparent:

1. More general analytic techniques such as those already developed by Frisch and others (References 6–11) will soon be needed to alleviate the computational tasks imposed by the theories under development.

2. The theory of Lindsey and Jefferies (1990)[2] needs to be extended to describe radiative transport in non-Markovian media properly. For simple multi-component models, this theory could be accomplished by the use of a quasi-continuous random variable for each species of medium that records the distance elapsed along the optical path since entry into the current medium. Quasi-continuous variables are also needed to represent atmospheres that are too complex to be described by convolutions of only a few discrete species of media.

3. A statistical theory for the spatial behavior of the two-intensity distribution, $f(I_1, I_2; \Delta y)$, and its extension to three-dimensional models needs to be developed to express spatial properties that are not intrinsic to the single-intensity distribution, $f(I)$.

4. An extension of the theory to accommodate non-LTE problems would be useful and should be practicable. In our opinion this will be facilitated by a formalism based on quasi-continuous random variables suited for non-Markovian models. However, Mitskevich *et al.* 1993[13] have already used the formalism of Lindsey and Jefferies 1990[2] to incorporate scattering in a study of Hα emission profiles from T-Tauri stars.

The possible applications of stochastic radiative transport theory are myriad; the solar atmosphere simply *is* stochastic. While the solar corona appears to be no less so than the photosphere, chromosphere and transition regions, it seems to be the latter for which stochastic analysis is substantially needed. This is because the consequences of stochastic structure are greatly complicated when individual stochastic elements can attain significant opacity, and the solar corona appears to be optically thin to the entire visible and IR spectrum.

We remind the reader that the infrared promises powerful applications in the development of the theory, because it forms in LTE in most lines as well as the continuum. We think that the techniques being developed at the present can already contribute to our understanding of the center-to-limb infrared contrast of plage and of the radiative dynamics of CO in the low chromosphere, and that this will soon be useful for diagnostics and modeling of stochastic structure in the solar atmosphere. One would hope that non-LTE applications will also become practical, particularly with increasing computational resources and experience. However the infrared, still in its infancy, already offers us a very promising window through which to acquire the experience. We expect infrared solar astronomy to play a major role in our understanding of how radiation behaves in stochastic atmospheres.

6. Acknowledgements

C. Lindsey and J. T. Jefferies hold visiting appointments at the National Solar Observatory. The authors have benefited greatly from cooperation between NSO and the Solar Physics Research Corporation. This work was done under the support of NSF Grants ATM-9311451 and ATM-9122073.

7. References

1. Gingerich, O., Noyes, R. W., Kalkofen, W. and Cuny, Y., *Solar Phys.* **18** (1971) 347.
2. C. Lindsey and J. T. Jefferies, *Ap. J.* **349** (1990) 286.
3. C. Lindsey, C. *Ap. J.* **320** (1987) 893.
4. D. Braun and C. Lindsey *Ap. J.* **320** (1987) 898.
5. J. T. Jefferies, C. Lindsey, *Ap. J.* **335** (1988) 372.
6. U. Frisch and A. Brissaud, *J. Quant. Spectrosc. Radiat. Transfer* **11** (1971*a*) 1753.
7. U. Frisch and A. Brissaud, *J. Quant. Spectrosc. Radiat. Transfer* **11** (1971*b*) 1767.
8. M. Auvergne, H. Frisch, U. Frisch, Ch. Froeschlé and A. Pouquet *Astron. & Astrophys.* **29** (1973) 93.
9. A. Brissaud and U. Frisch, *J. Math. Phys.* **15** (1974) 372.
10. C. D. Levermore, G. C. Pomraning, D. L. Sanzo and J. Wong, *J. Math. Phys.* **27** (1986) 2526.
11. P. Boissé , *Astron. & Astrophys.* **228** (1990) 483. **15** (1974) 372.
12. Y. Gu, C. Lindsey and J. T. Jefferies, *Ap. J.* (1995—submitted).
13. A. S. Mitskevich, A. Natta, and V. P. Grinin, *Ap. J.* **404** (1993) 751.

No Magnetic Field — No Chromosphere

M. Carlsson, R. Stein
Michigan State University

ABSTRACT

Enhanced chromospheric emission which corresponds to an outwardly increasing semiempirical temperature structure can be produced by wave motion without any increase in the mean gas temperature. Hence, the Sun may not have a chromosphere in magnetic field free internetwork regions.

A TURBULENT MODEL OF THE SOLAR GRANULATION

V. KRISHAN
Indian Institute of Astrophysics
Bangalore 560034, India

ABSTRACT

The turbulent nature of the motions on the solar photosphere is being emphasized more and more by the high resolution quality observations. The processes of inverse cascade of energy in a turbulent medium are being explored for modeling the hierarchal flow patterns on the solar surface. Here, the development of large scale flows is studied using the Kolmogorovic arguments for a statistically stationary turbulent fluid as well as through the solution of the Navier-Stokes equations including the Reynold's stresses generated by the small scale flows, in a manner akin to the dynamo problem. Some of the testable predictions of the model are also listed.

1. Introduction

We learn from the high resolution photographs that the solar photosphere is not of a uniform brightness. We see a pattern of bright irregular polygons separated by thin dark lanes, the bright regions representing the upward gas motion and the dark lanes, the downward. These cellular velocity patterns are believed to be the manifestations of the convective phenomenon occurring in the sub-photospheric layers. The cellular velocity fields are seen prominently on three scales: the granules with an average size of 1000 km and a lifetime of a few minutes; the supergranules with an average size of 30,000 km and a lifetime $\sim$ 20 hours; the mesogranules have properties intermediate to these two. Giant cells of the size of the convective zone are also suspected to be observed.

It is extremely important to understand the convective velocity fields, as they re-distribute angular momentum, causing the differential rotation of the Sun, which in turn affects the workings of the solar dynamo responsible for the magnetic activity seen in its myriad forms.

It has been concluded from the observations that the granules move in the flow fields of mesogranules which in turn are advected by the still larger supergranules.[10] The flows are turbulent and helical, the helicity being an essential ingredient of the dynamos, too.

The earlier attempts to model this phenomenon relied on the mixing length theory[12] and the linear instability description of the convection in the hydrogen and helium ionization zones of the sub-photospheric medium.[1,2] But these theories fail to account for the existence of a continuum of sizes, towards which the high quality observations tend to point.

We have proposed an inverse cascade model for the entire granulation phenomena. In particular, we have investigated the possibility of making large structures from small ones in a turbulent medium.[4]

The phenomenon of fluid turbulence is mainly approached by two different routes: (i) the study of statistically stationary states using Kolmogorovic arguments and (ii) through the solutions of the Navier-Stokes equations, which hopefully, will confirm the predictions of the former.

2. Helical turbulence

A turbulent fluid consists of motions on many spatial and temporal scales, within the constraints imposed by boundaries, buoyancy and dissipation. The cascading of energy from large scales to small ones is a familiar process. The inverse cascade, i.e. the energy flow from small scales to large scales can also occur under appropriate conditions e.g. in a two-dimensional hydrodynamic and a three- dimensional magnetohydrodynamic system. That the inverse cascade can occur in a three-dimensional hydrodynamic system is a recent revelation.[8] We have proposed that in the convection zone, most of the energy resides in small scales and its distribution, in the form of granules, meso and super-granules and probably giant cells, takes place through the process of inverse cascade. The essential ingredient for the formation of the large scale structures is the presence of helicity fluctuations in a turbulent fluid. In that case, a 3-D system supports helicity and helicity-helicity correlation as invariants, in addition to the total energy. This can be seen using the equations of mass and momentum conservation[9] as:

$$\frac{D\vec{U}}{Dt} = -\frac{1}{\rho}\ \vec{\nabla}p + \vec{F} \tag{1}$$

$$\frac{D\rho}{Dt} = -\rho(\vec{\nabla}\cdot\vec{U}) \tag{2}$$

from which, one can derive that:

$$\frac{D}{Dt}\left(\frac{\vec{\omega}}{\rho}\right) = \left(\frac{\vec{\omega}}{\rho}\cdot\vec{\nabla}\right)\vec{U} \tag{3}$$

$$\vec{\omega} = \vec{\nabla}\times\vec{U}$$

where, $\vec{U}$ and ρ are the fluid velocity and density respectively, p = p (ρ) the pressure, $\vec{F} = -\vec{\nabla}\phi$ the body force distribution and $\frac{D}{Dt} \equiv \frac{\partial}{\partial t} + (\vec{U}\cdot\vec{\nabla})$, the Lagrangian derivative. The fluid helicity H_s(t) in the volume V inside a closed surface S(t) moving with the fluid is defined as:

$$H_s(t) = \int_v h dV$$

where, h $= \vec{\omega} \cdot \vec{U}$

We can then show that

$$\frac{D}{Dt}(\frac{h}{\rho}) = (\frac{\vec{\omega} \cdot \vec{\nabla} B}{\rho}) = \frac{1}{\rho} \vec{\nabla} \cdot (\vec{\omega} B) \tag{5}$$

where B $= \frac{u^2}{2} - e - \psi$ is the Bernoulli function and e $\equiv \int \frac{dp}{\rho}$, the enthalpy per unit mass. Thus,

$$\begin{aligned} \frac{\partial}{\partial t} H_s(t) &= \int_V \frac{D}{Dt}(\frac{h}{\rho}) \rho dV \\ &= \int_V \vec{\nabla} \cdot (\vec{\omega} B) dV \\ &= \int_S (\hat{n} \cdot \vec{\omega}) dS = 0 \end{aligned} \tag{6}$$

for the normal component of the vorticity $\vec{\omega}$ being zero on the surface S. This condition remains valid for all times since the vortex lines are frozen in the fluid. Thus the helicity is a constant of the motion. The helicity conservation law can be written in the standard form as:

$$\frac{\partial h}{\partial t} + \vec{\nabla} \cdot \vec{F_h} = O \tag{7}$$

with the flux of helicity $\vec{F_h} = \vec{U} h - \vec{\omega} B$. If the vorticity is a stationary random function of position, as appropriate in the case of homogeneous turbulence, we can define the mean helicity

$$H = < \vec{U} \cdot \vec{\omega} > \tag{8}$$

where the angular brackets indicate either an ensemble or a space average. Again, H = constant under the three Kelvin conditions: (i) inviscid fluid, (ii) barotropic flow $p = p(\rho)$ and (iii) conservative body forces. If $H \neq O$, then the turbulence lacks reflectional symmetry and by a process analogous to the generation of large scale magnetic fields from small scale magnetic fields, large scale flows can be generated. If, on the other hand, in a situation where $< H > = O$, but the higher moments of the helicity distribution are constant[7] and influence the flow of energy, one can imagine the whole region to be divided into cells V_i bounded by surfaces S_i on each of which the condition $\vec{n} \cdot \vec{\omega} = O$ is satisfied, each h_i is then an inviscid invariant of the flow. In a large volume V containing many such cells V_i, the moments:

$$H_n = \frac{1}{V} \sum_i (h_i)^n \tag{9}$$

are then all inviscid invariants, such that H_1 is the mean helicity. If h_i are distributed in a random manner with equal probability of positive and negative values, then

$$\sum_i h_i \;\; \alpha \;\; V^{1/2} \tag{10}$$

and

$$H_1 = O$$

However, all even moments, specifically H_2 are finite and non-zero and although the mean helicity is zero, the fluctuations about the mean have constant variance. Thus, Levich and Tzvetkov[8] have used the 'I' invariant describing the correlations between helicity fluctuations to show that the conditions for the inverse cascade of energy obtain in a 3-D hydrodynamic system.

In this paper, we present the Kolmogorovic way of deriving the energy spectra in the inertial range of a turbulent flow possessing helicity fluctuations. We also illustrate the generation of a large scale flow from a small scale flow as an instability. Both of these approaches carry a lot of promise for modeling solar granulation. Several issues related to vorticity and helicity correlations in different parts of the energy spectrum lend themselves to observational verifications.

3. Inverse Cascade in 3-D, the Kolmogorovic Way

One learns from a 2-D system that it is the incompatibility in the inertial range spectra of the two invariants, that leads to the energy cascade towards large spatial scales.[3] So, if one had one more invariant in a 3-D system, perhaps an inverse cascade of energy could occur. Besides, the appearance of large scale structures in 3-D atmospheres of planets compels us to look for the possibility of an inverse cascade in 3-D.

Specifically, the observations of helical type of flow structures in circumstances varying from oceans to cloud complexes, brought to the fore the importance of helicity in turbulent fluids. The fundamental idea that needed to be appreciated was that the large helicity fluctuations always exist in a turbulent medium even if the average helicity vanishes. It was shown that the fluctuating topology of the vorticity field in turbulent flows can be characterized by the statistical helicity invariant I represented by conserved mean square helicity density[8]:

$$I = \int < h(\vec{x})h(\vec{x}+\vec{r}) > d^3r = \int I(\vec{r})d^3r \tag{11}$$

where

$$< h(\vec{x})h(\vec{x}+\vec{r}) >= \lim_{V\to\infty} \frac{1}{V} \int h(\vec{x})h(\vec{x}+\vec{r})d^3x$$
$$I(r=O) =< h(\vec{x})h(\vec{x}) >= \int I(k)d^3k \tag{12}$$

and

$$h = \vec{U} \cdot (\vec{\nabla} x \vec{U})$$

Here, $I(k)$ represents a density in wavenumber space of contributions to $< h(\vec{x})h(\vec{x}) >$ which is the mean square helicity density. Thus, energy E and I are the two invariants of a 3-D system. For a quasi-normal distribution of helicities, I can be written as:

$$I = A \int [W(k)]^2 dk \tag{13}$$

where A is a constant.

Using Kolmogorovic arguments for the I variant, one can determine the spectrum in the inertial range as: $(KU_k)K(W(k))^2$ = constant, so that

$$W(k) \alpha k^{-1} \tag{14}$$

and the total energy density is

$$E = \int W(k)dk \quad \alpha \quad log \frac{L(t)}{l} \tag{15}$$

Here, $L(t)$ is the transient largest scale excited at time t. One observes that the energy grows very slowly as the spatial scale $L(t)$ grows. So, practically, there is a little transfer of energy towards large scales. But, what happens is that as the correlation length of helicity fluctuations increases, the velocity and the vorticity become more and more aligned and consequently the nonlinear term $(\vec{U} \cdot \vec{\nabla})\vec{U}$ of the Navier-Stokes equation decreases and retards the flow of energy towards small scales. On the other hand, the growth of correlation length cannot go on indefinitely.

Especially if the medium is restricted in the vertical direction by gravity or buoyancy as is true of atmospheres of celestial objects, may it be a planet or a star. Under such circumstances, the correlation length continues to grow in the horizontal plane and the system becomes more and more anisotropic. In addition, since $I(k) = [W(k)]^2$ and $W(k)\alpha k^{-5/3}$, in analogy with the 2-D case one expects $I(k)$ to be dominant at small k while $W(k)$ will be larger at large k and this itself is a pointer to the inverse cascade of 'I'.

The growth of correlation length of helicity fluctuations in the horizontal direction makes the system anisotropic which can now be approximated to a quasi 2-D system. Here, the horizontal scale is much larger than the vertical scale and the vertical velocity U_z is much smaller than the horizontal velocities (Ux, Uy). This decouples the horizontal and vertical motions. U_z becomes independent of (x, y, z) and (U_x, U_y) independent of (z) leading to $\omega_{z,y} = (\vec{\nabla} x U)_{x,y} = O$. The invariant I becomes:

$$I = \int < U_z \omega_z)^2 > dxdydz$$
$$\simeq L_z < U_z^2 > U_k^2 \alpha U_k^2 = KW(k) \equiv I(k)k \tag{16}$$

and the constant decay rate of I gives $(KU_k)kW(k) =$ constant and $W(k)\alpha k^{-5/3}$.

Here L is the largest length scale in the horizontal plane. Thus $I(k)$ spectrum coincides with energy spectrum of 2-D turbulence $W(k)\alpha k^{-5/3}$, corresponding to the inverse cascade. One expects that an increasing fraction of energy is transferred to large spatial scales as the anisotropy in the system increases. This can go on until coriolis force begins to be effective. The length scale L_c where the nonlinear term of the Navier-Stokes equation becomes comparable to the coriolis force, can be determined from

$$(\vec{U} \cdot \vec{\nabla})\vec{U} = (\vec{U} \times \vec{\Omega}) \tag{17}$$

or

$$L_c \simeq \frac{U}{\Omega} \tag{18}$$

where $\vec{\Omega}$ is the angular velocity. Given sufficient energy, structures of size L_c must be form. At these large spatial scales, the system simulates 2-D behavior and entropy conservation begins to play its role. One may consider scales $L \geq L_c$ as a source of vorticity injection into the system. The entropy then cascades towards small scales with a power law spectrum given by

$$W(k)\alpha k^{-3} \text{ and E } \alpha \mathrm{L}^2 \tag{19}$$

Thus, there is a break in the energy spectrum as energy must cascade to large spatial scales as $L^{2/3}$ and to small scales as L^2. Therefore, the energy must accumulate at $L \sim L_c$ and pass on to the highest possible scales of the general circulation of the atmosphere. The complete energy spectrum of a hydrodynamic turbulent medium is given in Figure 1.

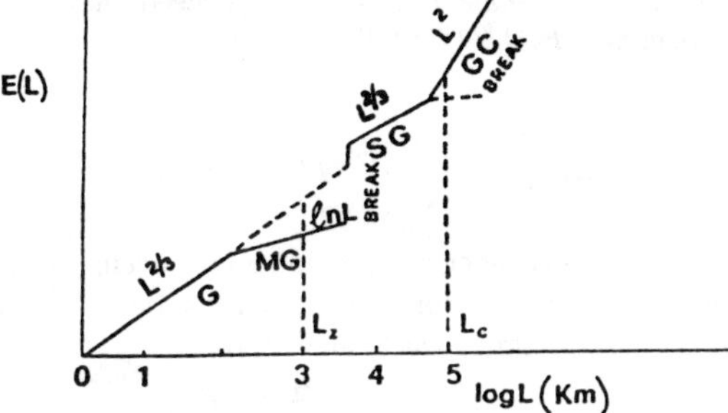

Figure 1: Turbulent energy spectrum. L_Z = scale of the first break due to anisotropy, L_C = scale of the second break due to the Coriolis force, G = granule, MG = mesogranule, SG = supergranule and GC = giant cell.

It is encouraging to know that the energy spectrum deduced from the observations (Fig. 2) does show the branches with $k^{-5/3}$ and $k^{-0.7}$ in good agreement with the predictions $k^{-5/3}(\equiv E(L)\alpha L^{2/3})$ and $k^{-1}(\equiv E(L)\alpha lnL)$ of our proposal. We may mention here that we have been able to model the flat rotation curves of galaxies using these ideas, in particular the log L branch of the energy spectrum without invoking the presence of dark matter,[11] in addition, we have proposed that the inverse cascade process may account for the large scale hierarchical structure of the universe.[5,6]

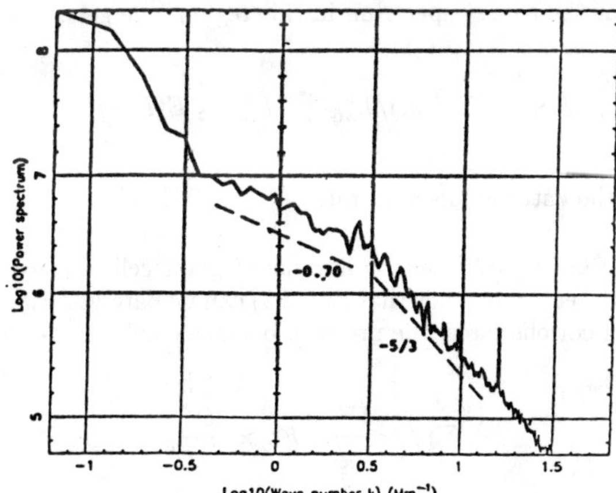

Figure 2: Power spectrum of the solar photospheric motions.[14]

4. Energetics

According to the picture[8] presented here, the energy in the larger structures has been inverse cascaded from the smaller structures, if so, then the energy density per unit gram $E(L)$ in the large scale L should not exceed that in the small scales(1). From the energy spectrum $E \alpha L^{2/3}$, it follows that,

$$E(L) = (\frac{E_o(l)}{\zeta})^{2/3} L^{2/3} \leq E_o(l) \tag{20}$$

where ζ is the time for which energy injection must occur. If we take ζ to be the lifetime of the larger structure then for $\zeta \sim 20$ hrs for supergranules and $E_o(l) \simeq (0.5)^2$ km^{2}sec^{-2} for energy density/gm in the granules, one gets,

$$L \sim L_{SG} \sim 36000 Km \tag{21}$$

which is the typical size of a supergranule. The size of a giant cell can be determined from Eq. (18) for $\Omega = (\frac{2\pi}{27})$ day^{-1} and assuming $U \sim 0.3$ km/sec for supergranular velocity (since they provide the stirring force for turbulence that organizes itself into giant cells) one gets:

$$L_c \simeq L_{GC} \sim 1.2 \times 10^5 Km \tag{22}$$

Again the energy content of giant cells should not exceed that of supergranules. Using Eq. (19) for the energy spectrum in this region, one gets:

$$E_{GC} = [E(L_{SG})/L_{SG}^2 \zeta]^{2/3} L_{GC}^2 \leq E(L_{SG}) \tag{23}$$

where $[\frac{E(L_{SG}}{L_{SG}^2 \zeta}]$ is the entropy injection rate.

For $L_{SG} \sim 10^5 km, \zeta = 30$ days = lifetime of giant cell, $L_{SG} \sim 30{,}000$ Km, and $E_{SG} \sim 90.3)^2$ km^2/sec^2. One finds that Eq. (23) can be barely satisfied. Furthermore, in the presence of coriolis force, the pressure balance condition becomes:

$$\frac{1}{\rho}/\nabla p/ \simeq \frac{U^2}{L_c} + Fc \simeq \frac{2U^2}{L_C} \tag{24}$$

in contrast to the case with no coriolis force where $\frac{1}{\rho}/\nabla p/ \simeq \frac{U^2}{L_c}$. Thus, one concludes that larger energy density is required to maintain structures at scale L_C. This may be the reason for their rare observability. The appearance of structures at L_C must be accompanied by a corresponding increase in the convective flux and therefore probably of total solar flux. Total solar luminosity changes of 0.1%, have been observed. If we attribute all of this 0.1% increase to the convective flux, Eq. (24) can be satisfied and structures of size Lc can get excited. The differential rotation of the Sun favors the formation of larger structures at the polar regions in comparison to the equatorial regions. This is further substantiated by the fact that the dominantly open magnetic fields in polar regions do not inhibit flow of convective flux. Thus, one may look for probable correlation between polar phenomena and solar luminosity enhancements with the appearance of giant cells. A very steep spectrum (Eq.19) practically forbids further organization of turbulence into structures larger than 1_c.

5. The Navier-Stokes Way

Analogous to the generation of large scale magnetic field by the α - effect, a large scale flow can develop from a small scale flow in a purely hydrodynamic system, essentially under the same conditions of anisotropy and/or parity violation. Following Sulem et al[13], we write the Navier-Stokes equations for the small scale as well as for the large scale flow including their interaction through the Reynold's stresses. We also restrict ourselves to the rotational and incompressible flows. The excitation of spatial scales much larger than the ones one starts with justifies the procedure of scale separation. The Navier-Stokes equations for a small scale incompressible flow $\vec{U}$ in the presence of a large scale flow $\vec{W}$ can be written as:

$$\frac{\partial U_i}{\partial t} + W_j \partial_j U_i + = \nu \nabla^2 U_i + f_i \tag{25}$$

$$\partial_j U_j = O \tag{26}$$

where ν is the kinematic velocity, f_i is the anisotropic force that produces the small scale flow $\vec{U}$ and may be periodic in space and time with space-time average $< f_i >= O$.

The Reynold's number is given by $R = \frac{l_o V_o}{\nu}$ for the small scale flow $\vec{U}$ characterized by the spatial scale l_o and a typical velocity V_o.

The Navier-Stokes equation for the large scale flow $\vec{W}$ can be written as:

$$\frac{\partial}{\partial t} W_i + \partial_j (W_i W_j + \alpha_{ij}) = \nu \nabla^2 W_i \tag{27}$$

$$\partial_j W_j = O, \quad \alpha_{ij} =< U_i U_j > \tag{28}$$

and the angular brackets represent the space-time or ensemble average. In order to calculate $\alpha_{i,j}$, we solve Eq. (25) under the assumptions: (i) the large scale flow $\vec{W}$

is independent of space-time (ii) the pressure force terms as well as the Reynold's stress term are small compared to the other terms. With these assumptions Eq. (25) describes a small scale forced flow $\vec{U}$ carried by the large scale flow $\vec{W}$. In order to simulate a helical and parity violating flow, the forcing term f_I, can be chosen to be:

$$\begin{aligned} f_x &= f_o Cos[y/l_o + t/t_o] \\ f_y &= f_o COS[x/l_o - t/t_o] \\ f_z &= \beta[fx + fy] \end{aligned} \tag{29}$$

where β is a measure of the anisotropy and $< fx >=< fy >= O$.

From Eq. (27), we find:

$$\begin{aligned} U_x &= \frac{V_o}{[1+(1+\frac{loWy}{\nu})^2]^{1/2}} Cos[\frac{y}{lo} + \frac{\nu t}{l_o^2} - y_o] \\ Sin^2 y_o &= \frac{(1+Wylo/\nu)^2}{I+(1+Wylo/\nu)^2}; f_o = \frac{\sqrt{2}\nu V_o}{l_o^2} \\ U_y &= \frac{V_o}{[1+(1\frac{loWx}{\nu})^2]^{1/2}} Cos[\frac{x}{lo} - \frac{\nu t}{l_o^2} - x_o] \\ Sin^2 x_o &= \frac{(1-Wxlo/\nu)^2}{1+(1-\frac{Wxlo}{\nu})^2} \\ U_z &= \beta(Ux + Uy) \end{aligned} \tag{30}$$

By substituting for $U'S$ in Eq. (28), one determines the components of the stress tensor α_{ij} as:

$$\begin{aligned} \alpha_{xx} &= < U_x^2 >= \frac{V_o^2}{1+(I+loWy/\nu)^2} \\ \alpha_{yy} &= < U_y^2 >= \frac{V_o^2}{1+(1-loWx/\nu)^2} \\ \alpha_{xy} &= \alpha_{yx} = 0; \alpha_{zz} = (\alpha_{xx} + \alpha_{yy})\beta^2 \\ \alpha_{xz} &= \alpha_{zx} = \beta\alpha_{xx}; \quad \alpha_{yz} = \beta\alpha_{yy} \end{aligned} \tag{31}$$

The linear characteristics of the large scale flow can be studied by retaining terms linear in Wx and Wy. The linearized large scale flow is governed by the equations:

$$\frac{\partial Wx}{\partial t} - \delta \frac{\partial Wy}{\partial Z} = \nu \frac{\partial^2}{\partial z^2} Wx \tag{32}$$

and

$$\frac{\partial Wy}{\partial t} + \delta \frac{\partial Wx}{\partial z} = \nu \frac{\partial^2}{\partial z^2} Wy \tag{33}$$

which admit solution of the form:

$$W_x + iWy \quad \alpha \quad e^{ikz} e^{t(\delta k - \nu k^2)} \tag{34}$$

where $\delta = \frac{\beta V_o^2 lo}{2\nu}$ and $k \sim \frac{1}{L}$ is the spatial scale of the large scale flow. It is clear that growing modes are excited for $\delta k > \nu K^2$ or $L > \frac{2lo}{BR^2}$. Thus the linear analysis points towards the excitation of large scale flow of dimension l at the expense of the small scale flow of dimension l_o.

Assuming the large scale flow W to be a function only of Z, one can attempt to look for steady state solutions of the nonlinear Eq. (27), which can be recast as:

$$\alpha_L \frac{\partial}{\partial z_L} [\frac{1}{1+P^2}] = \frac{\partial^2 Q}{\partial z_L^2} \tag{35}$$

and

$$\alpha_L \frac{\partial}{\partial z_L} [\frac{1}{1+Q^2}] = \frac{\partial^2 P}{\partial z_L^2} \tag{36}$$

where,

$$\alpha_L = \frac{L}{l_o} \beta R^2; \quad Z_L = Z/L$$

and

$$P = \frac{WyR}{V_o} + 1$$
$$Q = \frac{WxR}{V_o} - 1 \tag{37}$$

In order to understand a steady state with the flows on the largest possible scales, one takes the limit $\alpha_L \to \infty$. Integrating Eq. (35) and (36) once, one gets

$$\frac{\partial Q}{\partial z_L} = \alpha_L (\frac{1}{1+P^2} - A) \tag{38}$$

$$\frac{\partial P}{\partial Z_L} = \alpha_L(\frac{1}{1+Q^2} - A) \tag{39}$$

where

$$A = \langle\frac{1}{1+Q^2} = \langle\frac{1}{1+P^2}\rangle$$

and the $< >$ denotes the average over 2π period of Q and P in Z_L. Thus A represents the average Reynold's stress. Eliminating Z_L between Eq. (38) and (39), we find for Q and P

$$Q = \frac{1}{2A}[-S \pm \Delta^{1/2}], P = \frac{1}{2A}[S \pm \Delta^{1/2}] \tag{40}$$

$$\Delta = S^2 - 4A^2 + \frac{4AS}{tan(s+\epsilon)} \tag{41}$$

$$S = A(P - Q)$$

and ϵ is a constant of integration.

To determine the Z_L dependence of S, we take the difference of Eq. (38) and (39) to get:

$$\alpha_L dz_L = \pm\frac{1}{A}\frac{SdS}{\Delta^{1/2}Sin^2(S+\epsilon)} \tag{42}$$

After some manipulations and approximations,[13] find the asymptotic value of the flow speed W.

$$W = \sqrt{2}\frac{V_o}{R}[1 + (\frac{8L\beta R^2}{\pi lo})^{1/6}] \tag{43}$$

which could be larger or smaller than the small scale flow speed V_o.

The characteristic time ζ of the formation of the large scale flow is the inverse of the growth rate $(\delta k) \sim \delta/L$. If we further demand that $\zeta \leq$ life time ζ_F of the large scale flow L then:

the condition for the growth from Eq. (34) becomes

$$L > \frac{2\nu^2}{\beta V_o^2 l_o} = \frac{2lo}{\beta R^2} \tag{45}$$

Thus, combining Eq. (44) and (45) we find,

$$\frac{2lo}{\beta R^2} < L \leq \frac{\beta V_o R \zeta_F}{2} \tag{46}$$

In order to form a supergranular flow of size $L_S \sim$ 36,000 km, with a lifetime $\zeta_F \sim 20$ hours with a typical speed $W_s \sim 0.5$ km/sec (from Eq. 43) beginning with a granular flow of scale $l_o \sim 1000$ km with a typical speed $V_o \simeq 2$ km/sec, we need $\beta \sim 1/2$ and $R \simeq 24$, which is alright for the solar convection zone and the photosphere. The value of β indicates the degree of the required anisotropy in the force inducing the small scale flow.

For the mesogranular flow of size $L_M \simeq \frac{L_s}{5}$ and lifetime $\zeta_m \simeq \frac{\zeta_F}{6}$ with a typical speed $W_m \sim$ km/sec, we need $\beta R \sim 0.6$ and $R \sim 25$.

Flows on the giant cell scale $L \sim 2x10^5$ km can also be formed. But in this case, we must include the effect of the coriolis force which becomes influential at these large scales.[4] Besides, though the existence of the giant cells is perhaps not doubtful, more observations are required to establish their characteristics.

6. Some Testable Predictions

1. That the photospheric motions possess vorticity has been amply demonstrated. Not much attention has been paid to the helical nature of these flows. It will be worthwhile to measure helicity and helicity-helicity correlation of the photospheric motions, on various spatial scales. One could check if the helicity-helicity correlation length increases with height as the turbulent model predicts.
2. One may also check if the helicity increases towards larger spatial scales, since the helicity which represents the alignment of the velocity and vorticity retards the flow of energy towards small scales.
3. Much attention has been paid to the flow divergence, its relationship with helicity needs to be explored since the two are intimately connected through the equation.

$$\frac{\partial h}{\partial t} = \vec{\omega} \cdot \vec{\nabla} B - h(\vec{\nabla} \cdot \vec{U}) - \vec{U} \cdot \vec{\nabla} h$$

4. In steady state, the Bernoulli function B is related to the vorticity:

$$\vec{\nabla}B + \vec{\omega}x\vec{U} = O$$

and

$$\vec{U} \cdot \vec{\nabla}B = O$$

or B = constant on a streamline. If $\vec{\nabla}B = O$ then ω must also generally vanish in that region. Conversely, the existence of vorticity in the flow leads to the variation of B from streamline to streamline. Jumps in Bernoulli function are associated with the presence of shocks.

In conclusion, it may be a worthwhile exercise to incorporate some of the ideas related to turbulence while analyzing and interpreting the solar granulation data.

7. References

1. Antia, H.M., Chitre, S.M. and Narasimha, D.: 1984, *Ap. J.*, **282**, 574.
2. Bogart, Rs.S., Gierasch, P.J. and MacAuslan, J.M.: 1980, *Ap. J.*, **236**, 285.
3. Hasegawa, A.: 1985, *Advances in Physics*, **34**, 1.
4. Krishan, V.: 1991, *Mon. Not. R. Astro. Soc.*, **250**, 50.
5. Krishan, V. and Sivaram, C.: 1991, *Mon. Not. Roy. Astron. Soc.*, **250**, 157.
6. Krishan, V.: 1993, *Mon. Not. Roy. Astron. Soc.*, **250**, 257.
7. Levich, E. and Tsinober, A.: 1983, *Phys. Letters.*, **96A**, 292.
8. Levich, E. and Tzvetkov, E.: 1985, *Physics Reports*, **128**, No. 1, 1.
9. Moffatta, H.K. and Tsinober, A.: 1992, *Ann. Rev. Fluid Mech.*, **24**, 281.
10. Muller, R., *it el:* 1992, *Nature*, **356**, 322.
11. Prabhu, R.D. and Krishan, V.: 1994, *Ap. J.*, **428**, 483.
12. Schwartzschild, 1975, *Ap. J.*, **195**, 137.
13. Sulem, P.L., She, Z.S., Scholl, H., & Frisch, U.: 1989, *J. Fluid Mech.*, **205**, 341.
14. Zahn, J.P.: 1987, *Solar and Stellar Physics*, eds. Schroter and Schussler, 55 lecture notes in Physics, vol. 292; Springer Verlag.

MAGNETIC FIELD MEASUREMENTS IN THE INFRARED

SAMI K. SOLANKI
Institute of Astronomy, ETH-Zentrum
CH-8092 Zürich, Switzerland

ABSTRACT

Some results of infrared measurements of solar magnetic fields made and analyzed during the last few years are reviewed. These measurements have qualitatively affected our knowledge of small-scale magnetic features and of sunspots. For example, infrared observations have for the first time allowed us to quantitatively probe magnetic fields in the deepest observable layers, in the upper photosphere and the upper chromosphere. They have also provided us with more accurate field strengths, have made intrinsically weaker fields accessible to quantitative measurement and have enabled us to distinguish between different theoretical models and to identify hitherto unseen physical processes within magnetic features.

1. Introduction

The infrared has long been known to be an interesting spectral range for magnetometry, but not until the 1990s has it begun to realize its potential. A spate of publications by different groups has highlighted many of the possibilities of using spectral lines between 1 and 12 μm to measure solar magnetic fields. A major advantage of the infrared is that the Zeeman sensitivity of a spectral line increases roughly linearly with wavelength. Since for typical small-scale solar features the most Zeeman sensitive spectral lines in the visible are almost, but not quite completely split it is sufficient to go to a wavelength of 1.5 μm to observe complete splitting, which greatly simplifies the measurement of field strengths.

Another strength of the infrared is that it contains Zeeman sensitive lines with widely different properties, in particular with widely different formation heights. Thus, the Fe I 1.5648 μm ($g = 3$) line samples the lowest observable levels, the Mg I 12.32 μm line the upper photosphere and the He I 1.083 μm line the upper chromosphere. Recent efforts by Chang, Kuhn, Penn and coworkers (these proceedings) are aimed at extending this height range into the corona. In addition to the above lines other interesting lines are also present, but have been little used. For example, the Ti I 2.2310 μm line may well be the best umbral magnetic diagnostic currently available. More details on advantages and disadvantages of particular spectral lines for magnetometry are given by Solanki[1] and Rüedi et al.[2].

A number of reviews on the status in early 1992 of magnetic field measurements using infrared lines are to be found in the proceedings edited by Rabin et al.[3],[4],[5],[1], cf. the review by Solanki (1993)[6].

Here I first review magnetic elements followed by sunspots. Although visible observations have in many cases arrived at the same or similar results as the infrared, I concentrate on the latter in this review. Finally, I urge the reader to peruse the other papers in these proceedings on a related topic, since these often describe the most up-to-date work.

2. Magnetic Field of Magnetic Elements

Observations at 1.56 μm have provided the most convincing evidence so far that the small-scale magnetic features forming solar active region plages have kG fields. Figure 1, illustrates the difference between the 'field strength' obtained from a typical magnetogram (labelled 'NSO magnetograph') and directly from the splitting of $\lambda 1.5648$ μm. This difference is caused by the fact that the magnetic elements are not spatially resolved in the magnetogram, so that 'field strengths' derived therefrom, $\langle B \rangle$, are spatially averaged over the resolution element, while the $\lambda 1.5648$ μm splitting gives the intrinsic field strength, B. The two quantities are connected via α, the magnetic filling factor, which denotes the fraction of the solar surface (at a given height) covered by magnetic field: $\langle B \rangle = \alpha B$. Rabin[7,8] has determined B at a large number of spatial positions within two active regions and finds that B at the vast majority of them lies in the range 1000–1700 G in good agreement with optical results[76]. Within these limits Rabin sees field-strength variations on spatial scales of 5–10 arc s.

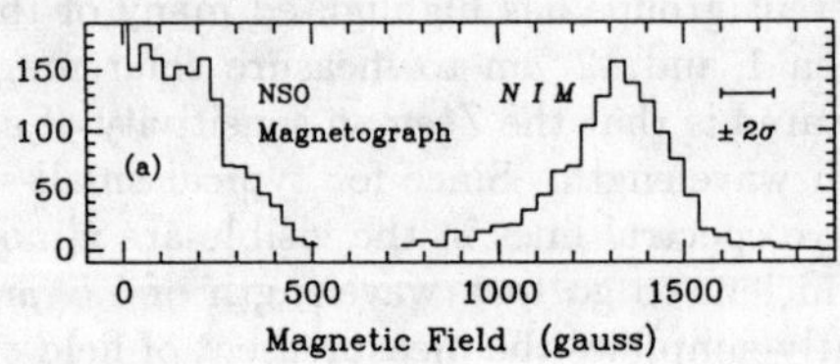

Fig. 1 Histograms of true field strength, as measured by the Near Infrared Magnetograph (NIM), and of the apparent field strength measured by the synoptic magnetograph on Kitt Peak (from Rabin[7]).

Rabin's results are supported and extended by the work of Rüedi et al.[9]. For a smaller data set they determined B and the plasma β within magnetic elements at a given geometrical height using basic flux-tube physics. The plasma $\beta(z) = 8\pi\, p(z)/B^2(z)$, where p is the gas pressure inside the magnetic element, is a measure of the ratio of the energy density in the gas to that in the magnetic field. A $\beta < 1$ implies that locally the magnetic field dominates energetically over the gas, while for $\beta > 1$ the opposite is the case.

Figure 2 shows $B(z = 0)$ vs. $\langle B \rangle$, where the height $z = 0$ corresponds to unit continuum optical depth at 500 nm in the quiet sun. The points above the dashed line have $\beta(z = 0) < 1$. Their average $\beta(z = 0) \approx 0.3$ and they carry roughly 90% of the magnetic flux in the analyzed data set. These kG elements are thus heavily

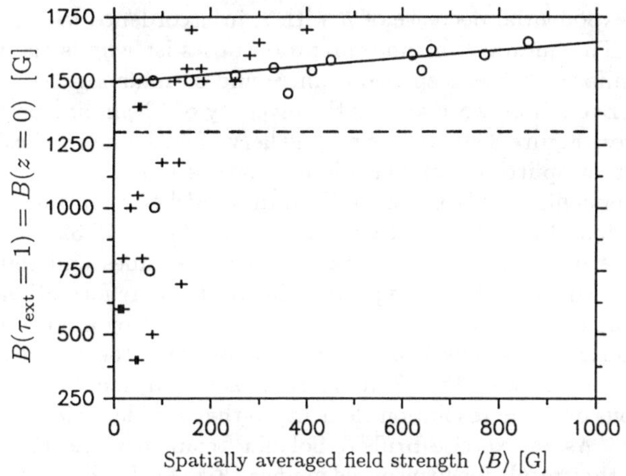

Fig. 2 The intrinsic field strength at the quiet-sun continuum forming level, $B(z = 0)$, vs. $\langle B \rangle$, the field strength averaged over the spatial resolution element. If more than one magnetic component was present in the resolution element, then $B(z = 0)$ for each component is plotted separately and denoted by a '+' [9].

evacuated with internal dynamics that are shaped by the magnetic field, although they may still be coupled to external gas motions. The spread in $B(z = 0)$ of these features is much smaller than the B obtained at the height of line formation, corresponding to a fixed continuum optical depth, since for small plasma β even a slight variation in $B(z = 0)$ produces a large change in gas pressure, which in turn affects the depth at which the spectral line samples B [10]. Since the field strength increases roughly exponentially with depth the sampled field strength varies significantly.

The weak-field, i.e. $\beta > 1$ elements below the dashed line represent a previously unseen component of the field. Note that all of these features have small fluxes. Recent results by various groups suggest that in the quiet sun weak fields may play an even more important role[77,78].

Theoretical calculations of the convective collapse of magnetic features predict a convectively stable end state whose field strength agrees very well with the measured value of the strong-field magnetic elements[11]. According to theory the weak-field features are in a convectively uncollapsed state and close to being in equipartition with the convective flows. The expected infrared observational signature of a convective collapse has been described by Steiner[12], who also gives an overview of some of the recent advances in the theoretical modelling of small-scale magnetic features.

If we now move from the 1.56 μm line to the Mg I emission line at 12.32 μm, i.e. from the bottom of the photosphere to near its top, a substantial decrease in the field strength of plage magnetic elements is seen (Chang[13] gives a brief introduction to this line in his review). $B \approx$ 200–500 G is observed[14,15,16], in agreement with the expected exponential decrease of B with z, in accordance with horizontal pressure balance. The question this interpretation raises is: why is the magnetic field strength so uniform in the deep photosphere and so inhomogeneous in the upper photosphere? Even more worrisome is the diversity of 12 μm line profile shapes observed in plages. Figure 3 shows a rogues' gallery of such line profiles[16]. They show not only different splittings, but also indications of multiple spatially unresolved magnetic components. Bruls & Solanki[17] synthesized 12.32 μm spectra in flux-tube models. They found that for sufficiently large $\alpha(z=0)$ the 12.32 μm line is formed at or above the merging height of neighbouring flux tubes and thus feels a field strength of $\alpha(z=0)\cdot B(z=0)$. Consequently, due to the large formation height of this line, its splitting is also sensitive to $\alpha(z=0)$, which varies by a much larger amount than the intrinsic field strength $B(z=0)$. This explains the broad range of plage field strengths measured by the 12 μm lines. In addition, the filling factor need not be homogeneous over the resolution element, so that a wide variety of line shapes is to be expected. As an exercise Bruls & Solanki[17] compared the 12.32 μm Stokes I and V profiles that result for two simple models (Fig. 4). In the first case flux tubes are evenly spaced over the whole spatial resolution element (solid profile), in the second case the flux tubes are uniformly distributed over only half the resolution element, with twice the $\alpha(z=0)$ as in the first case, while the other half is field-free. $B(z=0)$ and $\langle\alpha(z=0)\rangle$ are the same in both cases, where $\langle\ \rangle$ signifies averaging over the resolution element. Obviously, it is easily possible to produce a wide variety of line shapes in this manner. Conversely, this modelling implies that the observed 12 μm line shape gives us some information on the distribution of the magnetic field below the spatial resolution, a property unique to these lines.

In general, magnetic elements do not show signs of significant stationary flows when observed at 1.5–1.8 μm and averaged spatially and temporally[18]. This is in accordance with optical observations and theoretical considerations (mass conservation). Along small loops with a field strength imbalance between their footpoints, however, Meyer & Schmidt[19] proposed the presence of a siphon flow, with material flowing from the footpoint with smaller field strength to the other. Due to the requirements of total horizontal pressure balance the gas pressure is larger in the footpoint with lower field strength. It is this excess gas pressure that drives the siphon flow. Considerable theoretical work has been done on siphon flows[20,21,22,23,24,25,26,27,28]. Since the field-strength difference between the footpoints required to power a flow need not be particularly large, the high Zeeman sensitivity of infrared lines is of major advantage for siphon flow detection. Rüedi et al.[29], observing Stokes V profiles of 1.56 μm lines along a slit lying perpendicularly across the polarity inversion line of an active region plage, saw that the magnetic polarity with the smaller field strength exhibited an upflow, while the opposite polarity possessed a larger field strength and a downflow. This is exactly the observational signature expected for a siphon flow[26].

Subsequently, Degenhardt et al.[30] have studied 1.56 μm line profiles resulting from a set of MHD siphon-flow models differing in the maximum flow speed. As this increases so does the (consistently calculated) difference in B at the two footpoints.

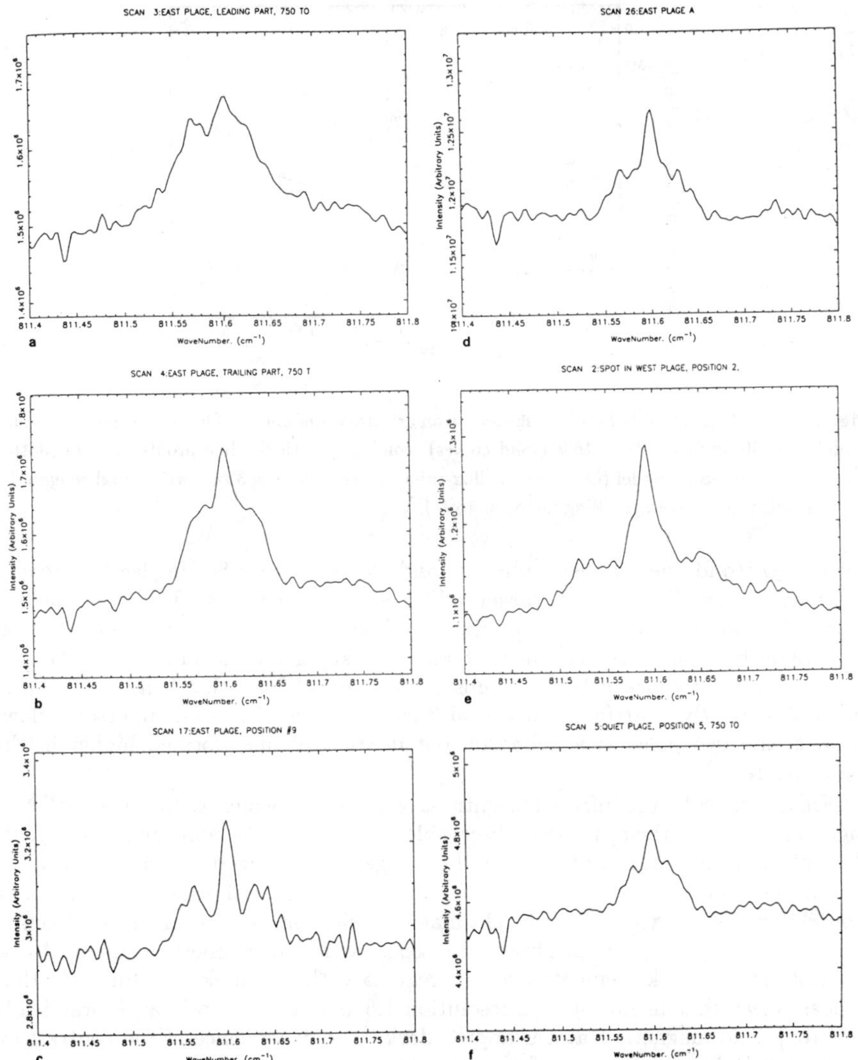

Fig. 3 Profiles of the Mg I 12.32 μm line in different plage and network regions. Note the variety of line profile shapes (from Zirin & Popp[16]).

Sufficiently strong upflows become supersonic near the apex of the loop and shock back to subsonic velocities in the downflowing leg.

Figure 5 shows simulated Stokes I profiles of λ 1.5648 μm formed along the up-

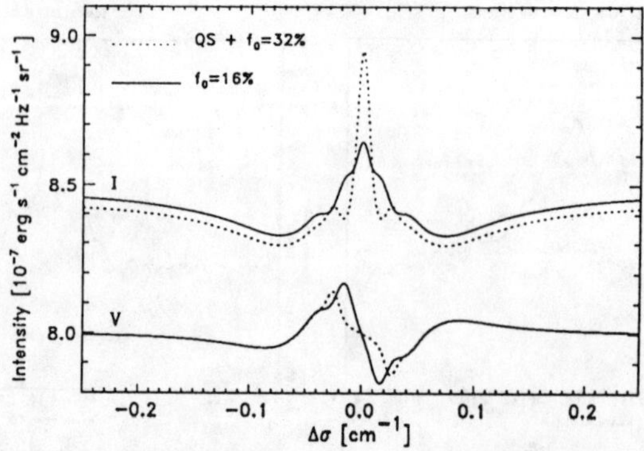

Fig. 4 Mg I 12.32 μm Stokes I and V (shifted by an arbitrary amount) profiles for a typical flux-tube model with filling factor $f_0 = 16\%$ (solid curves), compared with the line profile average (dotted curves) of a quiet-Sun model (QS) and the flux-tube model with $f_0 = 32\%$, with equal weights for both models (i.e. an average filling factor of 16%)[17].

flowing leg (solid line, smallest splitting) and along the downflowing leg for increasingly faster flows (increasingly larger splitting, increasing redshift). A comparison of the simulated with observed profiles led Degenhardt et al.[30] to conclude that the siphon flow observed by Rüedi et al.[29] was supersonic in and above the mid photosphere, but shocked back to subsonic above the formation height of λ1.5648 μm. Subsequently, Martínez Pillet et al.[31] saw direct evidence for supersonic flows between the two poles of a delta sunspot in spectral lines formed higher in the photosphere.

Finally, consider the infrared brightness of magnetic elements. Initial work[32,33,34] suggested that at their deepest observable layers (1.6 μm continuum) magnetic elements and their immediate surroundings together (averaged over a few arc s) are dark if the magnetic flux is sufficiently large. At higher spatial resolution Darvann & Koutchmy[35] found evidence for bright magnetic elements at 1.6 μm. Most recently Lin[36], from very sensitive brightness images, has found magnetic elements to be associated with dark elements even in regions with low magnetic flux. It thus appears clear that in low spatial resolution 1.6 μm continuum images practically all strong-field magnetic features appear dark. The dependence of the contrast on spatial resolution should repay further study.

3. The Magnetic Field of Sunspots

Sunspots, the oldest known solar magnetic features, hold a fascination all their own and still pose numerous puzzles. Nevertheless, much has recently been learnt about them, not least from infrared studies.

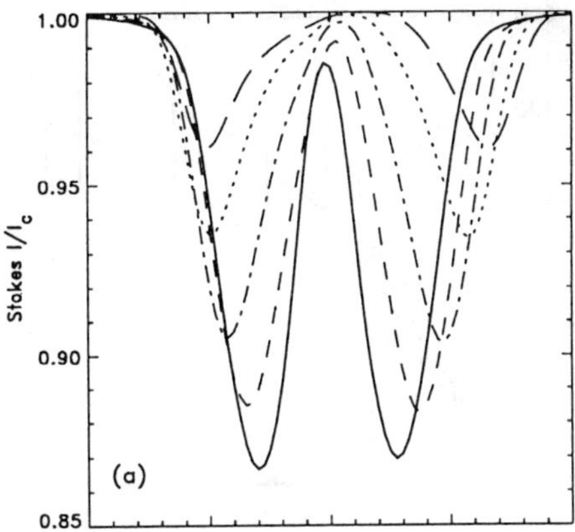

Fig. 5 Synthetic Stokes I profiles (normalized to the continuum intensity) emerging from a set of siphon flow models. The solid line represents the profiles emerging from the upstream leg, which are identical for all models. The other curves represent the downstream leg: subcritical downstream flow (dashed), shocked flows with shock positions at $z = 450$ km (dash-dotted), $z = 400$ km (dotted) and $z = 300$ km (long dashes). Note the strong increase of the line splitting with decreasing shock height[30].

The radial dependence of the magnetic vector of symmetric sunspots has been the subject of intensive recent investigation, mainly by the ASP[37,38,39] and other groups in the optical[40,41] and by Deming et al.[15], McPherson et al.[42], Solanki et al.[43] and Hewagama et al.[44] in the infrared. The last authors presented the first observations of the full magnetic vector in the infrared (at 12.3 μm).

In general the infrared provides good agreement with optical results. The largest B lies in the range 2100–3600 G and scales linearly with umbral radius[45]. It drops to 700–1000 G at the sunspot boundary. The field inclination to the vertical increases with radial distance, but the field is still inclined by 10–20° at the outer edge of the penumbra. Outside the white-light outline of the sunspot the field becomes almost horizontal and forms an extensive magnetic canopy, i.e. a layer of nearly horizontal field overlying a largely field-free atmosphere[43,46,47].

A low-lying magnetic canopy beyond the sunspot boundary is suggested by, for example, the much larger magnetic signal seen in the upper photospheric 12.32 μm line[44] than by the lower photospheric 1.56 μm line[43,48,49]. This is in contrast to the penumbra which both lines indicate to be completely magnetically permeated at all photospheric layers[43,44]. In addition to the height of hte canopy base (see Fig. 6), infrared lines also provide the field strength (down to values of only 200–300 G) and indicate interactions between canopy and underlying magnetic elements.

Now, the presence of a magnetic canopy implies that very little magnetic flux returns to the solar interior immediately outside the sunspot. In particular, infrared

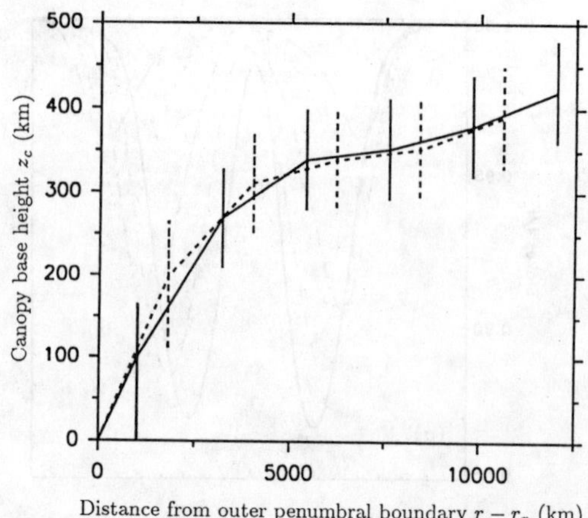

Fig. 6 Base height Z_c of the superpenumbral magnetic canopy vs. distance from the outer boundary of the penumbra, $r - r_p$. Solid and dashed curves represent Z_c derived from two different slices through the same large, symmetric sunspot. Error bars are estimates of systematic errors[43].

observations rule out the return-flux model of sunspots[50,51,52], although a small amount of return flux may still escape detection.

Infrared observations have also helped to distinguish between models of sunspots with shallow or deep penumbrae. In a shallow penumbra, the outermost field line outlines the $\tau = 1$ level, i.e. all the magnetic flux of the spot emerges within the umbra. On the other hand a significant fraction of the sunspot's magnetic flux emerges within a deep penumbra. From the measured $B(r)$ and $\gamma(r)$ dependences the fraction of the magnetic flux emerging in the penumbra can be determined (γ is the inclination of the magnetic vector to the vertical). In agreement with optical data 1.56 μm spectra indicate that 60–70% of the total magnetic flux of mature large sunspots emerges in the penumbra[53]. The 12 μm data of Hewagama et al.[44] give a similar number. Sunspot penumbrae are therefore deep.

The availability of magnetically sensitive lines formed at different levels makes the infrared an ideal wavelength range to determine vertical field-strength gradients $\Delta B/\Delta z$ in sunspots, although up to now most such investigations have compared two optical lines with each other. Within the umbra $\Delta B/\Delta z$ decreases from 1–2 G/km in the photosphere (obtained by comparing two photospheric optical lines[54,55] to 0.35–0.7 G/km averaged over the photosphere and chromosphere (obtained by comparing a photospheric with an upper chromospheric, transition zone or coronal base diagnostic [56,57,58,59,60]. In photospheric layers $\Delta B/\Delta z \approx 3$ G/km in the inner penumbra, and decreases to 0.7 G/km in the outer penumbra[61,49] (see Fig. 7). A combination of photospheric with upper chromospheric, transition zone or coronal base B-field, however, gives $\Delta B/\Delta z$ values of only 0.1–0.3 G/km in the outer penumbra[58,62,60].

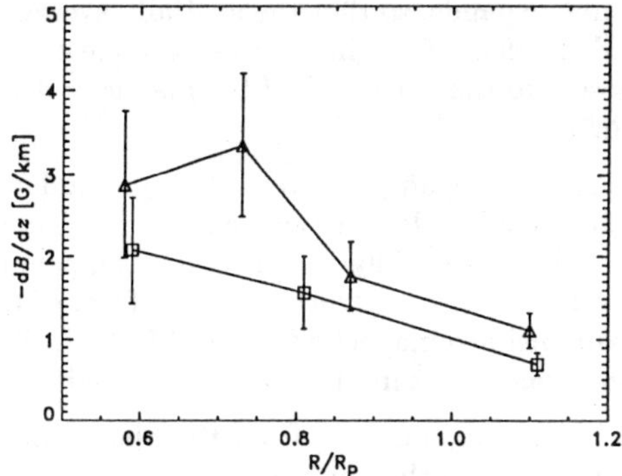

Fig. 7 Magnetic field strength gradients in the penumbra of a sunspot observed by Hewagama et al.[44]. Points on different sides of the sunspot are marked by triangles and squares, respectively[49].

The qualitative essence of these observations is that $\mathrm{d}B/\mathrm{d}z$ decreases with height and with radial distance from the sunspot axis, both of which dependences are in qualitative agreement with magnetohydrostatic (MHS) models of sunspots. There are, however, quantitative differences. The measured *photospheric* $\mathrm{d}B/\mathrm{d}z$ values are a factor of 2–4 times larger than predicted by MHS models (0.4–0.8 G/km in the umbral photosphere[50,51,63,67,73]).

There is also another technique which has been used to determine vertical field strength gradients. From photospheric vector magnetograms it is possible to obtain the horizontal gradients of the horizontal magnetic components, $\mathrm{d}B_x/\mathrm{d}x$ and $\mathrm{d}B_y/\mathrm{d}y$. Then $\mathrm{div}\,\mathbf{B} = 0$ should give us $\mathrm{d}B_z/\mathrm{d}z$ at the photospheric level. Near the centres of sunspots, where $B_z \approx B$, recent values for $\mathrm{d}B/\mathrm{d}z$ derived in this manner lie in the range 0.1–0.4 G/km[58,64], and are consequently almost an order of magnitude lower than values obtained directly from spectral lines formed at different heights. Maxwell's equation $\mathrm{div}\,\mathbf{B} = 0$ appears to be violated. How can this seemingly fundamental problem be resolved?

Consider now the two differnt types of observations whose results are being compared. The vector magnetograms typically have a spatial resolution of a couple of arc s. We are thus not simply deriving $\mathrm{d}B_z/\mathrm{d}z$ from them, but rather

$$\frac{\int\int \left(\frac{\mathrm{d}B_z}{\mathrm{d}z}\right)\mathrm{d}x\mathrm{d}y}{\int\int \mathrm{d}x\mathrm{d}y} = \frac{\mathrm{d}}{\mathrm{d}z}\left(\frac{\int\int B_z\,\mathrm{d}x\mathrm{d}y}{\int\int \mathrm{d}x\mathrm{d}y}\right) = \frac{\mathrm{d}}{\mathrm{d}z}\langle B_z\rangle\ , \tag{1}$$

where the integration is over the spatial resolution element. Now,

$$\frac{\mathrm{d}}{\mathrm{d}z}\langle B_z\rangle = \frac{\mathrm{d}}{\mathrm{d}z}B_z \tag{2}$$

if the magnetic field is homogeneous over the scale of the spatial resolution element. Otherwise Eq.(2) will in general not be fulfilled. As an example consider a small spatially unresolved flux tube. As we go up in height the true field strength B drops rapidly and the flux tube expands, so that the spatially averaged field strength $\langle B \rangle$ remains practically independent of height. In this example $d\langle B_z \rangle/dz$ and $d\langle B \rangle/dz$ are related much more closely to the derivative of the magnetic *flux* than to that of the intrinsic field strength.

In sunspots the large discrepancy between $d\langle B_z \rangle/dz$ and dB_z/dz suggests the presence of considerable small-scale structure and in particular that the amount of small-scale structure changes rapidly with height. The penumbra is well known to possess considerable magnetic fine structure in the form of uncombed fields, i.e. interlaced horizontal and inclined magnetic features[65,66,41,68]. There is also evidence that the amount of fine-scale structure decreases with height[60].

The most consistent picture emerging from the observations is of an inclined magnetic component threaded, in the photospheric layers, by horizontal flux tubes, whose number density decreases with height, so that at greater height the inclined component has expanded to fill the available space. Consequently, the *intrinsic* field strength of the inclined component (the only one contributing to B_z) decreases rapidly with height ($dB_z/dz \ll 0$), while the spatially averaged field strength of this component, like the flux, remains almost independent of height.

In the umbra no other evidence for significant magnetic inhomogeneity exists as yet, although the filamentary structure seen by Livingston[69] may provide an indication for its existence. The difference between dB_z/dz and $d\langle B_z \rangle/dz$ suggests that it might be worthwhile to look for direct evidence of magnetic inhomogeneity in sunspot umbrae as well, although it might be more difficult to detect than in penumbrae.

The large Zeeman sensitivity of infrared lines allowed for the first time the relationship between temperature and field strength to be studied throughout sunspots (visible work was restricted to the umbra[70,71,72]. Figure 8 shows B vs. I_c, the continuum intensity at 1.56 μm, for six sunspots[45]. Three regimes are visible. On the right is a regime in which I_c changes little, but B varies by a large amount. It corresponds to the penumbra. In the middle is a regime exhibiting an almost constant B and a rapidly varying I_c, which corresponds to the umbra-penumbra boundary. On the left is the umbral regime in which B and I_c are closely correlated. Remarkable is the seeming universality of the relationship between B and I_c (the six spots cover almost the whole observed range of sizes and maximum field strengths).

If we assume that the sunspot is in MHS equilibrium then such a relationship allows us to estimate the Wilson depression Z_W[72]. The infrared has the major advantage that it enables us to estimate Z_W throughout the sunspot. The Z_W derived assuming no magnetic curvature forces within the sunspot remains relatively constant at 100-150 km in the penumbra, but increases rapidly to 400-450 km at the umbral boundary[74]. Although this assumption can affect the results, the sudden change in Z_W at the umbral boundary is robust and partly reflects the change in continuum opacity due to the change in temperature at the umbral boundary. Wilson & Cannon[75] find roughly similar results from traditional Wilson-effect observations,

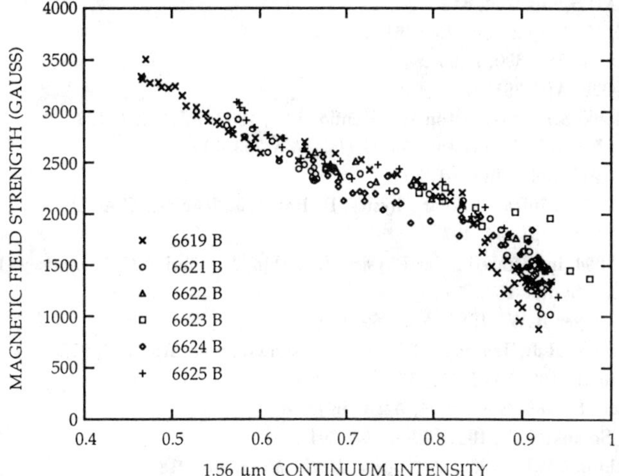

Fig. 8 Magnetic field strength vs. continuum intensity for six sunspots, each represented by a different symbol (from Kopp & Rabin[45]).

which are completely independent of the above technique.

4. Conclusions

I hope that this far from exhaustive review has provided a flavour of some of the many, often unexpected insights into solar magnetism that the infrared has provided us. The potential for more discoveries is immense and we can probably expect further surprises. Promising new developments are high spatial resolution and vector polarimetric observations, observations of lines formed in the upper chromosphere and the corona, observations with high polarimetric sensitivity and accuracy and the use of lines with unconventional properties, e.g. umbral lines.

References

1 Solanki S.K., 1994, in Infrared Solar Physics, D. Rabin, J. Jefferies, C.A. Lindsey (Eds.), Kluwer, Dordrecht, IAU Symp. 154, 393

2 Rüedi I., Solanki S.K., Livingston W., Harvey J.W., 1995, A&AS submitted

3 Rabin D., Jefferies J., Lindsey C.A., 1994, Infrared Solar Physics, Kluwer, Dordrecht, IAU Symp. 154

4 Deming D., Hewagama T., Jennings D., McCabe G., Wiedemann G., 1994, in Infrared Solar Physics, D. Rabin, J. Jefferies, C.A. Lindsey (Eds.), Kluwer, Dordrecht, IAU Symp. 154, 379

5 Rabin D., 1994, in Infrared Solar Physics, D. Rabin, J. Jefferies, C.A. Lindsey (Eds.), Kluwer,

Dordrecht, IAU Symp. 154, 449
6 Solanki S.K., 1993, Space Sci. Rev. 61, 1
7 Rabin D., 1992a, ApJ 390, L103
8 Rabin D., 1992b, ApJ 391, 832
9 Rüedi I., Solanki S.K., Livingston W., Stenflo, J.O., 1992, A&A 263, 323
10 Solanki S.K., Rüedi I., Livingston W., 1992a, A&A 263, 312
11 Spruit H.C., 1979, Sol. Phys. 61, 363
12 Steiner O., 1994, in Infrared Solar Physics, D. Rabin, J. Jefferies, C.A. Lindsey (Eds.), Kluwer, Dordrecht, IAU Symp. 154, 407
13 Chang E.S., 1994, in Infrared Solar Physics, D. Rabin, J. Jefferies, C.A. Lindsey (Eds.), Kluwer, Dordrecht, IAU Symp. 154, 297
14 Brault J.W., Noyes R.W., 1983, ApJ 269, L61
15 Deming D., Boyle R.J., Jennings D.E., Wiedemann G., 1988, ApJ 333, 978
16 Zirin H., Popp B., 1989, ApJ 340, 571
17 Bruls J.H.M.J., Solanki S.K., 1994, A&A in press
18 Muglach K., Solanki S.K., 1992, A&A 263, 301
19 Meyer F., Schmidt H.U., 1968, *Z. Angew. Math. Mech.* **48**, 218
20 Cargill P.J., Priest E.R., 1980, Sol. Phys. 65, 251
21 Cargill P.J., Priest E.R., 1982, Geophys. Astrophys. Fluid Dyn. 20, 227
22 Thomas J.H., 1988, ApJ 333, 407
23 Montesinos B., Thomas J.H., 1989, ApJ 337, 977
24 Montesinos B., Thomas J.H., 1993, ApJ 402, 314
25 Thomas J.H., Montesinos B., 1990, ApJ 359, 550
26 Thomas J.H., Montesinos B., 1991, ApJ 375, 404
27 Degenhardt D., 1989, A&A 222, 297
28 Degenhardt D., 1991, A&A 248, 637
29 Rüedi I., Solanki S.K., Rabin D., 1992b, A&A 261, L21
30 Degenhardt D., Solanki S.K., Montesinos B., Thomas J.H., 1993, A&A 279, L29
31 Martínez Pillet V., Lites B.W., Skumanich A., Degenhardt D., 1994, ApJ (Letters) in press
32 Foukal P., Little R., Mooney J., 1989, ApJ 336, L33
33 Foukal P., Little R., Graves J., Rabin D., Lynch D., 1990, ApJ 353, 712
34 Moran T., Foukal P., Rabin D., 1992, Sol. Phys. 142, 35
35 Darvann T.A., Koutchmy S., 1994, in Infrared Solar Physics, D. Rabin, J. Jefferies, C.A. Lindsey (Eds.), Kluwer, Dordrecht, IAU Symp. 154, 483
36 Lin H. 1995, These proceedings
37 Lites B.W., Skumanich A., 1990, ApJ 348, 747
38 Lites B.W., Elmore D.F., Seagraves P., Skumanich A., 1993, ApJ 418, 928
39 Skumanich A., Lites B.W., Martínez Pillet V., 1994, in Solar Surface Magnetism, R.J. Rutten, C.J. Schrijver (Eds.), Kluwer, Dordrecht, p. 99
40 Adam M.G., 1990, Sol. Phys. 125, 37
41 Title A.M., Frank Z.A., Shine R.A., Tarbell T.D., Topka K.P., Scharmer G., Schmidt W., 1993, ApJ 403, 780
42 McPherson M.R., Lin H., Kuhn J.R., 1992, Sol. Phys. 139, 255
43 Solanki S.K., Rüedi I., Livingston W., 1992b, A&A 263, 339

44 Hewagama T., Deming D., Jennings D.E., Osherovich V., Wiedemann G., Zipoy D., Mickey D.L., Garcia H., 1993, ApJS 86, 313
45 Kopp G., Rabin D., 1992, Sol. Phys. 141, 253
46 Giovanelli R.G., 1980, Sol. Phys. 68, 49
47 Giovanelli R.G., Jones H.P., 1982, Sol. Phys. 79, 267
48 Solanki S.K., Montavon C.A.P., Livingston W., 1994, A&A 283, 221
49 Bruls J.H.M.J., Solanki S.K., Carlsson M., Rutten R.J., 1994, A&A in press
50 Osherovich V.A., 1982, Sol. Phys. 77, 63
51 Flå T., Osherovich V.A., Skumanich A., 1982, ApJ 261, 700
52 Osherovich V.A., Garcia H.A., 1989, ApJ 336, 468
53 Solanki S.K., Schmidt H.U., 1993, A&A 267, 287
54 Wittmann A.D., 1974, Sol. Phys. 36, 29
55 Pahlke K.-D., 1988, Ph.D. Thesis, University of Göttingen
56 Abdussamatov H.I., 1971, Sol. Phys. 16, 384
57 Henze W., Jr., et al., 1982, Sol. Phys. 81, 231
58 Hagyard M.J., et al., 1983, Sol. Phys. 84, 13
59 Lee J.W., Gary D.E., Hurford G.J., 1993a, Sol. Phys. 144, 45
60 Rüedi I., Solanki S.K., Livingston W., 1994, A&A in press
61 Balthasar H., Schmidt W., 1993, A&A 279, 243
62 Lee J.W., Hurford G.J., Gary D.E., 1993b, Sol. Phys. 144, 349
63 Osherovich V.A., 1984, Sol. Phys. 90, 31
64 Hofmann, A., Rendtel, J., 1989, Astron. Nachr. 310, 61
65 Degenhardt D., Wiehr E., 1991, A&A 252, 821
66 Schmidt W., Hofmann A., Balthasar H., Tarbell T.D., Frank Z.A., 1992, A&A 264, L27
67 Osherovich V.A., 1980, Sol. Phys. 68, 297
68 Solanki S.K., Montavon C.A.P.: 1993, A&A 275, 283
69 Livingston W., 1991, Nat. 350, 45
70 Dicke R.H., 1970, ApJ 159, 25
71 Maltby P., 1977, Sol. Phys. 55, 335
72 Martínez Pillet V., Vázquez M., 1993, A&A 270, 494
73 Schlüter A., Temesvary S., 1958, in Electromagnetic Phenomena in Cosmical physics, B. Lehnert (Ed.), Reidel, Dordrecht, IAU Symp. 6, 263
74 Solanki S.K., Walther U., Livingston W., 1993, A&A 277, 639
75 Wilson P.R., Cannon C.J., 1968, Sol. Phys. 4, 3
76 Stenflo J.O., 1973, Sol. Phys. 32, 41
77 Keller C.U., Deubner F.-L., Egger U., Fleck B., Povel P., 1994, A&A 286, 626
78 Lin H., 1994, ApJ in press

44. Fleck B., Tarbell T., Jefferies P. [illegible], Dannenbaum [illegible] 1991, [illegible] 2 [illegible]
45. Bastian Tims T., Bogdan R., Jefferies D.M., Osterovich V., Wedemeyer G., Tippo I.D., Mitchell D.L., Fleck B. 199[illegible], ApJ [illegible] 13
45. [illegible] C. [illegible] 1992, Sol. Phys. 141, 223
46. [illegible] 1980, Sol. Phys. 68 [illegible]
47. [illegible] R.G., Jones H.P. 1981, Sol. Phys. 70, 2[illegible]
48. Solanki S.K., Montman C.A. [illegible] Livingston W. 1993, A&A 281, 221
49. Brandt P.N., Solanki S.K., [illegible] Rueden R., 1991, A&A in press
50. Osterovich V.A. 1968, Sol. Phys. 77, 62
51. [illegible] Osterovich V.A., [illegible] A. 1982, ApJ 261, 700
52. Osterovich V.A., Sacri Rao, 1989, ApJ [illegible] 106
53. Solanki S.K., Schmidt H.U. 1993, A&A 267, 287
54. Willmann [illegible] 1974, Sol. Phys. 36, 29
55. [illegible] K.D. 1985, Ph.D. Thesis, University of Göttingen
56. Abdussamatov H.I. 1971, Sol. Phys. 1[illegible], 284
57. [illegible] W. et al. 1982, Sol. Phys. 81, 231
58. Bogdan M. [illegible] 1983, Sol. Phys. [illegible] 1[illegible]
59. [illegible] W., [illegible] R.P., Turner [illegible] 1984a, Sol. Phys. 9[illegible], 10
60. [illegible] Solanki S.K., Livingston W. 1991, A&A in press
61. Balthasar H., Schmidt W. 1990, A&A 2[illegible]
62. [illegible] V., Bonfors [illegible] D.C., 1990b, Sol. Phys. 1[illegible]
63. Osterovich V.A. 1984, Sol. Phys. 9[illegible], 2[illegible]
64. [illegible] A., [illegible] 1989, Astroph. Space [illegible] 310, 67
65. [illegible] D., [illegible] A&A 25[illegible], 1[illegible]
66. Schmidt W., [illegible] A., Balthasar H., Tarbell T.D., Frank Z.A. 1992, A&A 264, [illegible]
67. Osterovich V.A. 1980, Sol. Phys. 65, 2[illegible]
68. Solanki S.K., Montagne C.A. P. 1993, A&A 272, 239
69. Livingston W. 1991, Na[illegible] 350, 45
70. [illegible] 1977, ApJ 1[illegible], 2[illegible]
71. [illegible] P. 1977, Sol. Phys. [illegible], 7[illegible]
72. Martinez Pillet V., Vázquez M. 1993, A&A 270, 494
73. Skumanich A., [illegible] 1993, in Electromagnetic Processes in Cosmic Magnetic Fields (eds.) (D. Reidel, Dordrecht, IAU Symp. [illegible]), 263
74. Solanki S.K., Walther U., Livingston W. 1993, A&A 277, 639
75. Wilson P.R., Cannon C.J. 1968, Sol. Phys. 4, [illegible]
76. [illegible] 1983, Sol. Phys. [illegible], 41
77. Keller C.U., Deubner F.-L., [illegible] 1994, A&A [illegible], 2[illegible]
78. Lin H. 1994, ApJ in press

IR Photometry of Magnetic Elements

H. Lin
Caltech, Pasadena, CA

ABSTRACT

Spectrographic observations of the Stokes I and V profiles with the IR Fe I 15648A and 15652A lines yield information about the magnetic field, filling factor, magnetic flux, and continuum contrast of the magnetic elements. These parameters are important for constructing realistic magnetic element models. We will present new data relating IR continuum contrast and the other three parameters of solar magnetic elements.

SUNSPOT UMBRAL STRUCTURES OBSERVED IN FE I 1027 NM

W. Schmidt and H. Balthasar
Kiepenheuer-Institut für Sonnenphysik, Schöneckstrasse 6, D-79104 Freiburg Germany

ABSTRACT

The magnetic field strength and velocity in bright umbral structures have been determined from time sequences of high resolution Stokes -I spectra using the Fe I line at 1027 nm. The umbral bright structures show velocity oscillations identical to the surrounding umbra and no systematic flow. The slightly reduced magnetic field strength in the bright dots may be due to a height effect.

1. Introduction

The dynamics and the magnetic field structure of the small scale phenomena within sunspots are essential for our understanding of the physics of sunspots. New instruments and techniques have stimulated a quite a number of corresponding sunspot observations in the last few years and significant progress has been made. The conference proceedings edited by Thomas and Weiss[1] provide a number of interesting articles and reviews on this topic. The nature of sunspot umbral dots is one of the open questions and there is little agreement between model predictions and observational results obtained to date (Parker, 1979[2], Degenhardt and Lites, 1993[3], Lites et al. 1991[4], Wiehr and Degenhardt, 1993[5], Schmidt and Balthasar, 1994[6]). Many, if not most of the spectral lines that are normally used for spectroscopic and polarimetric magnetic field measurements (lines with large Landé factors and simple splitting patterns) are strongly blended in sunspot umbrae or are of limited use because of their strong temperature dependence, or both. The still growing importance of the near infrared for magnetometry is thoroughly reviewed elsewhere in this book (Solanki[7]). The observations described below are made in the Fe I line at 1026.5 nm. At this wavelength the diffraction limited spatial resolution is still high (0.4 arcseconds at the 70 cm Vacuum Tower Telescope on Tenerife). This rather weak line is unblended in sunspots and has a triplet like splitting pattern; it is described in detail by Balthasar and Schmidt, 1994[8].

2. Observations

Observations have been made in October 1993 at the Vacuum Tower Telescope on Tenerife using a normal CCD camera (1024^2 pixels) in a 2x2 summing mode. A sunspot tracker was used to stabilize the image during the observations. The time sequence was obtained with a cycle time of 4 s and a total duration of 2.5 hours. Figure 1 shows one of the sunspot spectrograms. In the umbra (inserted in a different intensity scale) two bright structures are visible. The right hand part of the figure displays the same data, with each line profile normalized to the local continuum intensity. The splitting of the outer components corresponds to a Landé factor of 2.5 leading to complete splitting inside the spot. Some very weak interference fringes are also visible on the blue (left) side of the line. For the present analysis we used two subsets of the data, from 11:00 to 11:40 and from 12:00 to 13:00 UT.

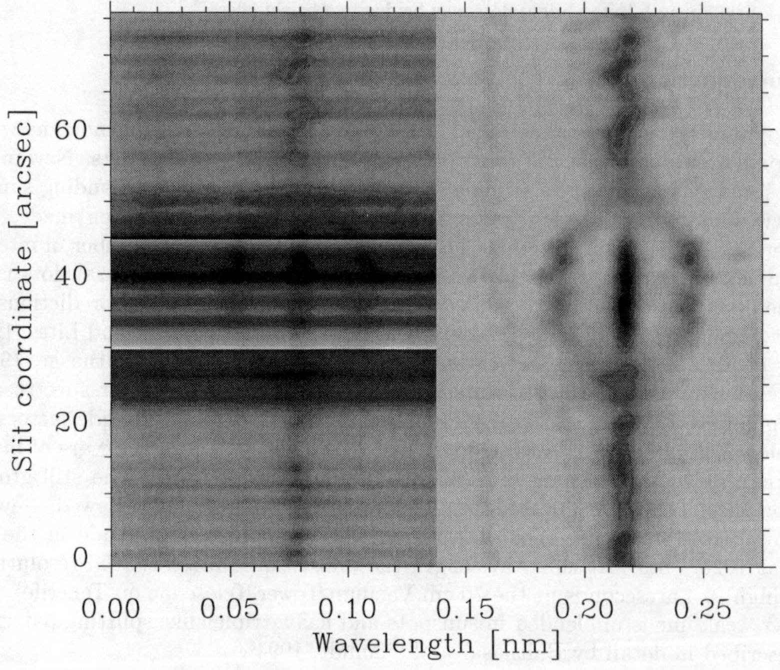

Fig. 1: Left part: spectrogram of the Fe I 1027 nm line, measured across a sunspot. The umbra is inserted in a different intensity scale to improve the visibility of the bright umbral structures. Right part: same data, but normalized to local continuum to remove spatial brightness variations. Some structures on the left side are caused by weak interference fringes.

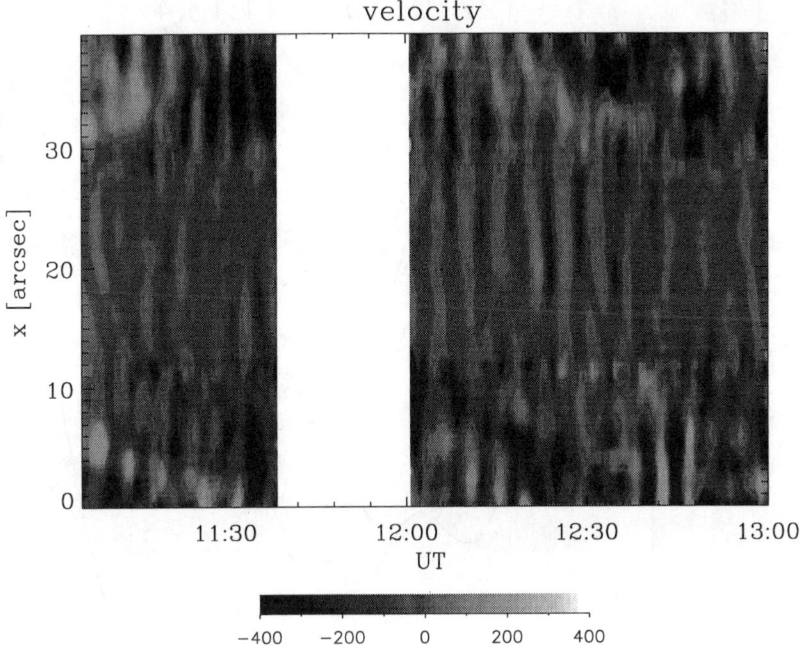

Fig. 2: Relative velocity across the sunspot vs. time. Two sequences taken at the same position in the sunspot have been analysed (11:00 to 11:40 and 12:00 to 13:00 UT). The five minute oscillations are the dominating feature. The mean time-averaged Doppler shift along the slit has been subtracted. The umbral bright features are located at x=17 and 19.

3. Results

The Doppler shift measurements confirm the absence of any systematic flow in or around bright umbral features found in earlier observations (e.g. Lites et al. 1991 [4], Schmidt and Balthasar 1994 [6]). After removal of the time average of the line-of-sight velocity only the five minute oscillations remain. Figure 2 readily demonstrates that the bright umbral features (located at x = 17 and 19) do not affect the velocity pattern at all, i.e. they have the same oscillatory behavior as the dark umbra. The Doppler shifts have been determined from the outer components of the magnetically split line and are therefore not affected by any scattered light from the photosphere outside the spot. Our findings are in qualitative agreement

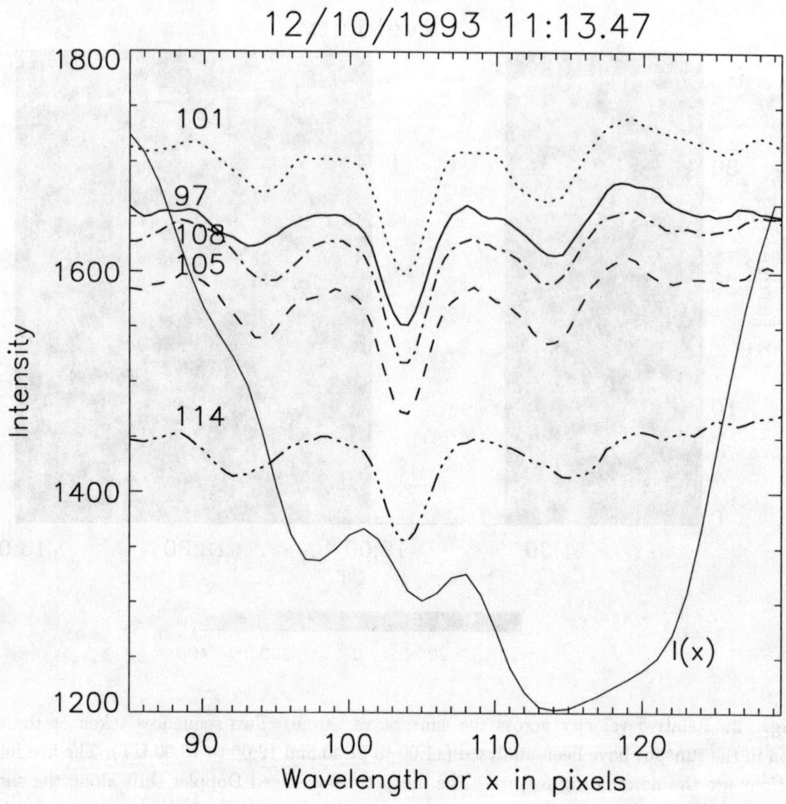

Fig. 3: Zeeman split line profiles a various spatial positions in the sunspot umbra. The thin full line shows the intensity profile across the umbra including two bright structures (located at x=101 and 108). The labels at the five line profiles refer to the spatial position x. The intensity scaling is different for the I(x) and the I(λ) profiles. The wavelength scale is 1.3 pm/pixel.

with measurements of Penn and LaBonte 1993[9]. There is no interaction between the oscillations and the magnetic field strength, i.e. the field strength at a given location in the umbra remains strictly constant during the measurements.

From the magnetic field analysis we find very little temporal changes and no correlation with the velocity oscillations. The thin full line in Figure 3 is the continuum intensity profile of the sunspot umbra. Zeeman split line profiles for selected positions are drawn in different styles. The labels correspond to the x positions of the intensity curve. The curve labeled "114" refers to the darkest part of the umbra and has a field strength of 2400 Gauss. The magnetic field in the bright structures (nos. 101 and 108) is slightly reduced compared to the surrounding dark umbra (97 and 105), but the effect is small and may be explained by the different height of observation in- and outside bright features and the vertical decrease of the magnetic field strength (Balthasar and Schmidt 1993[10]).

4. References

1. J. H. Thomas and N. O. Weiss (Eds.) *"Sunspots, Theory and Observations"* (1992).
2. E. N. Parker, *Astrophys. J.* **234** (1979) 333.
3. D. Degenhardt and B. W. Lites, *Astrophys. J.* **404** (1993) 383.
4. B. W. Lites, T. A. Bida, A. Johannesson, G. B. Scharmer, *Astrophys. J.* **373** (1991) 683.
5. E. Wiehr and D. Degenhardt *A&A* **278** (1993) 584.
6. W. Schmidt and H. Balthasar, *A&A* **283** (1994) 241.
7. S. K. Solanki, *in: "Infrared Tools for Solar Astrophysics: What's next?", J. Kuhn and M. Penn (Eds.)* 1995, ppp.
8. H. Balthasar and W. Schmidt, *A&A* **290** (1994) 649.
9. M. J. Penn and B. J. LaBonte, *Astrophys. J.* **415** (1993) 383.
10. H. Balthasar and W. Schmidt, *A&A* **279** (1993) 243.

OBSERVATIONS OF EVOLVING SOLAR MAGNETIC FEATURES IN THE INFRARED AND VISIBLE SPECTRUM

Thomas Rimmele
New Jersey Institute of Technology
National Solar Observatory, Sunspot, NM-88349, USA

and

Wolfgang Schmidt
Kiepenheuer Institut für Sonnenphysik, Freiburg, Germany

ABSTRACT

We observed several magnetic features simultaneously in the infrared and the visible spectrum. Time sequences of Fe I 1.56 μ spectra were recorded using a 256*256 NICMOS3-chip. We derived 2-dimensional maps of the magnetic field strength, polarization amplitude and Stokes-V zero crossing velocities by scanning a spectrograph slit across the magnetic features. The slit jaw image was fed to a UBF/FP narrow band filter, which periodically scanned the profile of the Fe I 5576 A line. Among the features observed was a pore, which disappeared during our observations, and a small spot positioned close to the solar limb. We present preliminary results from these observations.

1. 1 Introduction

Up to now most progress in observational solar physics has been made with observations in the visible. This is mainly due to the fact that optical detectors have been readily available since quite some time now. For example, large format, low noise optical detectors are standard equipment at solar observatories. On the other hand, infrared detectors, which are just becoming available to solar observers tend to have small format (256×256) and large background noise.

Most post focus instrumentation available to observers is limited or optimized for use in the visible. The UBF/FP filter combination that permits 2-dimensional spectroscopy and has been utilized so successfully in several recent observations including those resulting in an improved understanding of the penumbral fine structure or the excitation of solar p-modes, is limited in its use to the visible spectrum. Likewise existing vector magnetographs such as the Advanced Stokes Polarimeter (ASP), operated at the VTT/SP on a routine basis, are limited to the visible. Devices for spectroscopic imaging in the IR are under development at the NSO but are not yet available to observers.

Although the seeing improves considerably toward longer wavelengths, this advantage is in part offset by the lack of large aperture solar telescopes that provide spatial resolution comparable to the resolution achieved in the visible at almost any solar telescope in the world.

Considering this situation it is not surprising that most solar observers still focus on the visible. We think IR solar physics should not be considered as a standalone research branch that competes with visible observations or even replaces them. In our mind the best way to proceed is to extend our diagnostic capabilities by simultaneously observing in the visible **and** the IR. For example, the large Zeeman splitting in the IR allows us to easily measure the magnetic field strength whereas in the visible complex inversion techniques have to be applied to derive the field strength. The opacity minimum at 1.6 μ allows observations of the deepest photospheric layers. MHD-simulations presented during this meeting[1] show that in the 1.6 μ continuum individual flux tubes should be easily distinguishable as bright features against a low contrast granulation background. In the visible, flux tubes are not always seen as bright features in the continuum and can only be identified with certainty by means of polarimetry.

We performed simultaneous observations of several solar magnetic features in the infrared spectrum at 1.56 μ and in the visible. The objective was to evaluate the diagnostic capabilities of simultaneous measurements in the visible and the IR. We observed a pore which disappeared during the observing run. An active region was observed near the solar limb. We will show that for both observations the visible and IR data give complementary information resulting in an overall improvement of the diagnostic capabilities.

_. Observations

The observations were carried out at the VTT/Sacramento Peak. We observed the Fe I lines at 1.56μ using the Horizontal Spectrograph (HSG) and the NICMOS3 infrared camera available at the NSO/SP. The spectrum of the HSG was re-imaged onto the NICMOS3 chip in an attempt to optimize spectral and spatial pixel resolution, which were $41 m\AA/pixel$ and 0."$12/pixel$, respectively. The light reflected of the slit jaws was fed to the visible channel. In principle a dichroic beam splitter or a cold mirror could be used to split and direct the IR and visible light to the corresponding channels. In the visible channel narrow band ($20 m\AA$) filtergrams were recorded using the UBF/Fabry-Perot filter combination. Simultaneously with the each narrow band filtergram a white-light image was taken. The data were recorded using the MDA system. During this observing run we scanned the profile of the g=0 Fe I 5576 Å line in order to derive photospheric velocities.

While the UBF/FP filter was repeatedly scanning one or more spectral line profiles, the spectrograph slit was scanned across the solar feature of interest with a step size of 0.5". Left-circular (LCP) and right-circular polarized (RCP) spectra were recorded with the NICMOS3 camera. The exposure time was 60 ms. At each of the 20 slit positions five pairs of LCP and RCP spectra were recorded and subsequently averaged to reduce the noise level in the Stokes V spectra. In addition the feature was stabilized using the correlation tracker[2].

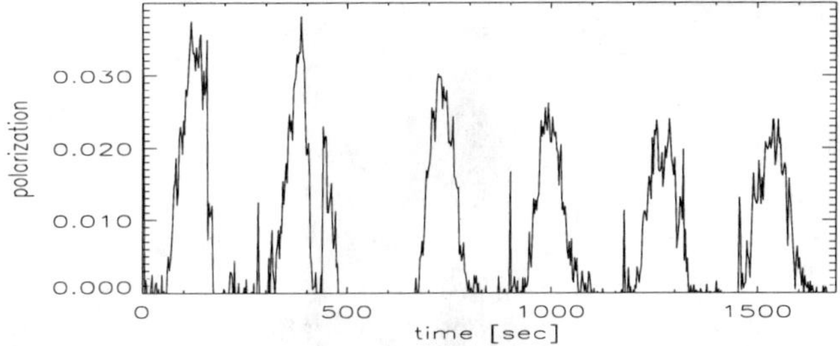

Figure 1: Polarization amplitude vs. time.

3. Results and Discussion

3.1. disappearing pore

On April 28, 1994 we observed a small pore close to disk center ($cos\theta = 0.88$). During the observing period of about 30 min the polarization signal measured in the pore gradually decreased. Figure 1 shows the polarization amplitude (LCP-RCP)/LCP+RCP) as a function of time. Each of the peaks of approximately 100 s width corresponds to a scan of the spectrograph slit across the pore. Figure 2 (right) shows a time sequence of magnetograms derived from the 1.56 μ Stokes V spectra. The magnetograms correspond to the last four scans shown in Figure 1. Also shown are cross sections through the center of the pore at different times. The intensity profiles, the corresponding magnetic field strength derived from the splitting of the σ-components, and the Stokes V amplitude, which is proportional to the magnetic flux are plotted. The intensity measured in the pore is increasing substantially in time to a point where the pore is hardly distinguishable from the granular background in the continuum images. During the same period the magnetic flux decreases by about 30% whereas the field strength shows little change.

The magnetic field drops from 1.9 kG in the center of the pore to about 1.5 kG at the edge of the pore. At the boundary of the pore the determination of the field strength becomes uncertain due to the small polarization amplitude. Compared to the field strength the flux decreases more rapid toward the edge of the pore which is consistent with the picture of an expanding flux tube where the field lines are more inclined (more horizontal) at the edge of the tube. The magnetic signal is mainly confined to the white light boundary of the pore within the spatial resolution achieved in our IR observations, which is of the order 1". Muglach et al.[3] observed a pore of 7" diameter with 3" resolution and, in contradiction to our observations, measured a magnetic profile that extends beyond the white light boundary of the pore by about 2-3". However, this discrepancy may be due to the different spatial

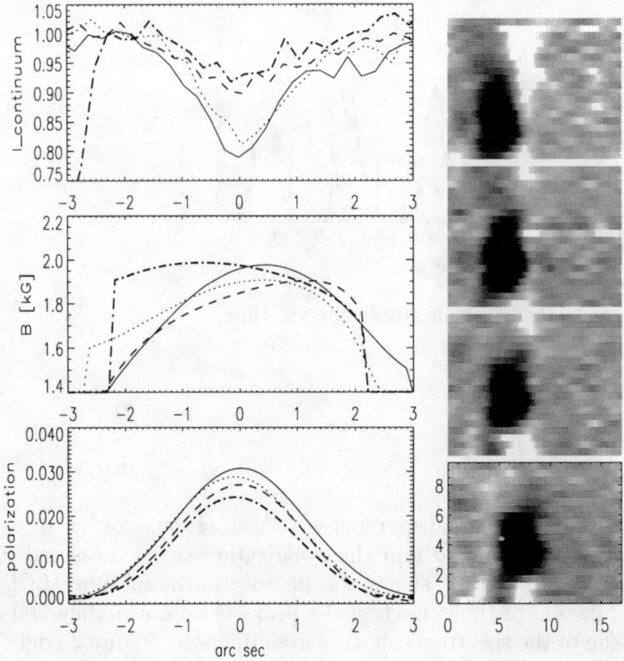

Figure 2: Panels to the right: Magnetograms of pore at different times. Left: cross sections through center of pore at different time steps. The time steps t_1 through t_4 correspond to the solid, dotted, dashed and dash-dotted line, respectively.

resolution achieved in both observations.

The 2-dimensional bisector-velocity maps derived from the Fe I 5576 Å profiles show no systematic flows in or around the pore.

What causes the magnetic flux in the pore to decrease? Several possible mechanism that can explain removal of flux from the solar surface have been discussed[4], including dispersion of flux, retraction of flux, escape of flux to the solar atmosphere and destruction of flux due to reconnection. Observational evidence for different cases has also been described in the literature [5,6,7,8]. In the magnetograms shown in Figure 2 flux of the opposite polarity is visible next to the pore. In particular in the magnetogram corresponding to the last scan a patch of opposite polarity seems to have emerged. We do not see any indication for flux spreading out into the surroundings of the pore, although there is the possibility that this flux is below the threshold for detection. The polarization signal measured in the opposite polarity is considerably smaller than the polarization measured in the pore and the field strength is 1.2 kG

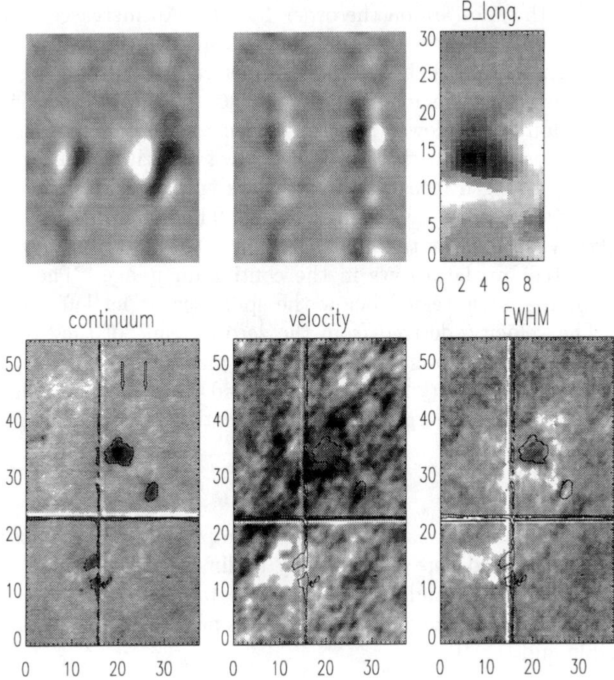

Figure 3: Active region near limb. Lower panels: data from the visible channel. Upper panels: infrared Stokes V data. Axis are labeled in arc sec.

compared to 1.9 kG in the center of the pore. At this point it is not clear if indeed flux of the opposite polarity (with its opposite polarity emerging somewhere out of the FOV) has emerged next to the pore causing flux cancellation or if at the outer boundary of the pore field lines start to return and maybe submerge.

3.2. active region near the limb

On Apr. 28, 1994 we also observed active region NOAA 7707 near the solar limb ($cos\theta = 0.45$). This region had started to emerge on Apr. 23 and on Apr. 28 was still undergoing rapid changes. Figure 3 shows a subfield of the recorded continuum slit jaw image and the corresponding maps for the velocity and the FWHM derived from the Fe I 5576 Å profiles. The solar limb is toward the lower right corner. We observe a ring of blue shift (dark) surrounding the spot close to the center of the FOV. The velocities measured around the spot are on the order 1 km/s. An increased FWHM is correlated with the blue shift, which points to enhanced turbulence in these areas. The corresponding IR magnetogram is shown in the upper right corner of Figure 3. The scan approximately covers the region from 16" to 26" in the x-direction and 20" to 50" in the y-direction in the lower panels. The magnetogram reveals that the blue shifted "ring" is correlated with magnetic field of polarity opposite from that of the spot. The field strength measured in the dark spot is 2 kG, and in the surrounding opposite polarity we measure a field of 1.3 kG. The two upper left panels show examples of Stokes V spectra at the 1.56 μ region. The spectra correspond to spatial positions indicated by the arrows in the continuum image. The Stokes V profile measured in the magnetic region below the spot (see upper left panel) is strongly blue shifted. The velocity, derived from the zero crossing of the Stokes V profiles and thus corresponding to flows confined to the magnetic feature(s), is of order 5 km/s. The ratio of the velocity amplitudes derived from the non-magnetic line FeI 5576 Å and the Stokes V spectra gives a rough indication for the filling factor.

4. Summary

We have hardly touched upon the wealth of information present in these observations and the first results presented here are still very preliminary. Nevertheless, we are encouraged by these first results and plan to continue to take advantage of the additional information and enhanced diagnostic capabilities provided by simultaneous observations in the visible and the IR.

5. References

1. M. Knölker, O. Steiner, M. Schüssler, *these proceedings*
2. T. Rimmele, von der Lühe, Wiborg P.H., Widener A. L., Spence G., Dunn R. B., in *Active and Adaptive Optical systems*, M. Ealey (ed.), SPIE **1542**, (1991), 186
3. K. Muglach, S.K. Solanki, W.C. Livingston, in Solar Surface Magnetism, R.J. Rutten and C.J. Schrijver (eds.), (1994) 127.
4. S. D'Silva, *ApJ* (1995) (in press).
5. S.G. Wallenhorst, R.F. Howard, *Sol. Phys.* **76** (1982) 203.
6. S.G. Wallenhorst, K.P. Topka, *Sol. Phys.* **81** (1982) 33.
7. D. Rabin, R. Moore, M.J. Hagyard, *ApJ* **287** (1984) 404.
8. G.W. Simon, P.R. Wilson, *ApJ* **295** (1985) 241.

MAGNETO-OPTIC EFFECTS ON FeI 1.56 MICRON LINE

K. S. BALASUBRAMANIAM*
National Solar Observatory/Sacramento Peak
National Optical Astronomy Observatories†
Sunspot, NM 88349, USA

and

CATHERINE E. PETRY‡
University of Arizona, Steward Observatory
Tucson, AZ 85721, USA.

ABSTRACT

The presence of magnetic fields in the solar atmosphere influences the polarization state of Zeeman induced spectral lines. We present results of calculations using polarized radiative transfer to identify the influence of magneto-optic effects on the FeI 1.56 micron spectral line about a sunspot. For a sunspot at the disc-center, the results of these calculation show that magneto-optic effects in the Stokes V profiles is strongest in the penumbral region, while for the Stokes Q and U profiles it shows an azimuthal variation.

1. Introduction

The presence of magnetic fields in solar active regions affects the polarization state of light emanating from them. The Zeeman effect causes the spectral line profiles to be polarized. The state of polarization depends on the orientation of the magnetic field and the thermo-dynamical character of the atmosphere where the magnetic fields are present. The character of polarization induced by the Zeeman effect can, in turn, be modified due to other magneto-optic effects such as the Faraday effect and the Voigt effect (FV effects)[1]. The Faraday effect is a rotation of the direction of the oscillation of a linearly polarized wave occuring when the wave traverses a medium along the direction of a magnetic field. The Voigt effect is the introduction of a phase difference in an optical beam, as it traverses a medium, perpendicular to the magnetic field.

The clarity of separation between the polarized components due to the Zeeman effect, increases with wavelength. The FeI 1.56 micron lines are formed in the deepest observable layers of the solar atmosphere[2]. The magnetic fields at the deepest observable layers are expected to be the strongest. The light originating from these

*Supported by Funding from the Air Force Office of Scientific Research.
†Operated by the Association of Universities Research in Astronomy, Inc.(AURA), under cooperative agreement with the National Science Foundation (NSF). Partial support for National Solar Observatory is provided by the United States Air Force under a Memorandum of Understanding with NSF.
‡NSO/NOAO/NSF 1992 REU Summer Research Associate.

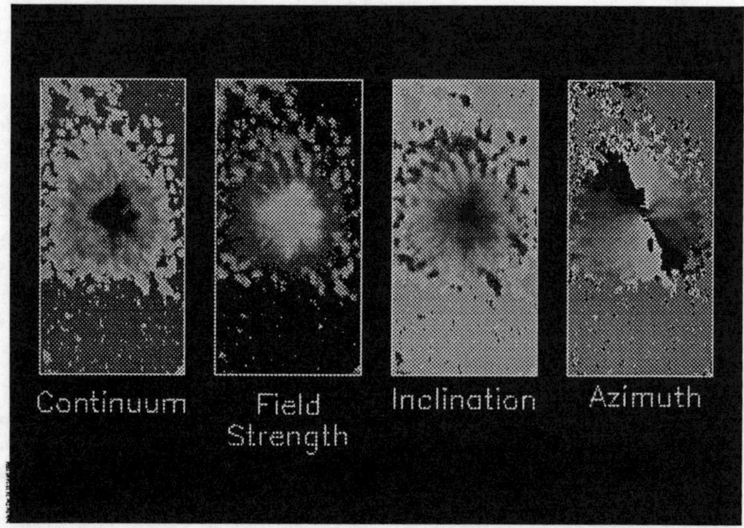

Figure 1: A model sunspot used in the calculation. This active region, covers an area of 25 Mm x 60 Mm, observed on March 25, 1992 (S9.5,W2.8) The magnetic field parameters were derived by Lites et al.[5], using the FeI 6302.5 Å line. The magnetic field strength is linearly scaled between 0 and 2500 Gauss, the inclination and azimuth from 0-180 degrees. We assume, for these calculations, the same magnetic field strength exists at height where FeI 1.56 micron line is formed.

deeper layers, in turn, traverse a longer path length. Hence, in understanding the polarization measured in the FeI 1.56 micron line, we would expect the Faraday and Voigt effects to be well discerned.

While the influence of the FV effects can be taken into account to infer vector magnetic fields using inversion procedures for the Zeeman effect, as is customarily done using optical spectral lines[3], it would be well worth to explore if they (FV effects) could independently offer any additional diagnostic capabilities of reconstructing the physical characteristics of the solar atmosphere, besides of the Zeeman effect. The infrared 1.56 micron line offers potential to exploring this diagnostic, with the construction of high spectral, spatial and polarization measurement devices.

In this paper we addresses the nature of variation of the FV effects about a typical sunspot, using methods of polarized radiative transfer.

2. The Method

Our approach to the problem was to calculate the polarized radiative transfer spectral line profiles in all four Stokes parameters, about a sunspot, with and without the magneto-optic effects, respectively and compare the differences in the profiles to understand the influence of the FV effects about a sunspot.

In order to calculate each of the Stokes profiles for the spectral line, a modified version of the polarized radiative transfer computer program MALIP, written by Landi Degl'Innocenti[4] was used. The spectral line studied is the Fe I line 15648.518 Å. It has a $5s^7D_1 - 5p^7D_1$ transition. Besides the transition levels, other inputs to the program included model sunspot magnetic field strength, its azimuth and inclination, model atmospheric parameters such macro-turbulent and micro-turbulent velocity, oscillator strength.

For a model sunspot, we chose a realistic realistic sunspot based on measurements. Fortunately, Bruce Lites was able to make available his measurements of an actual sunspot[5] near the solar disc center, using the HAO/NSO Advanced Stokes Polarimeter at Sacramento Peak Observatory[6] (see Figure 1).

The sunspot parameters included the magnetic field strength, its inclination and azimuth. We assume that the magnetic field quantities represented in this sunspot (observed using the FeI 6302.5 Å line) are the true values of the magnetic field quantities observed with the FeI 1.56 micron line. Two cases were considered, in one case, the magnetic field parameters were assumed to be a constant over the height of formation of the spectral line, while in the second case we used $B(h) = B_0e^{-h/300}$. Here B is the strength of the magnetic field and h is the height.

The next concern was the atmospheric models. Using the measured intensity of the sunspot, we identified, photosphere, penumbra and umbra regions based on the intensity discriminants. Atmospheric model parameters representing these features are characterized by temperature, gas pressure, partial pressures of abundant elements, micro-turbulent velocity, etc, at each optical depth in the atmosphere. The best available models for the photosphere and umbra were used. These are the Photospheric Reference Model and the Umbral Core Model M[7], respectively. The penumbral model of Kjeldseth Moe and Maltby[8] was used to define the penumbra.

Another realistic parameter is the macro-turbulent velocity. Steve Keil (personal communication) provided equations describing fitted functions to measured macro-velocities, as a function of height, for active regions. The vertical component of the macro-velocities used for the photosphere is given by $V_z(h) = 1.9e^{-h/200}$, while for the penumbra and umbra is given by $V_z(h) = 1.5e^{-h/350}$.

The published oscillator strength for FeI 1.56 micron line is, log(gf)= -0.54 (Zayer et al). However, spectral line profiles generated for for log(gf)=-0.54 did not match the atlas line profiles (Livingston et al). derived using the FTS. An updated value of log(gf)=-0.7 was provided by S. K. Solanki (private communication). Independently, observed spectral line profiles were provided by S. L. Keil and J. R. Kuhn (private communication) were used to match the fitted profiles, to derive an oscillator strength log(gf)=-1.4, however, the influence of scattered light in these observed profiles is unknown.

3. Data Analysis and Results

Four sets of polarized spectral line profile data (I, Q, U, V) were generated for the entire sunspot grid of 100 x 200 pixels. Each spectral line profile was sampled at 30 mÅ intervals, covering 50 spectral samples on either side of the spectral line center. Each set consisted of a pair of I, U, Q, V profiles, one with the FV effects accounted for and the other without taking into account the FV effects. Three of the data set pairs corresponded respectively to values of the oscillator strengths -0.54, -0.7 and -1.4. For each of these three sets the magnetic field was assumed a constant within the height of formation of the spectral line. The fourth data pair consisted of line profile pairs with an oscillator strength of -0.7 and the magnetic field having an exponential decay with height, of the form $B(h) = B_0 e^{-h/300}$.

For each data set pair, we considered the difference between the Stokes Q, U and V profiles, with and without the FV effects, respectively, as a measure of the FV effects. The difference between the Stokes, Q, U and V profiles, at -150 mÅ from the spectral line center, over the entire sunspot, is shown in Figure 2.

Cross-sectional plots of the polarization difference at -150 mÅ and -300 mÅ, are shown in Figure 3.

From Figures 2 and 3 it is clear that the (1) FV effects contribute to the polarization, of the order of 1-2%, or less. (2) In the Stokes V difference images, FV effects have a radial variation about a sunspot and the Stokes Q and V difference images show an azimuthal variation. (3) The FV effects increase with an increase in the oscillator strengths of the spectral line and (4) The FV effects cannot be ignored for high spectral and polarimetric resolution (5) FV effects increase with an increase in the gradient in the field strengths. These arguments suggest that the FV effects may themselves offer to be good diagnostic tools for probing the solar atmosphere, should one consider appropriate weighted differences in the Stokes Q, U and V profiles at the spectral line center (where the FV effects are strong), and the far wings of the spectral line (where the FV effects are weak and hence may be neglected). Further work is in progress.

4. Acknowledgements

We thank Dr. Bruce Lites for providing the Advanced Stokes Polarimeter vector magnetic field data. We thank Dr. Steve Keil for the observed infrared spectral line data as well as advice on using his velocity measurements, and several useful discussions. Finally, I would like to thank Drs. S.K. Solanki and J. Bruls for advice on the log(gf) data.

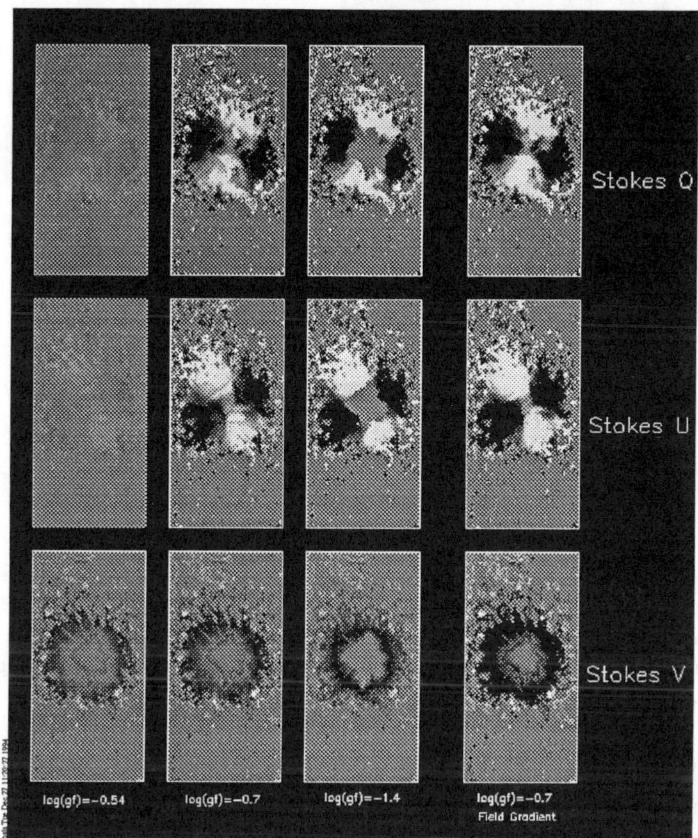

Figure 2: The difference between the Stokes Q, U and V profiles at -150 mÅ, with and without the magneto-optic effects, shown as grey scale images, for the sunspot shown in Figure 1. The first three columns represent the differences for log(gf)=-0.54, -0.7, -1.4, with constant magnetic field, and the fourth column represents the case for log(gf)=-0.7, with a gradient in the magnetic field, as described in text. The Stokes Q and U difference images have been linearly scaled between $\pm 7.\ 10^{-3}$. The Stokes V difference images have been scaled (moving from left to right)$\pm 2.\ 10^{-4}$, $\pm 8.\ 10^{-3}$, $\pm 8.\ 10^{-3}$ and $\pm 1.5\ 10^{-2}$, respectively.

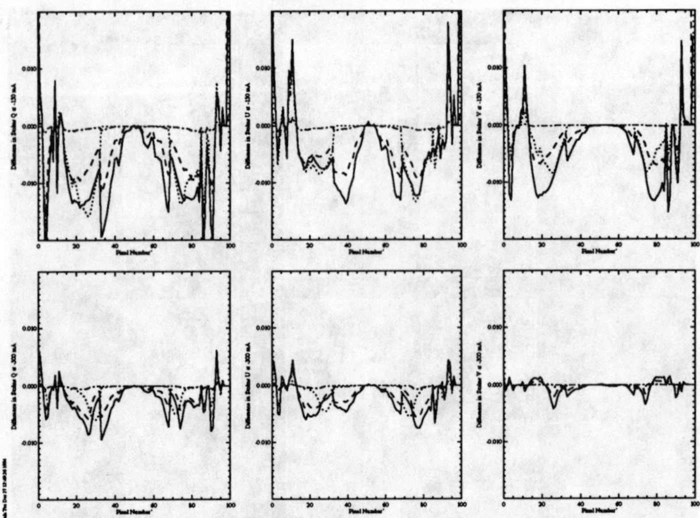

Figure 3: Cross-sectional plots of the Stokes Q, U and V difference intensities at -150 mÅ (top row) and -300 mÅ (bottom row). In each plot the different line-styles represent: continuous line for log(gf)=-0.7 with a gradient in the magnetic field (see text), dotted line log(gf)=-1.4, dashed line for log(gf)=-0.7 and dot-dashed for log(gf)=-0.54.

5. References

1. J.M. Beckers, *Solar Phys.* **9** (1969) 372.
2. J. H. M. J. Bruls, B. W. Lites and G. A. Murphy, in 1990, *Solar Polarimetry - Proceedings of the 11th NSO/SP Summer Workshop*, ed. Laurence J. November (1990) 468.
3. M. Landolfi and E. Landi Degl'Innocenti, *Solar Phys.* **78** (1982) 355.
4. E. Landi Degl'Innocenti, *Astron. Astrophys. Suppl.* **25** (1976) 479.
5. B. W. Lites, R. B. Dunn, D. F. Elmore, S. Tomczyk, A. Skumanich and K. V. Streander, *Bull. Amer. Astron. Soc.*, **24** (1992) 747.
6. S. Tomczyk, D. F. Elmore, B. W. Lites, R. B. Dunn A. Skumanich, J. A. Schuneke, K. V. Streander, T. W. Leach, C. W. Chambellan and L. B. Lacey, *Bull. Amer. Astron. Soc.*, **24** (1992) 814.
7. P. Maltby, P., et al. *Astrophys. J.* **306** (1986) 284.
8. O. Kjeldseth Moe and P. Maltby, *Solar Phys.* **8** (1969) 275.

SCALING OF SOLAR MAGNETIC FIELDS

A. RUZMAIKIN, C. CADAVID and J. LAWRENCE
Department of Physics and Astronomy, California State University, Northridge
18111 Nordhoff Street, Northridge, CA 91330, U.S.A.

DOUGLAS RABIN
*National Solar Observatory, National Optical Astronomy Observatories**
P. O. Box 26732, Tucson, AZ 85726, U.S.A.

and

HAOSHENG LIN
Solar Astronomy Department, California Institute of Technology
Pasadena, CA 91125, U.S.A.

ABSTRACT

Two types of singular behavior of the solar magnetic field on small scales are discussed. (1) Strong concentrations of the field into flux tubes, sheets or more complicated forms. (2) A sign variability of the field down to the smallest scale defined by diffusion. The role of infrared observations in studying these scaling properties is analysed.

1. Introduction

It was conjectured on an observational basis that most of the magnetic flux in the solar atmosphere exists as strong concentrations on a background of relatively weak field.[1,2] On the theoretical side, this idea has been modeled by a distribution of concentrated flux tubes in which the magnetic field strength exceeds 1000 Gauss at the photosphere.[3,4] Such flux tubes are now considered to be a basic constituent of solar and stellar atmospheres. However, most flux tubes are not directly detected. This is explained by the estimate that the flux elements should have sub-arcsecond size and thus lie beyond present observational abilities. The observations show a wide continuum spectrum of scales down to 0.5 arcsecond; a few high-resolution observations may have detected the upper end of the size distribution- –i.e., the largest discrete flux tubes.[5]

Here we discuss an approach which can help in the detection of the magnetic concentrations and in interpreting the small-scale field distribution. The approach is based on a study of scaling properties of the magnetic field.[6–8] The idea is to look not only at the smallest detectable structures but to study the field distribution at all available scales, at varying levels of coarseness. This study is used to demonstrate the existence of strong magnetic field concentrations (amplitude singularities), to estimate

*Operated for the National Science Foundation by the Association of Universities for Research in Astronomy

a set occupied by these concentrations and to discuss their forms. The singularities are expected not only in the form of flux tubes and flux sheets but also in the form of fractal structures.

The vector nature of the field opens the possibility of another type of singular behavior, sign-singularity, in which the magnetic field is variable in direction on all scales down to some small distance set by flux tube physics. Here we report preliminary results of a study of the magnetic singularities using, in addition to line-of-sight magnetograms, infrared magnetograms that measure intrinsic field strength, $|B|$, as well as magnetic flux. In particular we address the detection of sign-singular magnetic fields in the solar photosphere.

2. Multiscale Measures of Solar Magnetic Fields

Observed distributions of the solar magnetic field below the level of supergranules have no characteristic scales. This scale symmetry may be studied using a simple technique developed in recent years for other applications.[9–11] This technique allows us to characterize such distributions in a way that is independent of the observer—in particular, to calculate basic properties that distinguish between two images of irregularly distributed quantities such as the magnetic and velocity fields. The classical example is the Kolmogorov scaling of energy in hydrodynamic turbulence: $< v_l^2 > \propto l^{2/3}$, where $< v_l^2 >$ is the variance of velocity at scale l. In a case of Gaussian distribution one scaling exponent for second statistical moment characterizes the whole distribution. The observed distribution of the solar magnetic field is not Gaussian, and an infinite number of exponents associated with the scaling of all moments is needed. The distribution which produces the observed magnetic fields results from a dynamo process, and the set of scaling exponents is the main signature of this process.

Observations at visual wavelengths give line-of sight photospheric magnetic fields which are equivalent to the net magnetic flux integrated over an image pixel. The magnetogram can be considered as a set S of pixels. The natural measure on this set, $\mu(S)$, is a local magnetic flux Φ normalized by a total flux through the magnetogram. Let $\mu_+ = \Phi$ when the magnetic field is positive, and $\mu_- = |\Phi|$ when the field is negative. This measure is real, non negative and additive: $\mu(S_1+S_2) = \mu(S_1)+\mu(S_2)$, where S_1 and S_2 are two non-overlapping sets. It is zero on a zero set, $\mu(0) = 0$.

Measurements also exist of the magnetic field strength measured directly from the strong Zeeman effect exhibited by some near-infrared spectral lines.[12,13] In that case we can use the field energy per unit depth as a strictly positive measure.

3. On Two Types of Singularities

Let us assume that the flux measure is self-similar i.e. that it scales with a power law dependence on the dimensionless scale $\epsilon = l/L$: $\mu(\epsilon) \equiv \epsilon^\alpha$ where L is a characteristic size of the magnetogram. Then the line-of-sight magnetic field, which is the flux density Φ/S, must scale like $\epsilon^{\alpha-2}$ on the plane. Thus if $\alpha < 2$, the flux amplitude is singular on small scales.

The distribution density of the exponents for each coarse-graining scale ϵ can be found by counting the number $dN(\alpha,\epsilon) = n(\alpha,\epsilon)d\alpha$ of boxes with a particular value of α in bins of width $d\alpha$. The density $n(\alpha,\epsilon)$ is in general scale-dependent; it may be represented[10] in the form $n(\alpha,\epsilon) \propto \epsilon^{-f(\alpha,\epsilon)}$. Then, if the function $f(\alpha,\epsilon) \to f(\alpha)$ in the limit of small ϵ (i.e. f is scale independent in this limit) then the curve $f(\alpha)$ is an invariant characteristic of the distribution of the scaling exponents. Typically this function has a "cap" ($\cap$) form with a maximum value f_m at α_m (=2 if the magnetic field has nonfractal Euclidean support). The singularity spectrum functions $f(\alpha)$ for the measure $\mu_B = \epsilon^2 B^2$ constructed from the infrared magnetograms obtained by Rabin and Lin are shown in Figures 1 and 2. Note that weak fields are not resolved by the technique used, and corresponding pixels are set to zero. The resulting "holes" in the images lower the apparent dimension, $f(\alpha)$.

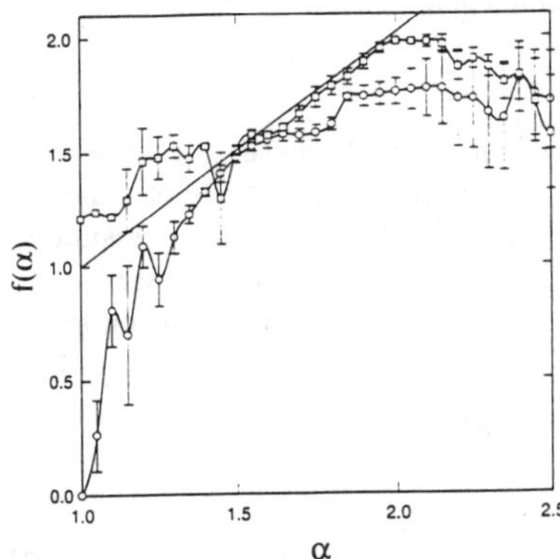

Fig. 1. The singularity spectrum for a magnetogram of active region NOAO 7525 obtained by the NSO Near Infrared Magnetograph on 21 June 1993. The squares correspond to the line-of-sight magnetic field; the circles correspond to the measure μ_B.

The measures considered above characterize the magnitude of the magnetic field. It also is interesting to look at alternations of direction of the field. In projection on some cross section such as the photospheric surface, one sees fields of both signs. By improving spatial resolution one might expect to observe a sufficiently small region as to contain only one sign of the field; this would be the case for a smooth magnetic field. *A priori* there is no reason to expect that a random field will be smooth on scales larger than that defined by diffusion. Thus there can be a situation where both signs are present at any resolved scale. For such a sign-singular field an improvement of resolution does not fix the sign of the field. In boxes of every scale, down to the practical limit of a single pixel, the field has both signs.

One needs a measure which can describe alternating-sign fields. Such a signed measure can be defined as, for example, the difference between the measures introduced above: $\mu = \mu_+ - \mu_-$.[14,6,8] Formally, a measure is sign-singular if, for any set

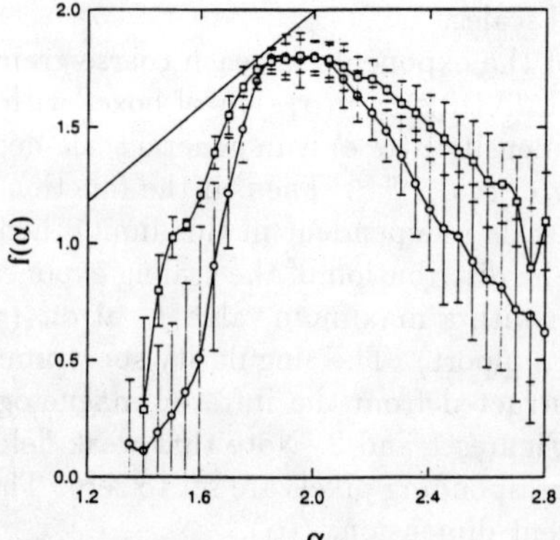

Fig. 2. The singularity spectrum for a plage magnetogram obtained by H. Lin (1994). The notation is the same as in Figure 1.

S (no matter how small) with $\mu(S) \neq 0$, there is a subset δS such that $\mu(\delta S)$ has the opposite sign from $\mu(S)$. Surprisingly, this type of singularity may occurs in the solar magnetic fields in quiet regions.[6] It is difficult at visual wavelengths to detect a signal which is the small difference of two adjacent, large, oppositely directed fields. Such a detection may be possible with infrared measurements.

4. How to Detect a Sign-singularity?

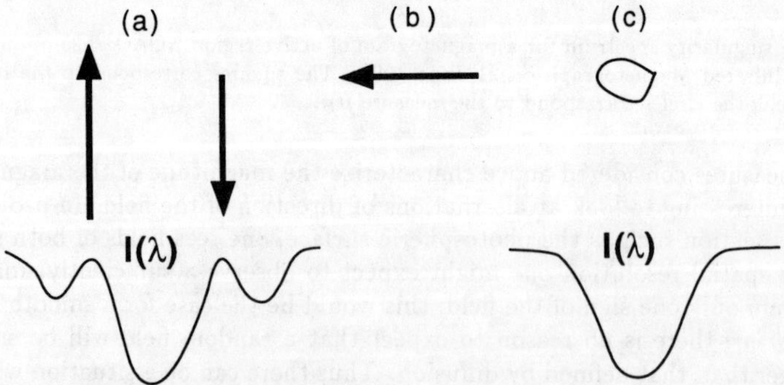

Fig. 3. (a) a magnetic sign-singularity; (b) the magnetic field is almost perpendicular to the line-of-sight; (c) the magnetic field is almost zero

To unambiguously detect a sign-singular magnetic field we need all four Stokes parameters. Figure 3 shows three possible configurations with almost zero line-of-sight magnetic field. Case b, with transverse field, can be distinguished by use of the Q or U Stokes profiles. Cases a and c have the same V profile and are thus indistinguishable in measurements of line-of-sight fields. However for true-field infrared observations, the I profiles will be different in these two cases. An extreme example is shown in Figure 3. In practice, this approach amounts to an application of the Zeeman broadening technique (introduced for stellar measurements by Robinson[15]) to small areas of the solar atmosphere, taking full advantage of the larger broadening of Zeeman-sensitive infrared lines.

The degree of cancellation can be described by the cancellation exponent[14,6], $\kappa = \lim_{\epsilon \to 0}(\ln \sum_i |\Phi(\epsilon)|/\ln(1/\epsilon))$, that characterizes the non-normalizibility of the sign-singular measure Φ.

5. Acknowledgments

We are grateful to Ch. Keller, S. Solanki, and I. Zayer for helpful discussions.

6. References

1. R. F. Howard and J. O. Stenflo, *Solar Phys.* **22** (1972) 402.
2. J. O. Stenflo, *Ann. Rev. Astron. Astrophys.* **1** (1989) 3.
3. H. Spruit, in *The Sun as a Star*, ed. S. Jordan, (NASA SP-450, 1981) p.385.
4. S. K. Solanki, *Space Sci. Reviews* **63** (1993) 1.
5. C. U. Keller, *Nature* **359** (1992) 307.
6. J. Lawrence, A. Ruzmaikin and C. Cadavid, *Astrophys. J.* **417** (1993) 805.
7. A. C. Cadavid, J. Lawrence, A. Ruzmaikin and A. Kayleng-Knight, *Astrophys. J.* **429** (1994) 391.
8. A. Ruzmaikin, C. Cadavid and J. Lawrence, in *Research Trends in Plasma Astrophysics, International Topical Conference*, ed. R. Kulsrud, G. Birbidge, and V. Stephan (La Jolla, AIP, 1994).
9. U. Frisch and Parisi, G., in *Turbulence and Predictability of Geophysical Flows and Climate Dynamics*, ed. M. Ghil, R. Benzi and G. Parisi (North-Holland: New York, 1985), p. 84.
10. C. Meneveau and Sreenivasan, K. R., *Phys. Lett. A* **137** (1989) 103; *J. Fluid Mech.* **224** 429.
11. C. J. G. Evertsz and B. B. Mandelbrot, in *Chaos and Fractals: New Frontiers of Science*, ed. H.- O. Peitgen, H. Jurgence, and D.Saupe (Springer Verlag: Berlin, 1992) p. 921.
12. D. Rabin, *Astrophys. J.* **390** (1992) L103.
13. H. Lin, these Proceedings (1994).
14. E. Ott, Du, Y., Sreenivasan, K. R., Juneja, A. and Suri, A. K., *Phys. Rev. Lett.* **69** (1992) 2654.
15. R. D. Robinson, Jr., *Astrophys. J.* **239** (1980), 961.

SOLUTIONS FOR THE REMOVAL OF THE 180° AMBIGUITY PROBLEM IN THE AZIMUTH OF TRANSVERSE MAGNETIC FIELD

Jing Li and Pascal Démoulin
DASOP, Observatoire de Meudon
France

1. Introduction

Three methods have been tested for the removal of the 180° ambiguity problem[1,2,4] with Low's[5] nonlinear Force-Free analytical magnetic field models. This study has three advantages: (1) it avoids the presence of noise in observational data; (2) it is independent of the magnetic field calibration; (3) the accuracy of methods can be obtained by comparing the ambiguity-solved fields with given field models.

In this poster, the published work is briefly reviewed. Recent works are described, which further test methods applied to Low's models with proper magnetic field noise.

2. Non-Linear Force-Free Field Models

A non-linear force free magnetic field model is produced by two systems of currents: first, an infinite straight line of current located below the photosphere; second, a system of currents distributed in the whole space.[5] A particular axisymmetric solution of the non-linear equations is given by the following:

$$\begin{aligned} B_x &= \frac{-B_0 a}{r} cos\varphi(r) \\ B_y &= \frac{B_0 axy}{r[y^2+(z+a)^2]} cos\varphi(r) - \frac{B_0 a(z+a)}{[y^2(z+a)^2]} sin\varphi(r) \\ B_z &= \frac{B_0 ax(z+a)}{r[y^2+(z+a)^2]} cos\varphi(r) + \frac{B_0 ay}{[y^2+(z+a)^2]} sin\varphi(r) \end{aligned}$$

$$r = \sqrt{x^2+y^2+(a+z)^2}$$

The solutions include a free "generating function", $\varphi(r)$. This function is related to force-free factor α by $\alpha(r) = -\frac{d\varphi(r)}{dr}$ defined by the equation $\nabla \times \vec{B} = \alpha(r)\vec{B}$.

In our work, five generating functions were used to obtain 5 models:

1. $\varphi(r) = f\ln$ (r) (α changes slowly with r);
2. $\varphi(r) = f\tanh$ (r) (α changes rapidly with r);
3. $\varphi(r) = f$ (r-l) for r ≤ 3;
   $\varphi(r) = 2f$ for r>3;
4. $\varphi(r) = f$(cos(2r)+sin(2r)) (α changes sign);
5. $\varphi(r) = f$ (analytical potential field).

We define $f = \pi/2 \mid \phi \mid_{max} / \phi_O$. Four values of f are considered for each model. They are $f = 0.5$, 1.0, 2.0 and 4.0, where $f = 0.5$ represents the simplest configuration of the field and $f = 4.0$ represents the most complex one. We have 5 models, and 4 cases for each model. It is 20 analytical fields in total.

3. Methods and Results

The summary of three methods are shown in the following table.

Method	Name	Criterion
I	Magnetic shear-based method	$B_{ref} \cdot B_{obs} \geq 0$
II	Magnetic Energy Gradient	$\frac{\partial B^2}{\partial z} \leq 0$
III	Divergency-Free Method	$\frac{\partial B_z}{\partial z}\left(\frac{\partial B_x}{\partial x} + \frac{\partial B_y}{\partial y}\right) \leq 0$

Table 1: Three methods are summarized here. In the method I, $\mathbf{B_{ref}}$ represents the transverse component model. $\mathbf{B_{obs}}$ represents the transverse component solved the ambiguity problem by method I. In the method II, the gradient of magnetic energy is given by $B_x \frac{\partial B_z}{\partial x} + B_y \frac{\partial B_z}{\partial y} - B_z(\frac{\partial B_x}{\partial x} + \frac{\partial B_y}{\partial y}) \leq 0$ which is only connected with magnetic field observed at the photosphere.

Figure 1 shows the accuracy of the three methods applied to 5 models. The Y-axis represents the accuracy of the methods. It is evaluated by counting the percentage of pixels where the 180° ambiguity was correctly resolved. From the application of the methods, it can be concluded that:

1. The methods II and III are much more accurate than the method I for the models 1, 3, and 4.

2. The accuracy of all three methods decreases as the complexity of fields increases;
3. The accuracy of all three methods increases as the array size increases. This shows the importance of high angular resolution of the magnetic field (this result is not seen in Fig. 1).

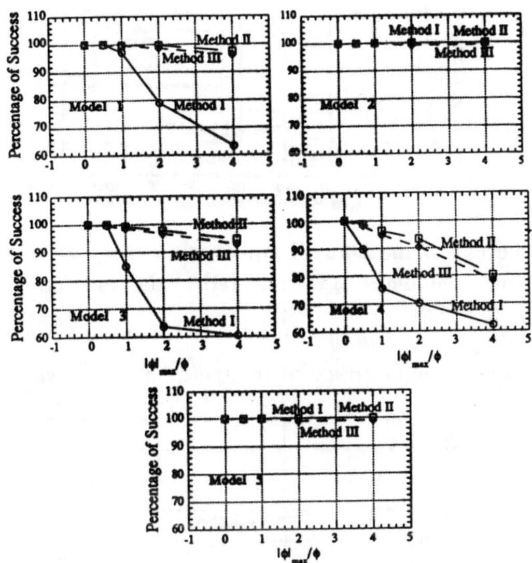

Figure 1: Method I is magnetic shear-based method. Method II is based on magnetic energy gradient decreasing as the height increases. Method III is magnetic divergent-free method. 5 figures represent 3 methods applied to 5 models, respectively. The percentage of success is obtained by counting pixels where the methods are accurate on solving the 180° ambiguity. The variation along X-axis represents the complexity of field structure.

4. Noise Work

Our recent work is to add proper noise to the FFF models so that the accuracy of the methods as they are applied to the observational data is evaluated. Noise of the longitudinal fields (B_Z) is set as 10 Gauss. The noise of the transverse field (B_t) varies from 0 to 300 Gauss.

Models 1, 3, and 5 are considered in this noise work, since they have closer configurations to real magnetic structures on the sun than do models 2 and 4. As methods II (magnetic energy gradient) and III (divergence-free) are more sensitive to observational noise than is method I (magnetic shear), we concentrate here on those two methods.

In Figure 2, models 1-1, 1-2, 1-3, and 1-4 represent 4 complex structures. The X-axis varies with the magnetic field noise added to the analytical transverse fields. It shows the accuracy of the methods is nearly independent of the complexity of the field structures. In addition, method II is more accurate than the method III when the noise is greater than about 10 Gauss.

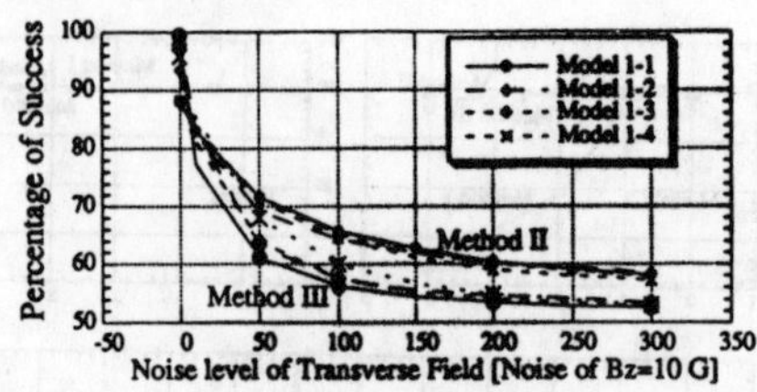

Figure 2: The success of methods II (magnetic energy gradient method) and III (magnetic divergence-free method) is indicated as the function of the transverse field. The array size is 128 x 128.

Figure 3 shows how the accuracy of the methods varies with the array sizes.

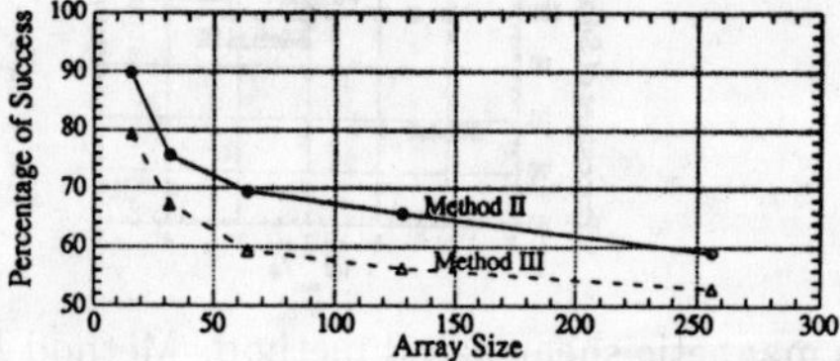

Figure 3: The accuracy of methods II and III is the function of array size. The noise level is 10G for the longitudinal field (B_Z) and 100G for the transverse field (B_t). The curve is made with model 1-1. The higher array size, i.e. higher angular resolution is, the less accuracy of methods is obtained.

5. Conclusions

1. The transverse field noise must be less than 50 Gauss if methods II and III are to successfully remove the 180° ambiguity problem.
2. The success of the "energy-gradient" and "divergence-free" methods is nearly independent of complex configurations of magnetic fields.
3. To benefit from the advantage of high angular resolution in magnetograms, the signal-to-noise ratio of the magnetic fields should be high.

6. References

1. Cuperman, S., Li, J., & Semel, M.: 1992, *A & A*, **265**, 296.
2. Cuperman, S., Li, J., & Semel, M.: 1993, *A & A*, **268**, 749.
3. Cuperman, S., Li, J., & Semel, M.: 1993, *A&A*, **278**, 279.
4. Li, J. Cuperman, S., & Semel, M.: 1993, *A&A*, **279**, 214.
5. Low, B.C.: 1982, *Sol. Phys.*, **77**, 43.

OSCILLATIONS IN ACTIVE PLAGE REGIONS AS OBSERVED IN 1.56 MICRON LINES

K. MUGLACH
Institut für Astronomie, Karl–Franzens Universität Graz
Universitätsplatz 5/2, A–8010 Graz, Austria
Kiepenheuer Institut für Sonnenphysik
Schöneckstraße 6, D–79104 Freiburg, Germany

S.K. SOLANKI
Institut für Astronomie, ETH–Zürich
ETH–Zentrum, CH–8092 Zürich, Switzerland

and

W.C. LIVINGSTON
National Solar Observatory, NOAO
P.O. Box 26732, Tucson AZ 85726, USA

ABSTRACT

We address problems of the dynamics of small–scale magnetic elements by analysing time series of spectra of two infrared spectral lines (Fe I at 1.5648 μm, $g = 3$, and Fe I at 1.5652 μm, $g_{\mathrm{eff}} = 1.53$). The data (Stokes I and V) were obtained with the McMath–Pierce telecope on Kitt Peak, together with the vertical spectrograph and single diode infrared detector. The dominant oscillatory signal in the velocity data is due to the 5 min oscillations. We also observe time variations of the magnetic field strength which exhibit a possible 5 min and 9–10 min oscillation.

1. Introduction

The problem of the dynamics of small–scale magnetic flux tubes can be addressed in various manners. The most direct information is obtained from the study of time series of spectra taken in polarized light (Stokes profiles).

So far the temporal behaviour of the Stokes V zero–crossing wavelength λ_V of photospheric lines in the visible have revealed velocity oscillations with a period of 5 minutes and amplitudes of 0.2 – 0.3 km s^{-1} [1,2,3,4,7,12]. This longitudinal velocity has been theoretically interpreted by Roberts [10] as representing radiatively damped sausage mode waves. Apart from its velocity signature the sausage mode produces another characteristic signature: it results in a small variation of the magnetic field strength. So far no periodic changes in the magnetic field strength have been observed.

The increased Zeeman sensitivity of lines in the infrared is an excellent tool for an accurate determination of the field strength, so the primary goal of this project was to look for periodic variations of the field in flux tubes of active plage regions. In addition, we will also study the velocity signal of the tube waves, since the IR lines are formed deeper in the solar atmosphere.

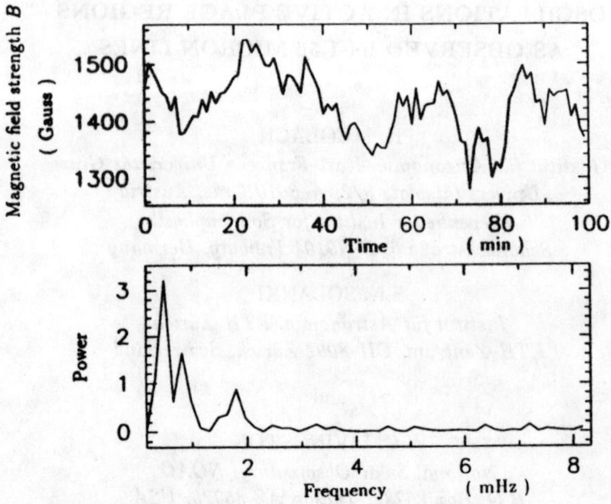

Figure 1: The upper figure gives the magnetic field strength B from the splitting of the $g = 3$ line versus time for the first time sequence, the lower figure shows the resulting power spectrum

2. Observations

A number of time series of Stokes I and V spectra where obtained using the McMath–Pierce telescope on Kitt Peak, together with the vertical spectrograph and the "Baboquivari" infrared detector [8]. We used two infrared spectral lines, namely Fe I at 1.5648 μm (with a Landé factor $g = 3$) and Fe I at 1.5652 μm ($g_{\text{eff}} = 1.53$).

This spectral combination allows a very accurate determination of field strengths of the deep photospheric layers of small–scale magnetic elements [11].

The aperture was centered on active plage regions near solar disk center. For the current analysis we selected the three time series with the best spectral quality which have a duration between 60 and 100 min and a time resolution between 30 and 60 sec. The spatial resolution was about 3 – 4″. A detailed description of the data can be found in Muglach [9].

3. Results

3.1. Magnetic Field

To get the field strength B we first determined the wavelength of the minimum and maximum of the red and blue Stokes V wing, λ_r and λ_b, of the $g = 3$ line. From their separation we calculate B.

The field strength (in Gauß) versus time (in minutes) and the corresponding power spectra of the three time series are shown in Figs. 1 – 3.

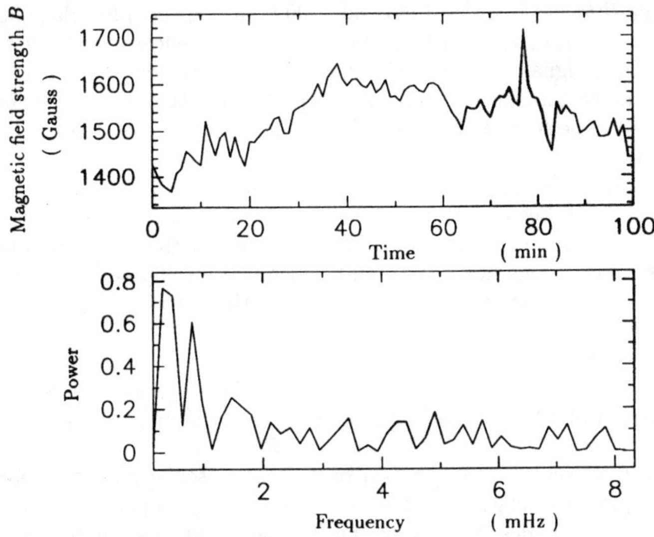

Figure 2: Same as Fig. 1, but for the second time sequence

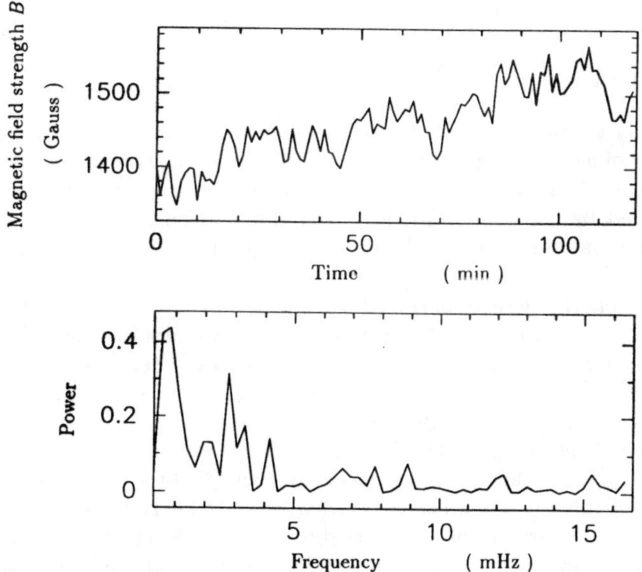

Figure 3: Same as Fig. 1, but for the third time sequence

We get B values between 1350 and 1700 G, as expected for plage flux tubes. The resulting power spectra show large peaks at low frequencies. On these time scales (50 to 15 min the signal can be due to a long term variation of the field, but the time series are much too short to consider this reliable. At somewhat higher frequencies we get smaller peaks at 9–10 min in Fig. 1 and 5 min in Fig. 3, while Fig. 2 shows no peak above noise level.

Up to now, the only observational analysis of the time behaviour of the field strength of flux tubes was done by Keller [7], using the ratio of the Stokes V amplitudes of Fe I 5247.1 Å and 5250.2 Å. He shows a plot of this magnetic line ratio versus time (his observations lasted 120 min) which exhibits fluctuations on time scales of minutes to one hour, but no significant peak in the power spectrum of the whole time sequence was found.

Theory predicts that magnetic elements are not in a stable configuration after forming in a convective collapse, but rather may be subject to (overstable) oscillations with a period of 10 – 20 min [5,6]. The 9–10 min peak in Fig. 1 fits quite nicely into this picture.

One other excitation mechanism of flux tube oscillation that theory proposes is strong coupling to the global p–modes which results in a periodical behaviour similar to that of the p–modes themselves. This would result in a 5 min oscillation as found in Fig. 3.

If the B oscillations are real, then we should see a corresponding velocity signature. In the following part we therefore examine the velocity oscillations.

3.2. Velocities

To study the mass motions the variation of the positions of Stokes I and V have to be determined. Although the observations were obtained in an active plage region, the influence of the polarized light on Stokes I is in the far wings, the central part is dominated by the non–magnetic surroundings. Consequently, by comparing the velocity fluctuations of I and V we are comparing flux tube and inter flux tube regions.

As wavelength reference of the intensity we simply calculate the wavelength of the line minimum, λ_I, to describe the wavelength of the V profile we use $\lambda_{br} = (\lambda_b + \lambda_r)/2$.

All of our Stokes I data reveal the 5 minute oscillation, an example is given in Fig. 4 for 1.5648 μm and 1.5652 μm respectively (for the same data set as in Fig. 2). The upper panel shows the time sequence of λ_I, the lower one its power spectrum. We get a single strong peak at 3.3 mHz which corresponds to a period of 5 minutes. Both lines show the same temporal behaviour and give the same power spectra. The amplitude of the oscillation has an RMS value of 100 – 200 m s^{-1}.

The Stokes V profiles of the same run also show 5 min oscillations (Fig. 5) but with reduced power at 3.3 mHz. The 5 min peak of the $g = 3$ line is even lower than the one of the $g_{\rm eff} = 1.5$ line. Nevertheless, the V profiles show roughly the same amount of total power as the I profiles, only some of the power is shifted to lower

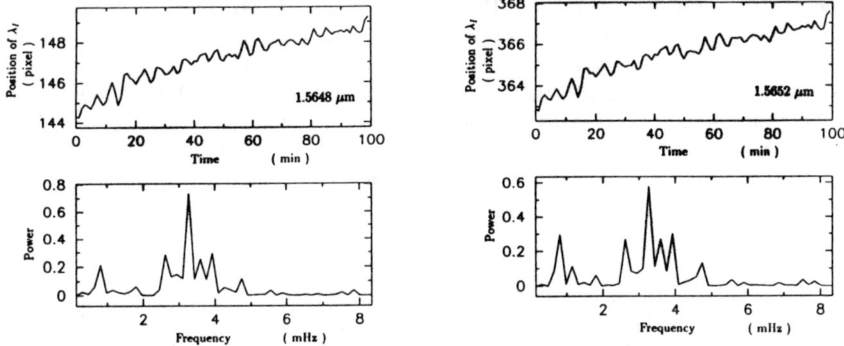

Figure 4: Position of the minimum wavelength of Stokes I λ_I versus time (in minutes) of 1.5648 μm and 1.5652 μm of the second time sequence and the corresponding power spectrum

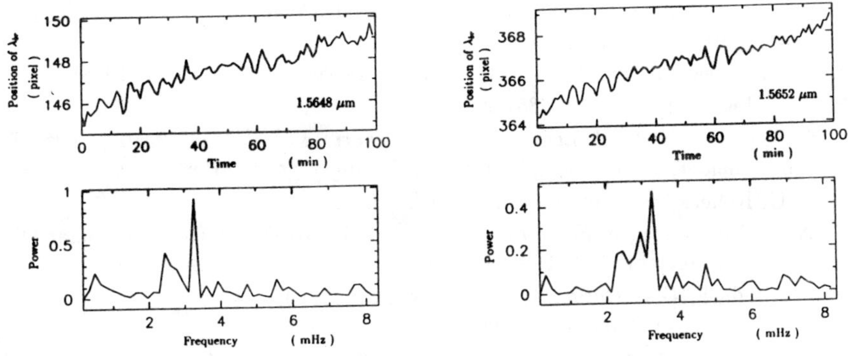

Figure 5: Position of λ_{br} versus time, (in minutes) of 1.5648 μm and 1.5652 μm and the corresponding power spectrum

frequencies (mainly 2.5 – 3 mHz). But these periods are still not sufficiently long to represent the velocity pertubation associated with a possible 9–10 min magnetic oscillation. We therefore think that this magnetic oscillation is not solar in origin. Since the magnetic features definitely exhibit 5 min velocity oscillations we cannot rule out that the 5 min B oscillations seen in one run are solar.

We hope that future investigations obtained with more advanced observational techniques answer the questions still left open by the current investigation.

4. Acknowledgements

K. M. wants to thank the organizers for their efforts to make a wonderful and successful workshop. Financial support from the "Österreichische Forschungsgemeinschaft" is gratefully acknowledged.

5. References

1. B. Fleck, *Rev. Mod. Astron.* **4** (1991) 90.
2. B. Fleck, F. L. Deubner, in *Mechanisms of Chromospheric and Coronal Heating*, eds. P. Ulmschneider, E. R. Priest, R. Rosner, (Springer, Berlin, 1991) p. 19.
3. B. Fleck, F. L. Deubner, W. Schmidt, in *The Magnetic and Velocity Fields of Solar Active Regions*, eds. H. Zirin, G. Ai, H. Wang, (Proc. IAU Coll. 141, Beijing, 1992) p. 522.
4. R. G. Giovanelli, W. C. Livingston, J. W. Harvey, *Sol. Phys.* **59** (1978) 49.
5. S. S. Hasan, *A&A* **143** (1985) 39.
6. S. S. Hasan, in *Mechanisms of Chromospheric and Coronal Heating*, eds. P. Ulmschneider, E. R. Priest, R. Rosner, (Springer Verlag, Berlin, 1991) p. 408.
7. C. U. Keller, *Dissertation*, ETH–Zürich (1992).
8. W. C. Livingston, in *Solar Polarimetry*, ed. L. November (National Solar Observatory, Sunspot, NM, 1991) p. 365.
9. K. Muglach, *Dissertation*, University of Graz (1994).
10. B. Roberts, *Sol. Phys.* **87** (1983) 77.
11. S. K. Solanki, I. Rüedi, W. C. Livingston, *A&A* **263** (1992) 312.
12. E. Wiehr, *A&A* **149** (1985) 217.

Polarimetric IR Array Observations of a Flare

M.J. Penn and J.R. Kuhn
National Solar Observatory/Sacramento Peak
National Optical Astronomy Observatories†
PO Box 62, Sunspot NM 88349 USA

ABSTRACT

We take slit spectra simultaneously in left and right circular polarization of 8 Å of the solar spectrum centered at 1083nm. Moving the slit after a pair of exposures, we scan the solar active region NOAA 7629 with a cadence of about four minutes from 1724 to 1900 UT on 06 Dec 1993. The region was in the decay phase of a C class flare. We compute a simple proxy for the magnetic field strength in the photosphere and chromosphere by fitting Gaussians to the Si I and He I absorption features in the left and right circulary polarized images. We show a movie of the intensities, velocities and magnetic fields and find the following morphological items: (1) we measure the magnetic field *inside* three decaying flare kernals with He I 1083nm emission, (2) we observe small magnetic bipoles in the photospheric magnetic field (both emerging AR bipoles and MMF bipoles associated with a small sunspot) but do not see these bipole features at the chromospheric level, suggesting that they are small loops with heights less than 1500km, (3) we find He I AR filaments avoid strong line-of-sight magnetic fields in the chromosphere, (4) we see strong shear in the line-of-sight velocity in some AR filaments in the chromosphere, (5) we see a line-of-sight velocity flow correlated with the line-of-sight magnetic field in sunspots in the photosphere in rough agreement with recent theorectical work.

1. The Data

A 128x128 pixel HgCdTe near-IR array was used at the NSO/Sac Peak Vacuum Tower Telescope to observe the solar spectrum near He I 1083 nm. A liquid crystal polarization analyser was used with the horizontal spectrograph and a Wollaston prism to obtain simultaneous left and right circular polarization spectral images at each slit position. Figure 1 shows three sample intensity and Stokes V images obtained from these observations. In the upper section of the figure are the intensity spectra; wavelength runs from blue to red (left to right) and spatial position slit along the slit runs vertically. Three dark absorption features are seen, the Si I line at 1082.6 nm, the He I line at 1083.0 nm and an atmospheric line at 1083.2 nm. The left intensity spectrum shows a quiescent section of the active region, the center shows a flare kernel and the right shows a dark filament showing a large redshift. The lower sections of the figure show the Stokes V images derived from the data. The observations scan the entire active region with a cadence of four minutes by stepping the slit.

† Operated by the Association of Universities for Research in Astronomy, Inc. (AURA) under cooperative agreement with the National Science Foundation

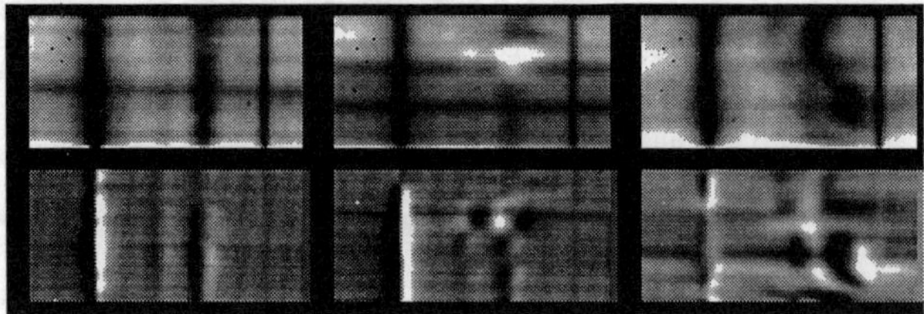

Figure 1: Sample Intensity and Stokes V Spectra for Three Slit Positions.

2. Magnetic Field Inside Flare Kernels

In the middle panels of Figure 1 the spectrograph slit crosses a flare kernel in the active region. This flare kernel shows strong He I emission in both spectral components of the line with the theorectical ratio of 8:1 in intensity. The Stokes V spectra of the flare kernel clearly shows a V signal in both spectral components of the line. The sign of the V spectra appears to reverse in the flare kernel compared to other regions of the slit; this is simply due to the subtraction technique used. The magnetic field in this and two other flare kernels appears to be larger than the magnetic field observed in the same spatial position at another time. We feel that this may be due to a lower formation height of the emission line combined with a vertical decrease in the magnetic field, but we will discuss other possibilities in another paper [1].

3. MMF and Emerging AR Bipoles Not Seen in Chromosphere

Figure 2 shows the measured photospheric (Si I) and chromospheric (He I) LOS magnetic fields plotted versus continuum contrast. This provides a illustration that while the photospheric magnetic field appears bipolar, the chromospheric field is more unipolar. Examining a time-series of the data reveals that many photospheric magnetic bipoles (such as some moving magnetic features seen near a sunspot and emerging bipolar active-region features) are not observed in the chromospheric magnetic field. We suggest that these features represent low loops which do not extend to the height of formation of the He I line, about 1500 km. This is consistant with previous examination of MMFs by Harvey and Harvey [2] who saw little evidence for MMFs at the height of Hα.

4. Filaments: Magnetic and Velocity Fields

The mean magnetic field in the dark He I filaments at the chromospheric level is about -170 $\pm$ 120 G. The filaments appear to avoid regions of strong LOS magnetic field, appearing at the edges of such regions instead. This is consistant with a more horizontal field in the filament which provides support against solar gravity. In some filaments strong shear in the vertical velocity is seen as if the filament were twisting.

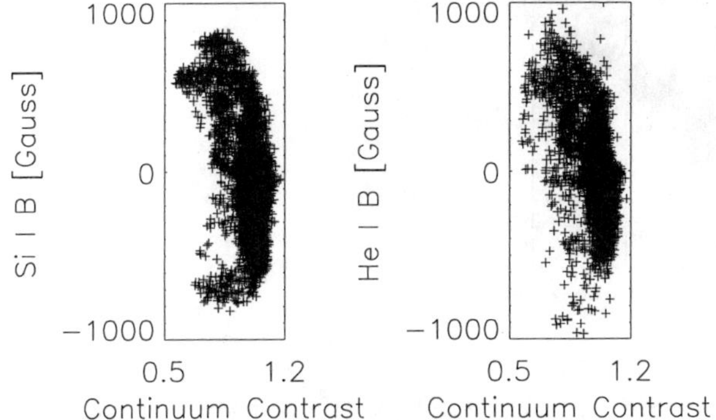

Figure 2: Dark Photospheric Bipoles Not Seen in Chromosphere.

5. Photospheric Velocity in Sunspots

A correlation is found with the photospheric LOS velocities and the photospheric LOS magnetic fields in agreement with recent theory about flows in rising magnetic flux tubes [3]. An upflow is observed in the leading polarity and a downflow is observed in the following polarity. The magnitude of the observed flow is about 200 m s^{-1}.

6. Magnetic Field at Two Heights

A simple proxy for the line-of-sight (LOS) magnetic fields are derived from the left and right polarized spectra by fitting Gaussians to the absorption lines. In Figure 3 we plot the LOS magnetic field derived from the Si I line (photospheric) versus the LOS magnetic field derived from the He I line (chromospheric). In some of the field-of-view the magnetic fields show different structures and this causes the spread in the diagram. In the positive polarity regions two sunspot umbrae are the dominant the magnetic structures, and we can see that the correlation between photospheric and chromospheric LOS magnetic fields is much better. Deriving a vertical magnetic field gradient from this data is difficult, but fitting the positive polarity regions we find that $B_{HE} = 0.84 \times B_{SI}$.

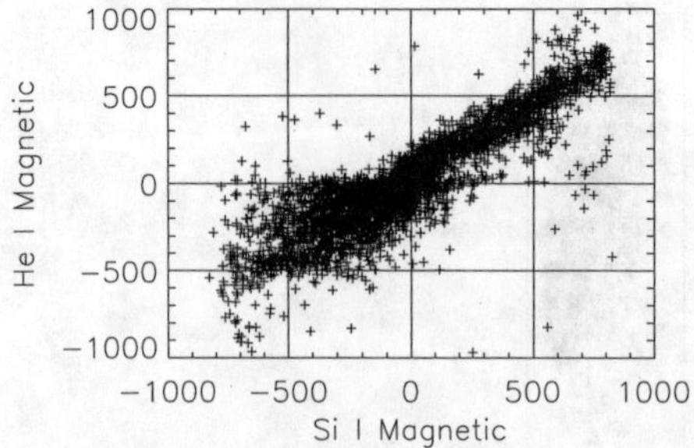

Figure 3: Photospheric (Si I) vs. Chromospheric (He I) magnetic fields.

7. Acknowledgements

We would like to acknowledge the help of the observing and support staff at the Vacuum Tower Telescope.

1. M. J. Penn and J. R. Kuhn, *ApJ* **in press** (1995)
2. K. Harvey and J. Harvey, *Sol. Phys.* **28** (1973) 61.
3. Y. Fan, G. H. Fisher and E. E. DeLuca, *ApJ* **405** (1993) 390.

ON THE DISTRIBUTION OF THE SOLAR MAGNETIC FIELDS

HAOSHENG LIN
Big Bear Solar Observatory, California Institute of Technology
Pasadena, CA 91125, USA

ABSTRACT

We use the two magnetically sensitive infrared Fe I lines at We use the two magnetically sensitive infrared Fe I lines at 15648.5 Åand 15652.9 Å to study the distribution of the solar magnetic fields. We found that the intra-network elements have an average field strength of about 500 gauss. Furthermore the solar surface magnetic field distribution can be described by a 2-component mode. The first component is the strong-field component. It includes the magnetic elements found in sunspots, plages, and quiet region network, and has an average field strength of 1400 gauss. The second component is the weak-field component comprised mostly of the intra-network elements. We also found that the average size of the weak-field elements is approximately 70 km in diameter.

We suggest that the clear distinction between the strong- and weak-field components imply that the strong-field represent the large-scale, deeply rooted solar-cycle field supported by the thermal pressure of the solar plasma; the weak-field component may be generated by the turbulent velocity of the solar convection zone, and is supported by the turbulent pressure of the solar convective motion.

1. Introduction

The magnetic flux on the surface of the sun is distributed non-uniformly in various structures with different size and field strength. It is intriguing to note that while the total flux of the surface magnetic structures varies by several order of magnitude, from 10^{22} Mx in sunspots to 10^{16} Mx in IntraNetwork (IN) fields, the mean magnetic field strength only changes by a factor of 2, from close to 3000 gauss in sunspots down to about 1500 gauss in the quiet region network elements. The intrinsic field strength of the IN fields was not well determined; however, most researchers agreed that they are also of kilogauss field strength[1].

Why do most of the magnetic fluxes exist in the form of strong (kilogauss) fields? If strong fields are preferred over weak fields, then it has strong implication about the physical conditions in the solar photosphere. The magnetic field strength B is an important parameter that determines the dynamics of the magnetic flux tubes in the solar plasma. The very existence of the magnetic flux tubes stipulate the Magneto-HydroDynamical (MHD) equilibrium between the magnetized and n on-magnetized solar plasma. Thus, the number distribution of magnetic field $N(B)$ also bears information about the thermodynamical properties of the solar plasma in which they reside. The importance of knowledge of B and $N(B)$ cannot be overstated.

Rabin[2] measured the magnetic field distribution $N(B)$ in a plage region and found that $N(B)$ peaked around 1400 gauss, which is close to the thermal equipartition field strength of 1440 gauss[3] of the solar photosphere. In quiet regions, most authors obtained kilogauss fields[4,5,6]; however, the sample of the measurements is small, and we don't have a reliable measurement of the intrinsic magnetic field strength and magnetic field distribution of the IN fields to date. Thus, we carried out an experiment mapping the solar magnetic field configuration in both active and quiet regions using the Fe I 15648.5 Å and Fe I 15652.9 Å lines with an infrared array camera. We will present new results of the measurements of the magnetic field strength of the IN fields and $N(B)$ in the active and quiet regions. The implications of these results will be discussed as well.

2. Observations and Data Reductions

2.1. *The Least-Squares Fitting*

The data described in this paper were obtained at the National Solar Observatory/Sacramento Peak, New Mexico in February 1994, and at Big Bear Solar Observatory in May 1994. Stokes V profiles of several regions were obtained by scanning the spectrograph slit across the regions. The Stokes V profiles of the Fe I 15648.5 Å g=3, and Fe I 15652.9 Å g=1.53 lines were fitted simultaneously by

$$\begin{aligned} V(\lambda) &= \frac{1}{2}A[(p(\lambda-\lambda_1+\triangle\lambda_B)-p(\lambda-\lambda_1-\triangle\lambda_B)) \\ &+ 0.77(p(\lambda_1-\lambda_2+0.51\triangle\lambda_B)-p(\lambda-\lambda_2-0.51\triangle\lambda_B))] \end{aligned} \quad (1)$$

where A is the amplitude factor of the Stokes V profile, λ_1 and λ_2 are the line center wavelengths of the $g=3$ and $g=1.53$ lines, respectively, $\triangle\lambda_B$ is the Zeeman splitting of the $g=3$ line, and $p(\lambda)$ is the line profile. The factor of 0.77 is the ratio of quiet-sun line-depth D_q of the $g=1.53$ to the $g=3$ line. We assumed a gaussian profile for $p(\lambda)$,

$$p(\lambda) = 1 - D_q exp(-\frac{\lambda^2}{2\sigma_E^2}). \quad (2)$$

For the usefulness and limitations of this formulation, the readers are referred to the discussion in Rabin[7].

2.2. *Fitting Criteria*

Noise, systematic errors and contributions of unresolved magnetic elements within our imaging pixels may result in anomalous Stokes V profiles. Thus, we adopted two criteria to be satisfied before we can perform the least-squares fitting algorithm. The first criterion is that the apparent flux ϕ_A of the $g = 3$ line must be greater than 0.02. That is,

$$\phi_A = \mid \sum_{\lambda=\infty}^{0} V(\lambda) - \sum_{\lambda=0}^{+\infty} V(\lambda) \mid \geq 0.02. \tag{3}$$

The apparent flux ϕ_A from a non-magnetic spectrum is typically a few times 10^{-3}. The 0.02 threshold was set to assure that the fitted spectra have enough signal-to-noise ratio to obtain a good fit. Using the definition of $V(\lambda)$ in Eq. (1) and (2), we can express the magnetic flux ϕ in terms of ϕ_A in our $0.48'' \times 1''$ pixel by $\phi = 7.56 \times 10^{17} \phi_A \mu^{-1}$ Mx. Thus the threshold of $\phi_A \geq 0.02$ corresponds to $\phi \geq 1.5 \times 10^{16}$ Mx.

The second criterion is that the asymmetry p of the Stokes V profiles, defined as the absolute value of the sum of the V profile over all wavelength range, must be less than 0.02. That is,

$$p = \mid \sum_{\lambda=-\infty}^{+\infty} V(\lambda) \mid \leq 0.02. \tag{4}$$

A typical non-magnetic spectrum usually yields $p \approx 0.002$, and we allowed for some distortion of the V profiles by setting a higher threshold for p. Both the thresholds were determined by inspecting the fitted results of many weak-field elements.

3. Results

3.1. The Magnetic Field Distribution

Figure 1 and Figure 2 show the magnetic flux ϕ and true-field B maps of active region NOAA 7675, and a disk center region, respectively. In active region, most of the flux are in strong-field form. In quiet region, however, we found that most of the magnetic elements outside the network elements are of strength below 1000 gauss. Comparison with BBSO magnetogram shows good correlation between the weak-field elements and the IN fields. Thus, we conclude that most of the IN fields are of sub-kilogauss field strength. Figure 3 shows magnetic field distribution $N(B)$ of active region and quiet region. The active region distribution peaks around 1400 gauss. This is similar to the distribution obtained by Rabin[4]. However, while Rabin's result showed sharp turn-off for field below 1000 gauss, we found a substantial fraction of the magnetic pixels are of sub-kilogauss field strength. About 33% of the magnetic pixels have sub-kilogauss field in the plage. Inspection of the true-field maps showed that most of these weak-field magnetic pixels appear at the edge of regions with strong-field. Only a small fraction of these weak-field pixels show up as small, isolated magnetic elements. This is consistent with the finding of Ruëdi et al.[8].

In quiet regions, we find that, in addition to the kilogauss component, the magnetic field distribution $N(B)$ has another component that peaks around 500 gauss. This is the dominant component in the quiet sun. About 65%, 75%, and 59% of the magnetic pixels are of sub-kilogauss field in the three quiet regions. Comparison between the BBSO magnetograms and the IR flux and true-field maps shows that almost all the kilogauss pixels are in the high-flux network elements, most of the weak-field pixels are distributed more or less uniformly in the regions between the strong network elements. Comparison showed that there is good one-to-one correspondence between the weak-field elements in the IR measurement and the BBSO IN elements. Thus, we conclude that the weak-field component observed in the quiet regions is associated with the intra-network magnetic elements.

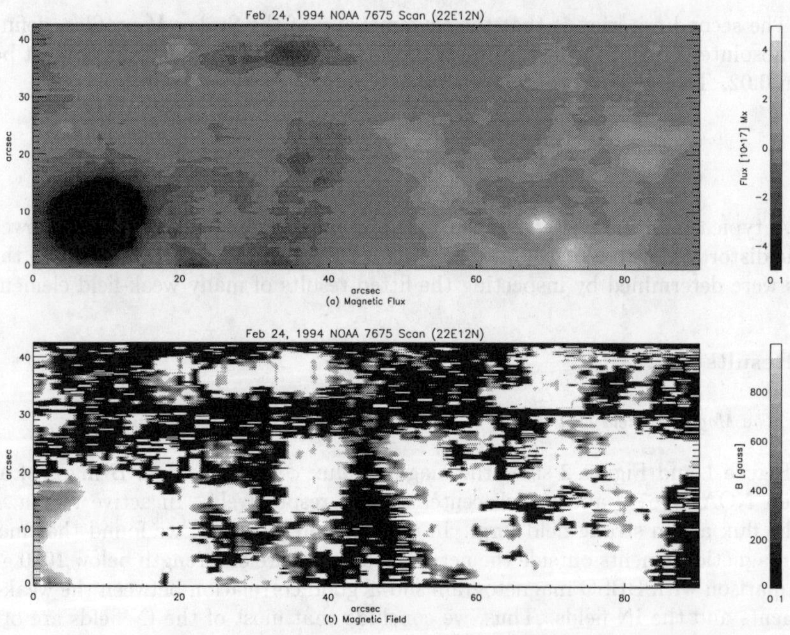

Figure 1: The (a) magnetic flux σ, and (b) true-field B maps of NOAA 7675.

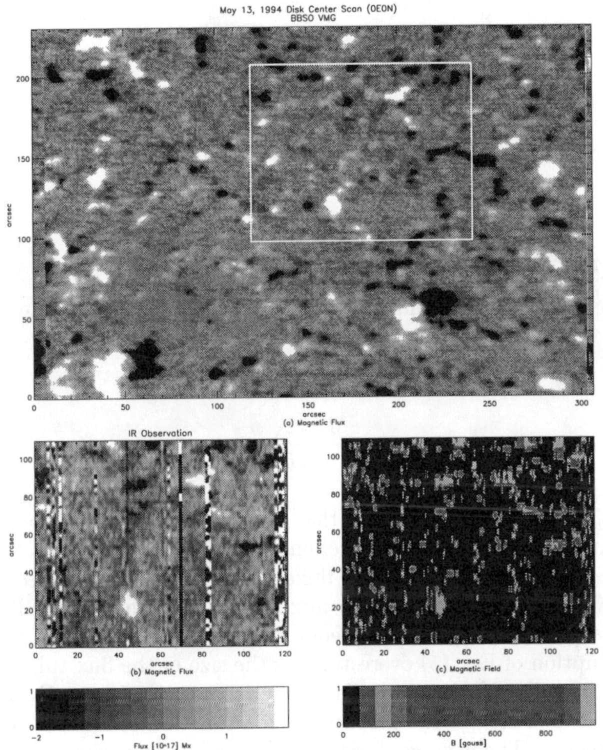

Figure 2: The (a) BBSO magnetogram, (b) IR magnetic flux, and (c) IR true-field maps of a disk center quiet region.

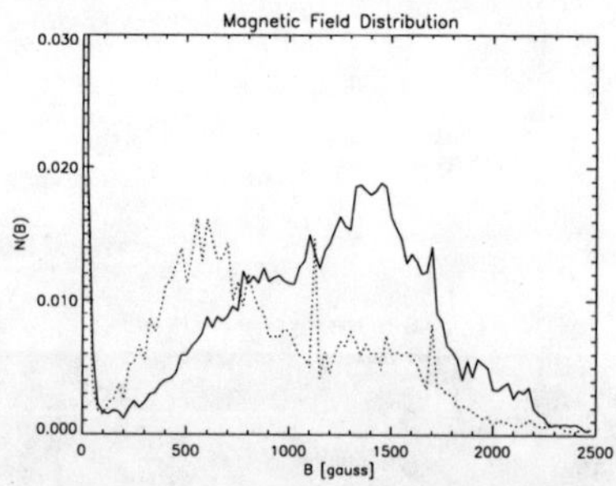

Figure 3: The magnetic field distribution of active region (solid line) and quiet region (dashed line).

3.2. The Size of the Magnetic Elements

Given the magnetic flux and field strength, we can calculate the size (diameter) of a flux tube by

$$\frac{d = 2\sqrt{\phi}}{\pi B} \tag{5}$$

However, Eq. (5) implicitly assumes that we can measure the true flux of a single magnetic element. We know from experience that the detection of IN fields depends highly on seeing conditions. Thus, the fluxes of IN fields we measured in our imaging pixels usually underestimate the true fluxes. On the other hand, it is most likely that there are multiple magnetic elements in each of the IN pixels. Thus, the single-element assumption of Eq. (5) overestimates the size of the flux tube. Therefore, the size derived using Eq. (5) is more likely the *average* size of the IN fields. Figure 4 shows the measured size of the flux tubes as a function of B of a plage and a disk center quiet region. The average diameter of the weak-field component (between 400 and 1000 gauss) has a well-defined value of about 70 km with very small spread in both datasets. For the region below 400 gauss, the size measurement may be biased by our 1.5×10^{16} Mx/pixel flux limit, and we do not have a strong confidence here. In the strong-field regime $B \geq 1000$ gauss, the sizes derived from the two datasets differ significantly. This suggests that the size of the strong-field component varies over a wide range. This is not unexpected. When the flux increases, the flux tube must expand in order to maintain MHD equilibrium.

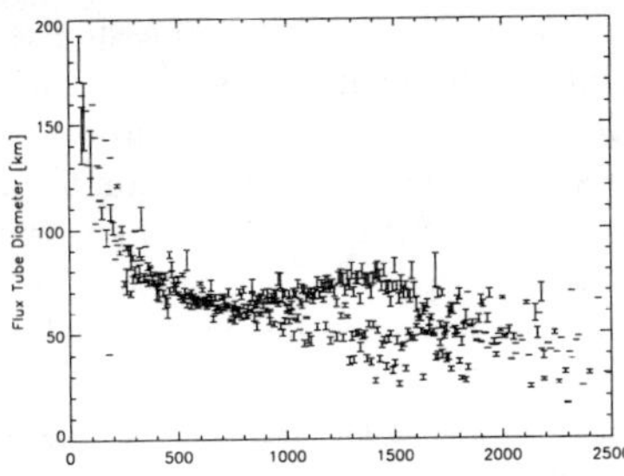

Figure 4: The diameter of flux tubes calculated from a plage and a quiet region. Error bars represent a 3σ error.

4. Discussions and Conclusions

The clear distinction between the weak- and strong-field components is a somewhat surprising result. The natural question to ask is: How does the sun generate and support two distinct components of magnetic field in the photosphere?

To try to answer this question, we first note that the elements of the strong-field component have very different characteristics from the elements of the weak-field component. In general, the strong-field elements are long-lived, stable (against solar plasma motion) structures. Their numbers exhibit strong correlation with the solar cycle. These characteristics imply that the strong-field component is the solar-cycle field that originates at the base of the solar convection zone. Comparing with the strong-field component, the weak intra-network fields are short-lived ($\approx$ hours), they move with the supergranulation flow, and they do not show any sign of solar cycle dependence[9]. Thus, it is unlikely that they are also linked with the solar-cycle field. Spruit et al.[10] proposed that 'U-loops' associated with the formation of active regions will disperse through the SCZ and form a weak field that covers a significant fraction of the solar surface. Petrovay & Szakály[11] pointed out the need for a cycle-independent small-scale solar dynamo operating in the solar convection zone in order to explain the observed flux density of intra-network field. Durney, De Young, & Roxburgh[12] suggested that, in addition to the large-scale cycle-dependent magnetic field, there is another component of magnetic field generated by the turbulent motion of the SCZ. Numerical simulations[13,14] also showed indications of the existence of a small-scale dynamo. Although most of these calculations predicted a very weak field - in the range of 50 to 150 gauss - they all pointed out that the intra-network fields are generated in the SCZ.

In a large kilogauss flux tube, the magnetic pressure is balanced by the thermal pressure difference in and outside of the flux tube. This suggests that the average field strength of the strong-field component should be close to the thermal equipartition field strength of 1400 gauss. Both Rabin's[2] results and our measurement confirmed this. In a weak-field flux tube the average size of 70 km is comparable to the photon diffusion length of the solar photosphere. This means that radiative energy transfer can efficiently heat the magnetized plasma inside the weak-field flux tube. Thus there should be little thermal pressure difference in and outside the flux tube. The 500-gauss field strength can still suppress the convective energy transport into the flux tube. Thus, the weaker turbulent pressure term may be responsible for balancing the magnetic pressure. If we take the granular velocity field of $v_t \approx 2$ km/sec as the turbulent velocity, then we can calculate the energy density of the turbulent field,

$$\epsilon_{turb} = \frac{nm_H v_t^2}{2} \approx 10^{17} \times 1.67 \times 10^{-24} \times (2 \times 10^5)^2 = 6.6 \times 10^3 \mathrm{ergcm}^{-3} \quad (6)$$

where m_H is the mass of hydrogen atom. In comparison with the thermal energy density of the photosphere, $\epsilon_{therm} = nkT = 8.28^{-24}$ erg cm-3, ϵ_{turb} is one order of magnitude smaller, thus it is not important in the strong-field elements. We can find the turbulent equipartition field strength by

$$\bar{B}_{turb} = \sqrt{8\pi\epsilon_{turb}} = 400 gauss. \quad (7)$$

This is in good agreement with the observed average field strength of the weak-field component.

For the turbulent pressure term to be important, the turbulent eddies must be smaller than the flux tube size; otherwise, the flux tubes would only see a uniform velocity field. In this case, the weak-field flux tubes should move with the plasma flow, and will not experience any turbulent pressure. However, the lifetime of the turbulent eddies ($\approx$ 20 min, the lifetime of granule) is much shorter than the lifetime of the IN fields; hence, on time average, the flux tubes should still experience a random velocity field. Thus, the turbulent convective motion of the SCZ is indeed exerting a turbulent pressure on the flux tubes.

Thus, we can conclude that the strong-field component is generated by the large-scale cycle-dependent dynamo operating at the base of the solar convection zone, and is supported by the thermal pressure of the photosphere. The weak-field component is generated by a small-scale dynamo associated with the turbulent motion of the solar convection zone, and is supported by the turbulent pressure of the turbulent velocity field.

5. Acknowledgments

The author would like to thank the observers at Big Bear Solar Observatory and at the Vacuum Tower Telescope of NSO/Sac Peak for many helps during the observations. The author is grateful to the scientists of the Caltech Solar Astronomy group, and the colleagues attending this workshop for many inspiring discussions. Dr. Jeff Kuhn generously lent the infrared camera for use at BBSO. This research was supported by a grant from the Office of Naval Research, grant number N00014-89-J-1069.

6. References

1. J. O. Stenflo, in *IAU Symp. 141, The Magnetic and Velocity Fields of Solar Active Regions*, ed. H. Zirin, G. Ai, and H. Wang (ASP, San Francisco, 1992).
2. D. Rabin, *Ap. J.*, **390** (1992) L103.
3. H. Zirin, *Astrophysics of the Sun*, (Cambridge University Press, Cambridge, 1988).
4. R. W. Howard and J. O. Stenflo, *Solar Phys.* **22** (1972) 402.
5. E. N. Frazier and J. O. Stenflo, *Solar Phys.* **27** (1972) 300.
6. J. O. Stenflo, *Solar Phys.* **32** (1973) 41.
7. D. Rabin, *Ap. J.*, **391** (1992) 832.
8. I. Ruëdi, S. K. Solanki, W. Livingston and J. O. Stenflo, *A&A* **263** (1992) 323.
9. S. F. Martin, *Solar Phys.* **117** (1988) 243.
10. H. C. Spruit, A. M. Title and A. A. Ballegooijen, *Solar Phys.* **110** (1987) 115.
11. K. Petrovay and G. Szakály, *A&A* **274** (1993) 543.
12. B. R. Durney, D. S. De Young and I. W. Roxburgh, *Solar Phys.* **145** (1993) 207.
13. M. Meneguzzi, U. Frisch and A. Pouquet, *Phys. Rev. Lett.* **47** (1981) 1060.
14. M. Meneguzzi and A. Pouquet, *J. Fluid Mech.* **205** (1989) 297.

SOLAR OBSERVATIONS NEAR 1.6μM WITH THE GERMAN TELESCOPES AT TENERIFE

DIRK SOLTAU
Kiepenheuer-Institut für Sonnenphysik
Freiburg, Germany

1. Introduction

Observations in the near-infrared at wavelengths around 1.6μm are of particular interest for several reasons:

- Because of the opacity minimum continuum observations near 1.6μm wavelength allow for investigations of the deepest directly accessible layers of the solar atmosphere.
- Since the Zeeman splitting grows with the square of the wavelength, whereas since the Doppler width increases only linearly, there is a gain in magnetic sensitivity by a factor of 3 compared with visible observations.
- Effects of instrumental polarization and seeing tend to be smaller in the infrared.

In order to take advantage of these properties, the Kiepenheuer Institut purchased a two-dimensional infrared camera system. The first examples of observations made at the German Vacuum Tower Telescope and the Gregory Coudeé Telescope at Tenerife are shown demonstrating the potential of the equipment.

2. The Camera

The Kiepenheuer-Institut bought an Amber AE4256 camera system. The cooled (liquid nitrogen) detector (InSb) is sensitive from 1 to 5μm and consists of a 256 x 256 pixel array, where the pixel size is 38μm. The system is controlled by a 486 PC and the data are digitized with 12 bit. The integration time can be varied and allows frame rates from 1 to 60 frames per second. Data are written in FITS format on exabyte tapes. The software allows for previewing the images as a real time image as well as for basic image arithmetic.

3. Infrared Imaging at the VTT

The VTT allows for simultaneous observations in up to four different wavelength channels (white light, CaK, Hα, and a wavelength selected with a tunable filter). Figures 1 and 2 show continuum images of granulation including two pores. The exposures were taken simultaneously but with different exposure times. The time of observation was May 3, 1994 at 10:33 UT.

Figure 1 is a white light image (an interference filter around 500 nm was used). The exposure time was 0.1s. The uncorrected normalized contrast of this image is 2.3% (rms) corresponding to medium class seeing.

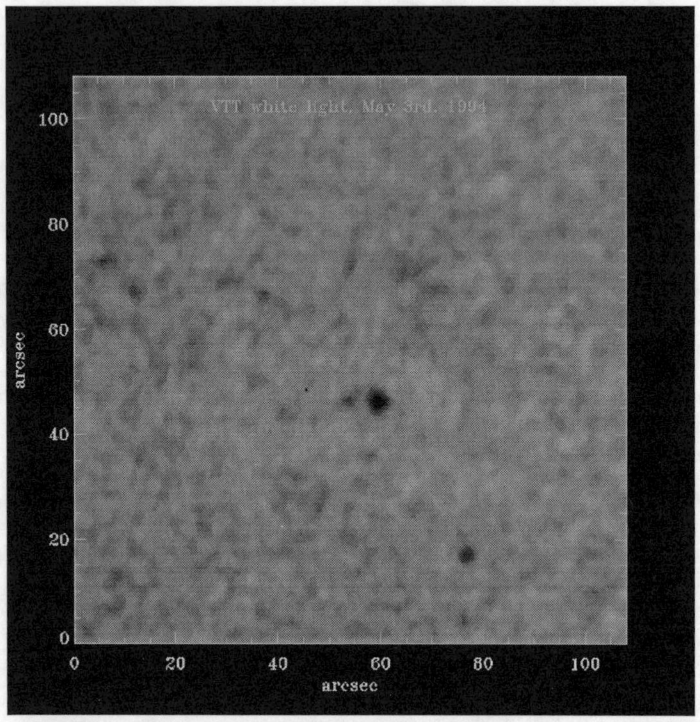

Figure 1: Continuum image taken with an interference filter near 500 nm. The bandwidth was about 6 nm.

Figure 2 shows the same structures in infrared light using the wavelength interval between 1.4μm and 1.7μm.

Here the exposure time was 5 ms. In that case the seeing was practically frozen in, which is one reason for the better appearance of this image. Koutchmy[2] and Keil et al[3] pointed out that near 1.6μm the seeing is expected to be better. This may be another reason for the better quality of the IR image compared with Fig. 1. The smallest structures are close to the resolution limit of the telescope at this wavelength (0.7 arcsec). The (uncorrected) rms contrast is 1.1%.

Image pairs like those in Figures 1 and 2 can be used to investigate e.g. convection effects in the deeper layers of the photosphere by comparing their size and intensity distributions.[1,3,2] Preliminary results confirm the findings of these authors that the average granule size is smaller when observed near 4.6μm. Figure 3 shows the two-dimensional power spectrum of the data shown in Figure 1 whereas Figure 4 shows the infrared power spectrum. Obviously there is more power at the higher spatial frequencies for the infrared image.

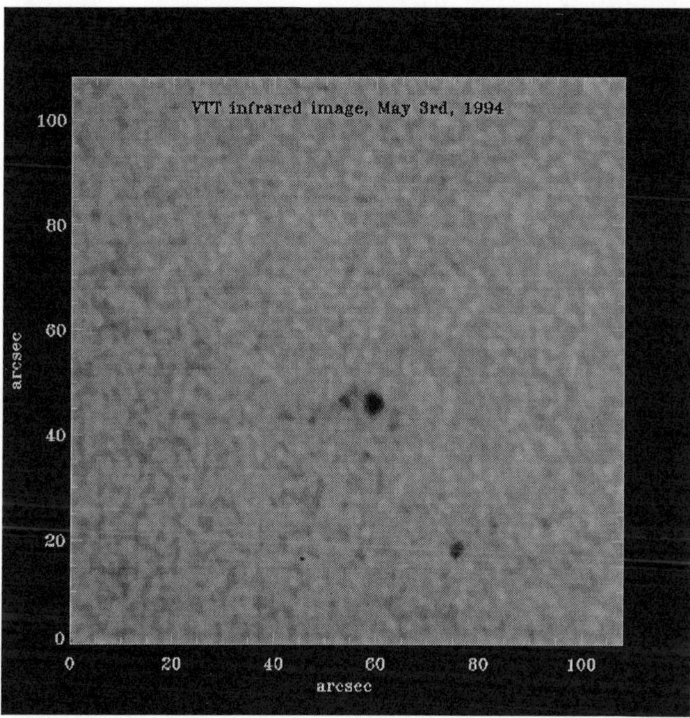

Figure 2: Continuum image taken in the near infrared using a broad band filter transmitting between 1.4μm and 1.7μm.

Figures 1 and 2 are also examples for measuring the contrast of pores in two very different wavelengths. In the visible, the (uncorrected) relative intensity of the bigger pore was determined to be 67% of the surrounding quiet region. In the infrared image, the same feature shows a relative intensity of 83%. The diameter of this feature (FWHM) is 1.75 arcsec. For several reasons, such IR intensity variations are easier to interpret in terms of temperature differences than observations in the visible continuum.[2] A simple estimation using the Planck formula leads to a temperature decrease of 550K as compared to the surroundings. These numbers shall only demonstrate possible future observations which will have to take into account corrections for the MTF properly.

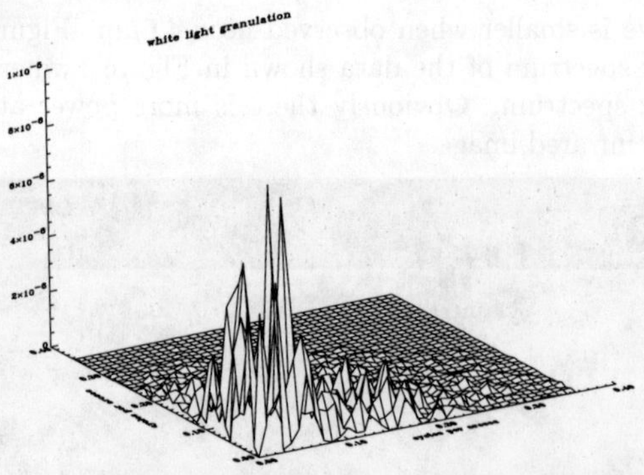

Figure 3: Two-dimensional power spectrum of the white-light image.

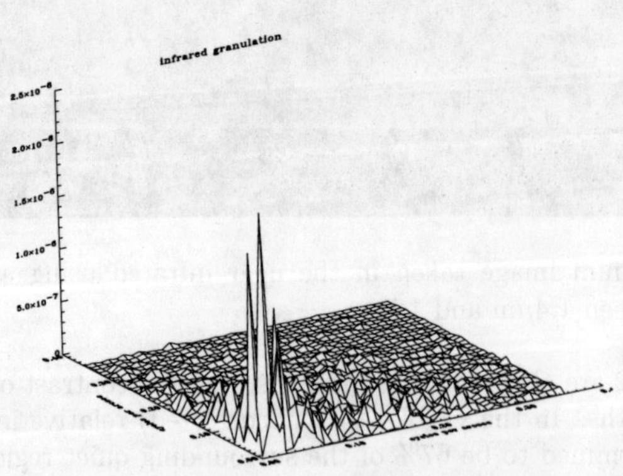

Figure 4: Two-dimensional power spectrum of the infrared image.

4. Infrared Spectroscopy

Infrared spectroscopy can be done either at the Gregory Coudé Telescope (GCT) or at the VTT. At the GCT, we have a dispersion of about 0.0029 nm per pixel. Although the intensity level is low, we have to use short exposure times to avoid the thermal noise to saturate the camera. This results in relatively low signals. Figure 5 shows a spectrum of a pore for the line at 1564.86 nm. This line is of particular interest because of its large Landè factor (g=3). The diagnostic value of this line has been described by Solanki et al.[4] In our example, the Zeeman splitting is about 0.14 nm leading to a field strength of 205mT (2050 Gauss). The signal appears rather noisy because of the necessary short exposure time. To increase the signal-to-noise ratio, it probably will be necessary to average over several exposures. The gratings at the observatory are not very suitable for infrared observations. This is why a new grating (200 gr/mm, blaze angle 52°) will be provided by the Astrophysikalisches Institute Potsdam from next year on. We hope to increase the efficiency of the spectrograph by a factor of 2.

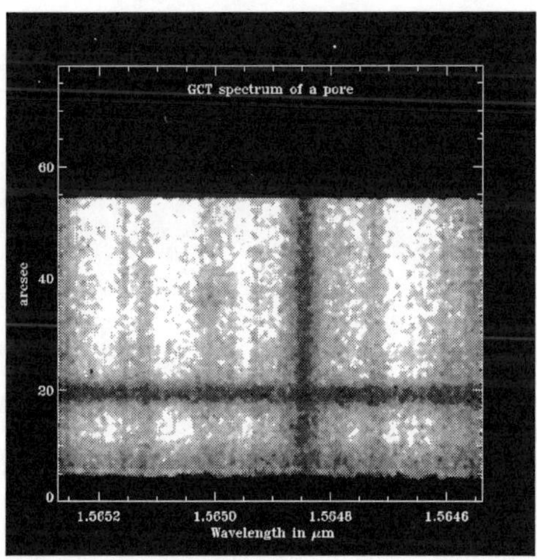

Figure 5: Spectrum of a pore obtained with the Gregory Coudé Telescope.

5. References

1. S. Koutchmy in "Solar Photosphere: Structure, Convection and Magnetic Fields", J. Stenflo (ed.) IAU Symp. 138, p. 81 (1990).
2. S. Koutchmy in "Infrared Solar Physics", D.M. Rabin, J.T. Jefferies and C. Lindsey (eds.) IAU Symp. 154, p. 239 (1994).
3. S. Keil, J. Kuhn, H. Lin, and K. Reardon, in "Infrared Solar Physics", D.M. Rabin, J.T. Jefferies and C. Lindsey (eds.) IAU Symp. 154, p. 251 (1994).
4. S.K. Solanki, I. Rüedi, and W. Livingston *Astron & Astrophys.* **263** 312 (1992).

He I λ1083nm OBSERVATIONS AND CHROMOSPHERIC AND CORONAL ACTIVITIES

Y. Suematsu, K.Ichimoto, and T.Sakurai
National Astronomical Observatory
Mitaka, Tokyo 181, Japan

ABSTRACT

We present some results on studies of line profile fitting analysis of He I 1083nm line spectra which were obtained with a 25-cm aperture coronagraph at the Norikura Solar Observatory.

1. Introduction

The He I 1083nm line is a good indicator of coronal as well as chromospheric structures and activities. Since 1989, we have obtained a number of spectra and spectroheliograms of this useful line at the Norikura Solar Observatory. From these data, we have found: (a) a close relationship between the steady structures in He I 1083nm intensity, velocity fields, and photospheric magnetic fields, (b) many dark points in active regions which vary with a time scale of 10–20 min, and (c) various manifestations of flares in this line[1,2]. Here we present some results of line-profile fitting analysis of this line.

2. Observations and Method of Line Profile Fitting

The spectra were obtained with a CCD camera (512×480 pixels), mounted on one of the exits of the Littrow-type spectrograph. The solar image was scanned so as to give a full disk spectroheliogram; it consists of four scans in east-west direction. One pixel corresponds to 1.08 arcsec in spatial direction and 9.7pm in dispersion direction. Two wavelength bands were usually recorded: One is the band of 0.175nm width centered at 1083.03nm and the other is the band of 0.068nm width at its nearby continuum of 1083.4nm.

We assumed that the He I λ1083nm line profiles are given by[3],

$$I(\Delta\lambda) = I_c\ exp[-\tau(\Delta\lambda)] + S(1 - exp[-\tau(\Delta\lambda)]), \tag{1}$$

where $I(\Delta\lambda)$ is the intensity in He I λ1083nm line, I_c is the continuum intensity near the line, S is the source function of the line, and

$$\tau(\Delta\lambda) = \tau_0\ exp[-(\Delta\lambda - \Delta\lambda_0)^2/\Delta\lambda_D^2], \tag{2}$$

where τ_0 is the optical thickness at line center, $\Delta\lambda_0$ is the Doppler shift, and $\Delta\lambda_D$ is the Doppler width. For simplicity, S was assumed to be constant which is 0.4 times

I_c[3 4]. It should be noted that this number may not be suitable for high-lying features such as filaments; it could be smaller than 0.4.

The observed profile was normalized by the nearby continuum as

$$C(\Delta\lambda) = (I(\Delta\lambda) - I_c)/I_c. \tag{3}$$

The normalized profile $C(\Delta\lambda)$ was finally expressed as:

$$C(\Delta\lambda) = (S/I_c - 1)(1 - exp[-\tau(\Delta\lambda)]). \tag{4}$$

We obtained, with a nonlinear least square fitting procedure, three unknown parameters τ_0, $\Delta\lambda_0$, and $\Delta\lambda_D$, which specify physical and dynamical properties of the He I λ1083nm chromosphere. The equivalent width was calculated using the fitted line-profile, because the observed one did not cover a wide enough wavelength range.

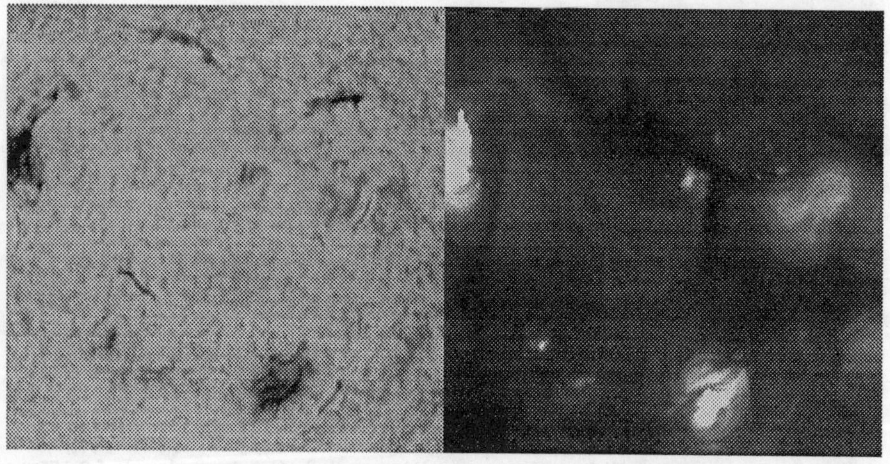

Fig. 1. Comparison of He I λ1083nm (left) and YOHKOH/Soft X-ray (right) images. These images were obtained simultaneously on 14 Aug. 1992. The field of view is 22'×22' at the central disk.

3. Results

We studied the relationships between line-profile parameters obtained. Generally, the He I λ1083nm line was weak and the observed profiles were noisy, and so we selected good fittings in which the standard deviation of fittings divided by the line depth is less than 5%. Examples of scatter plots between the line-profile parameters for the data of Sep. 4, 1994 are shown in Figures 3 and 4; corresponding spectroheliograms are given in Figure 2. Results for other dates were almost the same as those presented here.

(1) It has been well known that there exists an anticorrelation between the He I 1083nm and EUV/soft X-ray intensities. We roughly confirm this between the He I 1083nm and soft X-rays from YOHKOH/SXT; the deviation from the relation seems to be due to the formation height difference between them (Figure 1). Hence the comparison of line-profile parameters with YOHKOH/SXT was not straightforward and is not given here.
(2) The equivalent widths (EW) are in the range of 0.02 to 0.2Å. We tentatively classify the region by two parameters, namely EW and optical thickness to (or line depth)[5]; the quiet region where τ_0 is less than 0.1 has EW less than 0.1, the active region has EW between 0.1 and 0.2, and the filament has EW between 0.05 and 0.15 where the optical depth is large (0.15–0.3) and the Doppler width is small (about 0.5Å).
(3) We have large scatters in EW–Doppler width (DW) relations (Figure3). Especially it is large in quiet regions. This is partly because of poor fittings there and could be partly because of random motion of spicules[6]. It is interesting to suggest that large DWs are associated with the heating of the overlying corona[7], although we need further information in order to confirm this.
(4) Large Doppler shifts up to 10 km s^{-1} (0.4Å) are seen in quiet regions, i.e. in small EWs (Figure 4), which might indicate the spicule motion of large vertical velocity.

1. E. Hiei, K. Ichimoto, and G. Fang, *Flare Physics in Solar Activity Maximum 22, Lecture Note in Phys.* **387**, (1991), P. 67.
2. K. Ichimoto, G. Fang, and E. Hiei, *Proc. of the First China-Japan Seminar on Solar Physics* (Kunming TonDar Institute, China, 1993), p. 158.
3. E.H. Avrett, J.M. Fontenla, and R. Loeser, *Infrared Solar Physics, IAU Symp.* **154**, (Kluwer Academic Publ., 1994) p. 35.
4. R.G. Giovanelli and D. Hall, *Solar Phys.* **52** (1977) 211.
5. H.P. Jones, *Infrared Solar Physics, IAU Symp.* **154**, (Kluwer Academic Publ., 1994) p. 49.
6. P. Venkatakrishnan, S.K. Jain, J. Singh, F. Recely, and W.C. Livingston, *Solar Phys.* **138** (1992) 107.
7. P. Venkatakrishnan, *Solar Phys.* **148** (1993) 233.

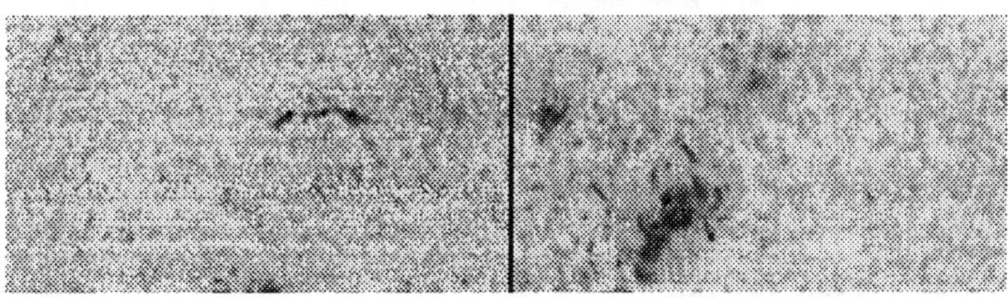

Fig. 2. He I λ1083nm spectroheliograms on 4 Sep. 1994. The left panel is for north (quiet and filament region) and the right for south hemisphere (quiet and active region). Image size is 11.5 arcmin in vertical and 21 arcmin in horizontal.

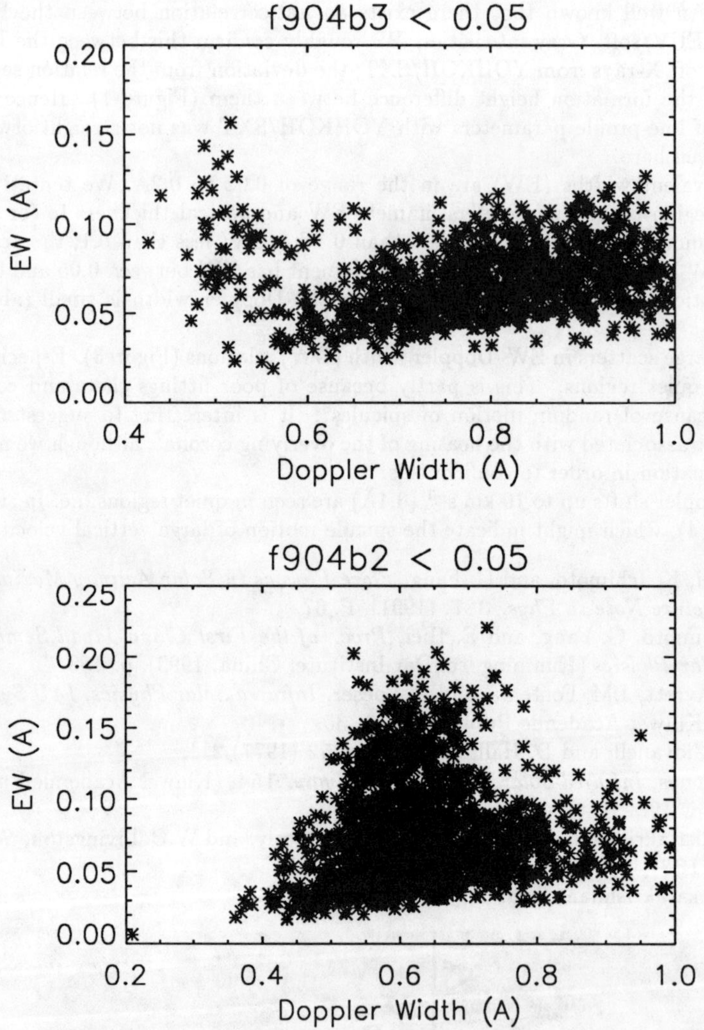

Fig. 3. Scatter plots of equivalent widths vs. Doppler widths of He I λ1083nm line for the data of Sep. 4, 1994. The upper panel is for north (quiet and filament region) and the lower for south hemisphere (quiet and active region).

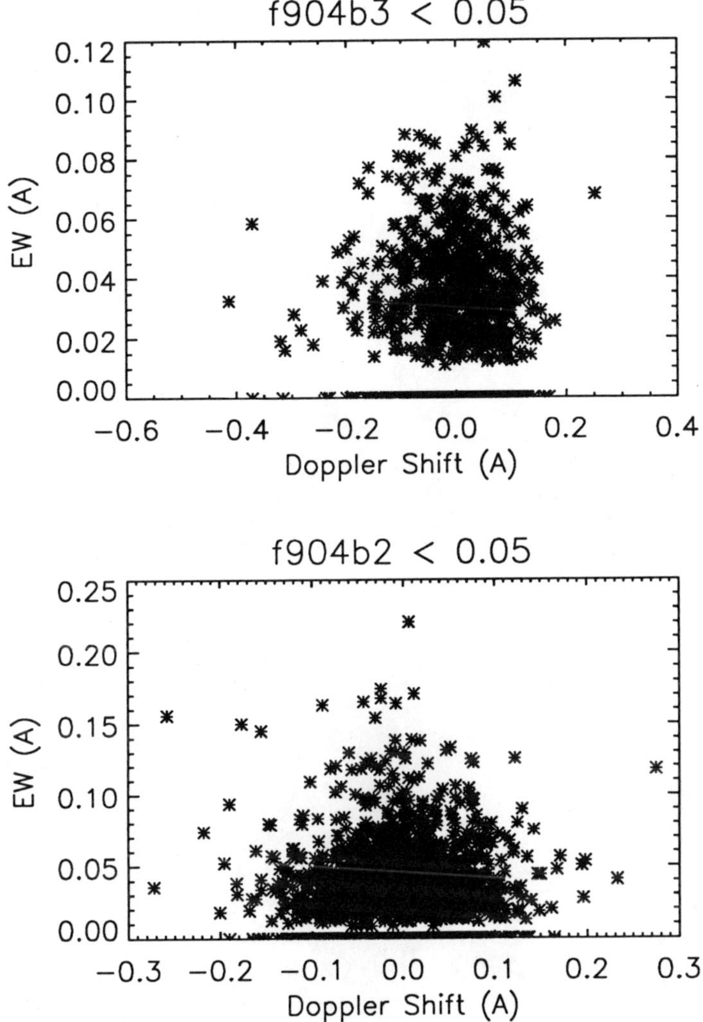

Fig. 4. Scatter plots of equivalent widths vs. Doppler shifts of He I λ1083nm line for the data of Sep. 4, 1994. The upper panel is for north (quiet and filament region) and the lower for south hemisphere (quiet and active region).

Limb observations of the He 1083 line

W. Schmidt
Kiepenheuer-Institut für Sonnenphysik
Schöneckstrasse 6, D-79104 Freiburg, Germany

ABSTRACT

We have observed the He I 1083 emission line outside the solar limb. The observations give direct indications of the height of formation of that line. This line is a very useful and powerful diagnostics for chromospheric magnetism. Direct information about the layers that contribute to the line is essential for the interpretation of polarimetric measurements of magnetic fields in the chromosphere above sunspots and Active regions using the He line.

Observations of Prominences in the Light of He I 10830 A Line

Iraida S. Kim
Sternberg State Astronomical Institute, Moscow State University,
13 Universitetsky Pr., 119899, Moscow, Russia

ABSTRACT

Coronagraphic observations of solar prominences in the light of the near infrared He I, 10830A line made recently through narrow band interference filters as well as a Fabry-Perot filter are discussed. "I-1060 B" high speed infrared film and video camera were used as detectors. Prominence images in the He line are compared with the H-alpha ones. Intensity ratios of He (10830)/ H-alpha are analysed for different classes of prominences. This ratio seems to be an index of prominence dynamics and probably stage of evolution.

WAVE PROPERTIES OF THE CHROMOSPHERE IN HE I 1083 NM AND CA II K

K. Bocchialini and F. Baudin
Institut d'Astrophysique Spatiale, Bât. 121, Université Paris XI Orsay F-91405 France

and

S. Koutchmy
Institut d'Astrophysique de Paris - CNRS, 98 bd Arago Paris F-75014 France

ABSTRACT

From observations simultaneously performed in He I 1083 and Ca II K lines, with the Horizontal Spectrograph of the Vacuum Tower Telescope of NSO/SP, we derived some new dynamical properties of the quiet chromosphere, using an 83 nm long sequence. For both the magnetic network and the intra-network, amplitude spectra of the Doppler velocities are presented. From a time/frequency analysis using the wavelet transform of the Doppler velocity in the two lines, the temporal behavior of oscillations at different frequencies is analyzed. Propagating waves are found in the magnetic network.

1. Introduction

As it has been discussed by J. Zirker[7,8], the diagnostic of dynamical phenomena in the solar atmosphere is necessary in order to understand the coronal heating and the acceleration of the solar wind. It is essential to identify and to study mhd waves which are good candidates to heat the chromosphere and the corona.

Therefore, we performed observations of the quiet chromosphere, at the Vacuum Tower Telescope of the NSO/SP on March 22, 1993. We observed a magnetic network element and an intra-network or cell element (without strong magnetic field) near the disk center, and simultaneously in two lines: He I line at 1083 nm and Ca II K line at 393.4 nm. These two lines being formed at different altitudes, the analysis of the wave propagation is possible.

We used the horizontal spectrograph and slit-jaw images; the same structures were followed during the whole sequence of observations thanks to the computer controlled guiding system of the VTT. Spectra (256x403 px) in the two lines were recorded during 83 mn, at a 5 sec rate. More details concerning the observations are given in Bocchialini[4,5].

2. Illustration of the dynamic al phenomena

After the usual data reduction (dark current effects and flat-field removing), Bocchialini[3], we performed a spatial and spectral averaging over several pixels, in order to improve the signal/noise ratio and to reduce the seeing noise. The spatial resolution of the final spectra (32x45) is 1.2 arcsec/px in the two lines, and the spectral dispersions are 5.8 pm/px in Ca II K and 17.2 pm/px in He I.

We recorded a short movie using a 910 sec long sequence of spectra, to illustrate the dynamical motions of the chromosphere. In order to enhance He I line, the spectral images have been divided by a reference spectrum which is the time average of the spectral images. The movie of the accelerated spectra shows dynamical motions as evidenced by the shifts of the line, which easily translate in velocities. Oscillations are well apparent both in the network and the intra-network. The Si I line at 1082.71 nm evidences the photospheric motions; the leaving trace of the telluric H_2O line at 1083.21 nm does not move: it indicates that the "motions" on the spectral images are not due to the atmospheric motion. They are predominantly of solar origin.

In the case of Ca II K line, oscillations near the center of the line from blue to red are even better apparent; the motions and the period of the oscillations in the network and intra-network elements appear well different.

In the next section, we concentrate our study on the analysis of the Doppler velocity.

3. Temporal analysis of the Doppler velocity

We determined the Doppler velocity assuming that this velocity is proportional to the difference of the intensities in the blue and in the red wings, situated close to the line center. We computed the temporal variations of the velocity in the two lines and for the two regions, for the whole sequence (Fig. 1).

We observe an oscillating signal, with a constant period. This period is significantly longer for the network than for the intra-network, in both lines. The amplitude of the velocity varies from -6 km/s to +6 km/s, in the case of the intra-network in both lines, and in the case of the network in Ca II K line. The amplitude varies from -2 km/s to +2 km/s in the case of the network in He I line. This small amplitude is also shown on the amplitude spectra that we discuss in the next section.

4. Fourier analysis of the Doppler velocity

In order to determine the specific frequencies of these oscillating signals, we performed a Fourier analysis and we computed the amplitude spectra in the two lines for the two regions, per unit of passband (Fig. 2).

In both lines, the dominant frequency is 3.5 mHz for the network; in the intra-network, the 3.5 mHz frequency is present but the peak is now shifted to 6 mHz. The frequency equal to 3.5 mHz corresponds to the 5 min period oscillations of the photosphere, and the frequency equal to 6 mHz (3 min) corresponds to the cut-off frequency of the chromosphere.

The amplitude detected near 3.5 mHz in He I line for the network is small compared to the amplitude detected for the same structure in Ca II K line; it could

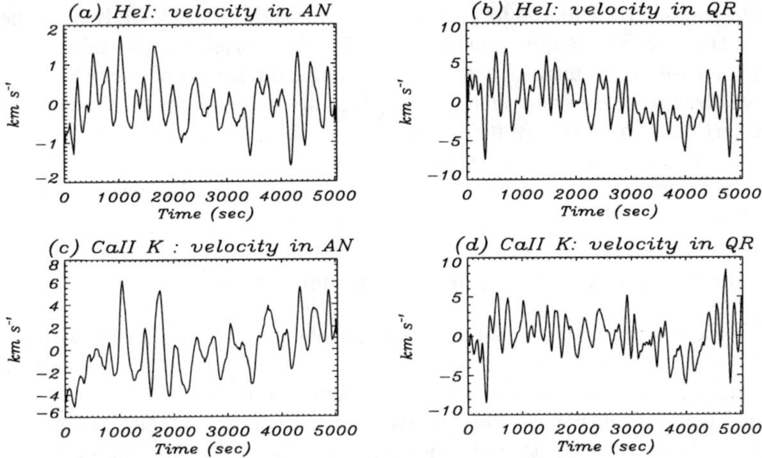

Figure 1: Temporal variations of the Doppler velocity over the 83 min sequence; (a) in He I for the network, (b) in He I for the intra-network, (c) in Ca II K for the same network as in (a), (d) in Ca II K for the same intra-network as in (b).

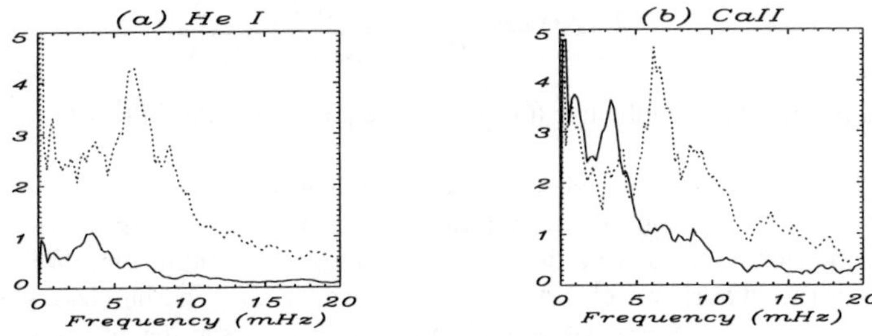

Figure 2: Amplitude spectra of the Doppler velocity per unit of passband; (a) in He I, (b) in Ca II K; full line: network, dotted line: intra-network.

indicate that, if He I line is formed at higher altitude than Ca II K line, the upper limit of the resonant cavity is located near He I line formation altitude.

In the case of the intra-network, the predominant period of 3 min could indicate that we observed here transverse waves, which are detected where the magnetic field is not vertical. These transverse waves could be mhd (Alfvén) waves.

In order to obtain information on the temporal behaviour of the Doppler velocity, and on the duration of the oscillations, we performed a time/frequency analysis, using the wavelet transform.

5. Time/frequency analysis of the Doppler velocity

The wavelet transform consists of the convolution of a signal with a wavelet; this wavelet is a sinusoidal signal modulated by a Gaussian envelope. This is the first application made to study the solar chromospheric oscillations[2,6].

On Fig. 3 is displayed the squared modulus of the wavelet transform of the Doppler velocity as a function of time and frequency, in the two lines and for the two regions. It corresponds to the energy of the signal. We first observe that the signal is not continuous in time: several wavetrains are well apparent, and they seem to be correlated in the two lines. The duration of these wavetrains is longer for the network than the intra-network. Around the specific frequency of 3.5 mHz in the network, we detect 3 wavetrains during the 83 min of the sequence; around the specific frequency of 6 mHz in the intra-network, we detect 4 well-separated wavetrains.

In order to check more accurately if the wavetrains are correlated or not, we computed the correlation coefficient $C(\Delta t)$ as a function of time, between the two lines:

$$C(\Delta t) = \frac{\Sigma_t(v_K(t).v_{He}(t+\Delta t))}{\sqrt{\Sigma_t(v_K(t)^2).\Sigma_t(v_{He}(t)^2)}} \tag{1}$$

and we plotted this coefficient for the two regions and the different wavetrains (Fig. 4).

In the case of the intra-network, there is no lag between the two lines, while in the case of the network, the best correlation occurs after a 10 s lag for the first and the third impulsions, and after a 30 sec lag for the second impulsion. We conclude that for the intra-network we observe stationary waves, and in the case of the network, the signal in He I preceeds the signal in Ca II K, by a significant amount.

6. Conclusion

Thanks to the time/frequency analysis of the Doppler velocity performed in the network and the intra-network, simultaneously in the He I and Ca II K lines, we demonstrate the presence of stationary waves in the intra-network and of running waves in the network. The direction of these propagating waves depends on the formation altitudes of the two lines. From the 1D model of Avrett and Loeser[1],

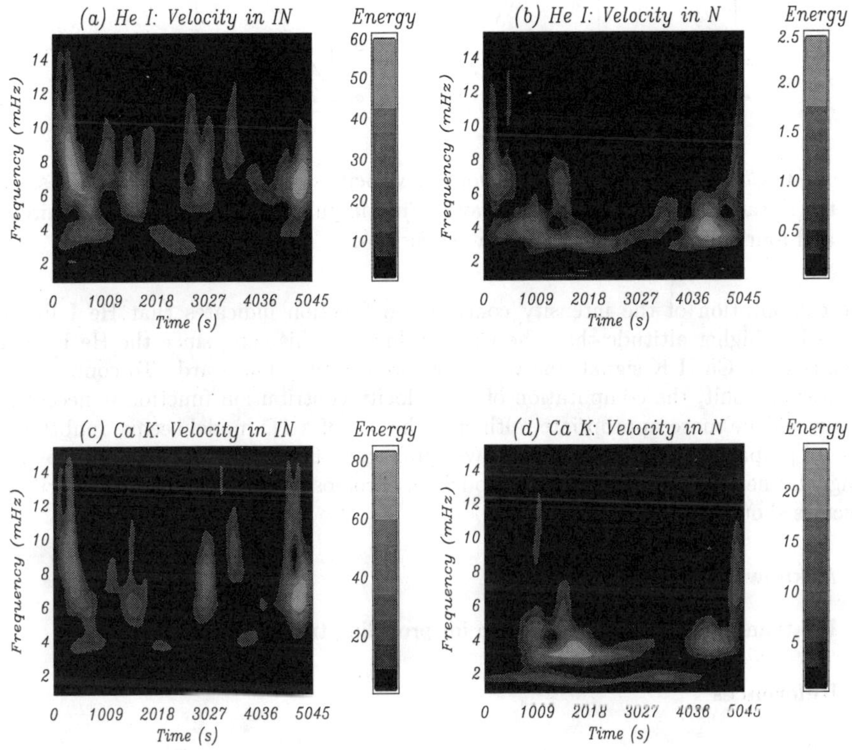

Figure 3: Time/frequency variations of the Doppler-velocity (a) in He I for the intra-network (IN), (b) in He I for the network (N), (c) in Ca II K for the intra-network (IN), (d) in Ca II K for the network (N). Energy is in arbitrary units.

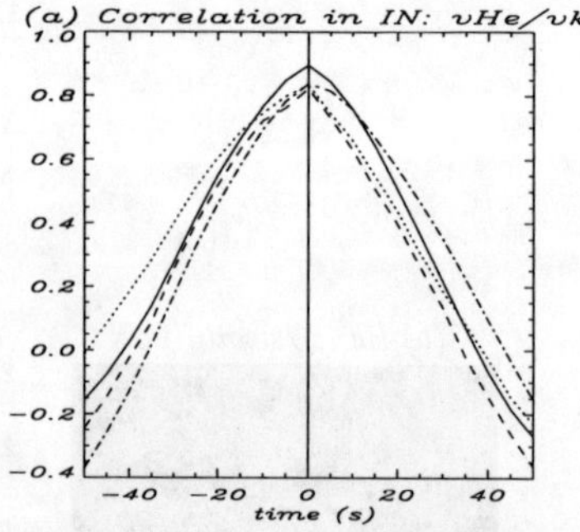

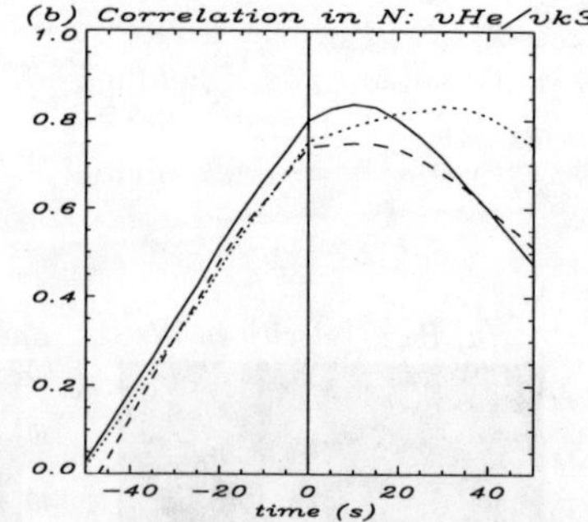

Figure 4: Correlations between the Doppler velocities in He I and Ca II K lines, (a) for the intra-network, (b) for the network. First impulsion: —, second: . . ., third: - - -, and fourth in the case of the intra-network:-.-.

the computation of the intensity contribution function indicates that He I line is formed at higher altitude than the Ca II K line. In this case, since the He I signal preceeds the Ca II K signal, the waves are propagating downward. To confirm this surprising result, the computation of the velocity contribution function is necessary and would be more appropriate, although the use of a 1D model seems doubtful in case of propagating waves. Indeed, waves propagate along structures confined by the magnetic field of the network and models of chromospheric structures like fibrils or spicules should be considered.

7. Acknowledgments

KB thanks the NSO/SP workshop for providing travel funds.

8. References

1. Avrett E.H., Loeser R., 1992, 7th Cambridge Workshop on Cool Stars, Stellar systems and the Sun, ASP Conference Series, vol.26, M. Giampapa & J. Bookbinder Eds
2. Baudin F., 1994, Proceedings of the Conference "Helio- and Astero-Seismology from the Earth and Space", ASP Conference Series, R.K. Ulrich & E.J. Rhodes Eds
3. Bocchialini, K., 1994, Thesis, Paris XI University.
4. Bocchialini, K., Vial, J.-C., Koutchmy. S., Zirker J.B., 1994a, 14th NSO/SP International Summer Workshop, in "Solar Active Region Evolution - Comparing Models w ith Observations". Eds. K. S. Balasubramaniam and George W. Simon. Publications of the Astronomical Society of Pacific, 68, 389.

5. Bocchialini, K., Koutchmy, S. Vial, J.-C 1994b, *Astrophys. J.*, **423**, L-67
6. Bocchialini, K., Baudin F., 1994, *Astron. & Astrophys*, submitted
7. Zirker, J. B. 1993, *Sol. Phys.*, **148**, 43
8. Zirker, J. B. 1994, Invited Review Paper, see these proceedings.

CHROMOSPHERIC MAGNETIC FIELD MEASUREMENTS IN HE I 10830 Å

I. Rüedi and S.K. Solanki
Institute of Astronomy, ETH-Zentrum, CH-8092 Zürich, Switzerland

H. Balthasar
Kiepenheuer-Institut für Sonnenphysik, Schöneckstrasse 6, D-79104 Freiburg Germany

W. Livingston
National Solar Observatory, NOAO, P.O. Box 26732, Tucson, AZ 85726, USA

and

W. Schmidt
Kiepenheuer-Institut für Sonnenphysik, Schöneckstrasse 6, D-79104 Freiburg Germany

ABSTRACT

We present upper chromospheric magnetic field measurements carried out using the He I 10830 Å line. The observations sample active regions located at disk centre, as well as at the solar limb. At the limb projection effects between the chromospheric line and a nearby photospheric line are seen. Above the limb, at least over active regions, the line strongly saturates, while in the quiet sun layers with much larger horizontal velocities are observed than on the disk close to the limb.

1. Introduction

Magnetic fields dominate the structure and the energetics of the outer solar atmosphere, but in the vast majority of cases they have been measured only in the photosphere. A number of diagnostics do exist for the measurement of chromospheric and low-coronal magnetic fields, but have been only moderately used, in particular for quantitative studies because they are generally affected by significant drawbacks.

In this paper we present chromospheric magnetic field measurements carried out with the He I 10830 Å line. This line is the best currently available diagnostic of chromospheric magnetic fields since it overcomes most of the problems affecting the other potential diagnostics (see Rüedi et al. [12] for details).

2. Observations

The observations presented in this paper are of two types. The first data set, obtained at disk centre, was recorded using the McMath-Pierce telescope on Kitt Peak, the main spectrograph and a single diode. The second data set, observed at the extreme limb, was recorded using the Gregory-Coudé telescope on Tenerife with a CCD array detector. The spatial resolution is estimated to be 3″–5″ for the Kitt Peak and 1-1.5″ for the Tenerife data. At Kitt Peak Stokes $I \pm V$ were recorded consecutively at each wavelength and 12 wavelength scans were coadded in order to average out seeing-induced distortions of the spectrum. At Tenerife, both polarities were observed simultaneously on the CCD.

In addition to the chromospheric He I 10830 Å line, all spectra show the photospheric Si I 10827 Å line which provides simultaneous longitudinal magnetic field values in the photosphere.

3. Results from Disk Centre Observations

3.1 General Features and Plages

The most striking feature of the measurements is the homogeneity of the magnetic field in the chromosphere relative to the photosphere. Above an active region containing a sunspot the chromospheric $\alpha B \cos\gamma$ changes only by a factor of 4 while the photospheric $\alpha B \cos\gamma$ varies by a factor of 30.

On the other hand, in plages we observe a close correspondence between $\alpha B \cos\gamma$ at both heights in the atmosphere. This is what is expected for spatially unresolved flux tubes expanding with height and filling all the space at the formation level of the helium line. Typical observed values for $\alpha B \cos\gamma$ are 150 – 300 G.

3.2 Sunspot Magnetic Field Properties

The umbral magnetic field is approximately 700–1000 G stronger in the photosphere than in the chromosphere. This difference can be explained by the divergence of the umbral field with height inherent to all MHD models of sunspots [9,11]. We estimate the height difference between the two layers to be around 1500–2000 km. The resulting gradient of the field strength dB/dz lies between 0.35 and 0.67 G km^{-1}, which is in good agreement with gradients determined over a similar or larger height range by other authors [1,8,10]. In the outer penumbra we get $dB/dz \approx$ 0.15–0.35 G km^{-1} which is smaller than the dB/dz obtained in the umbra. In the inner penumbra dB/dz is closer to the umbral value.

The trend of decreasing dB/dz with increasing distance from the centre of the sunspot is in good agreement with the results of Lee et al. [10] and Bruls et al. [3]. Although the magnitude of our gradients agrees well with dB/dz derived over a similar height range [7] in both the umbra and the penumbra the dB/dz derived here is much smaller than the gradients obtained from purely photospheric lines [2,3,15]. This comparison suggests that throughout sunspots dB/dz decreases with height.

3.3 Sunspot Canopies

At some positions outside, but close to a penumbral boundary the chromospheric $\alpha B \cos\gamma$ is actually larger than in the photosphere. These positions are further marked by small $\alpha B \cos\gamma$ values and Si I Stokes V profiles with particularly narrow peaks and a small peak separation. Good fits to such profiles require the presence of a magnetic canopy located roughly 300 km above the unit optical depth, $\tau_c = 1$, surface. This also explains the larger He I magnetic flux value, since this line is formed entirely above the canopy while the photospheric line obtains its main contribution from the magnetic field free region located under the canopy.

3.4 He I Line Width

Interestingly, the Stokes I and V profiles of the He I line are particularly broad there where the Stokes V profile of the Si I line is particularly narrow, i.e. at the location of the superpenumbral canopy. The He I line is also most strongly shifted relative to the Si I profile at these positions. The shifts, reaching 300–400 m s^{-1}, are consistent with the signal of the inverse Evershed effect seen in the He I line. The largest width that the He I line reaches in the (disk centre) data corresponds to an enhanced broadening velocity of 9.1 km s^{-1} relative to the width at the neutral line. This broadening velocity is a lower limit, since the He I line width at the neutral line may partly be due to a velocity broadening as well.

This difference in line width suggests that the major part of the velocity broadening is due to field aligned motions. If the large excess broadening velocity is related to the chromospheric inverse Evershed effect then there are sizable small-scale velocities that are much larger than the large-scale inverse Evershed flow, or these small-scale velocities are much more inclined to the horizontal. In either case, the Evershed effect would be highly inhomogeneous. The extra broadening is unlikely to be purely thermal, since it would correspond to temperatures much higher than those expected at the formation height of this line.

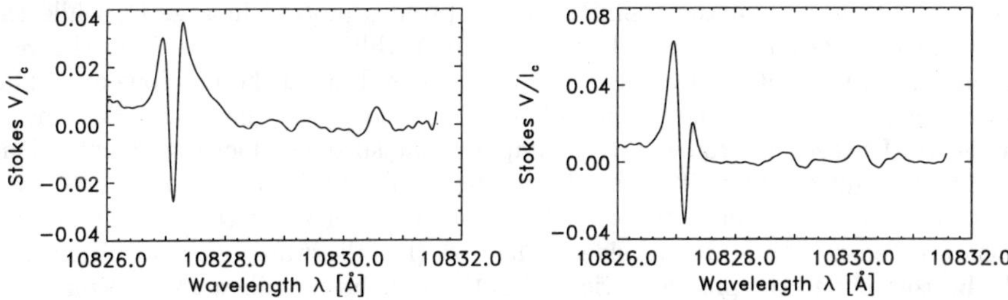

Fig. 1: Complex Si I and He I Stokes V profiles probably exhibiting the crossover effect. These profiles were observed near the neutral line of a sunspot penumbra.

3.5 Complex Profile Shapes at the Neutral Line

At the neutral line next to the limbward penumbra of the following sunspot we observe complex Si I profiles suggestive of the crossover effect [4,5], which is best explained by a combination of magnetic and velocity gradients along the line of sight [13,14]. Two such Stokes V spectra are shown in Fig. 1. We tested if instrumental cross-talk could produce such asymmetric profiles by assuming the worst possible cross-talk situation expectable from the telescope at the time of the observations.

Although the final Stokes V profile can be quite asymmetric, it never shows the three-lobed structure seen in Fig. 1. We conclude that the observed complex shapes of the Si I Stokes V profiles near the neutral line are mainly due to the solar crossover effect. The corresponding chromospheric line profiles cannot be sufficiently distorted by cross-talk either. Consequently, the He I line profile also shows the crossover effect, although to a lesser extent. We conclude that the small-scale inhomogeneity of the penumbral field (uncombed field), which best explains the crossover effect [14], is reduced in the upper chromosphere but apparently not absent.

4. Results from Limb Profiles

In this section we mainly point out a few general features that can be recognized on the limb spectra.

4.1 Projection Effects

Figure 2 displays the Stokes I (left) and V (right) profiles of the He I 10830 Å and Si I 10827 Å lines. All four spectra were observed simultaneously above an active region containing two sunspots. The top spectrum was recorded closest to disk centre (in front of the sunspots) while the bottom spectrum was recorded closest to the limb.

We can see in the top panel that the photospheric line is already distinctly split, while no significant Stokes V signal is exhibited by the chromospheric line. In the second panel the He line shows a normal Stokes V signal while the Si I Stokes V profile is somewhat distorted. In the third row, the photospheric Stokes V profile has a complicated structure (similar to the profiles plotted in Fig. 1), while the He I Stokes V profile is relatively normal, but highly asymmetric. Note the very extended red wing of the He I line in this row as well as in the previous one. This is evidence for the existence of large horizontal velocities. The strong asymmetry of the He I Stokes V profiles further implies that sizable velocity gradients along the line of sight are present in the chromosphere, i.e. horizontal velocity gradients. The complex Si I profile probably reflects the crossover effect (cf. Fig. 1), since it was observed on the limbward side of the first (i.e. diskward) sunspot. Finally in the bottom panel, the photospheric Stokes V profile has regained an antisymmetric shape (with opposite polarity), while the chromospheric line now shows a complex profile. In some of the spectra located between the third and fourth panel both lines showed two-component profiles. The He I line consistently repeats the behaviour of the Si I line, but at smaller distances to the limb due to the different heights of formation of the two lines. A more or less vertical magnetic feature will be observed closer to disk centre in a photospheric line than in a chromospheric line such as the He I 10830 Å.

4.2 Line Saturation

Above the limb, the He I line goes into emission. Over active regions it becomes extremely saturated as can be seen on the left panel of Fig. 3. In this case the blue

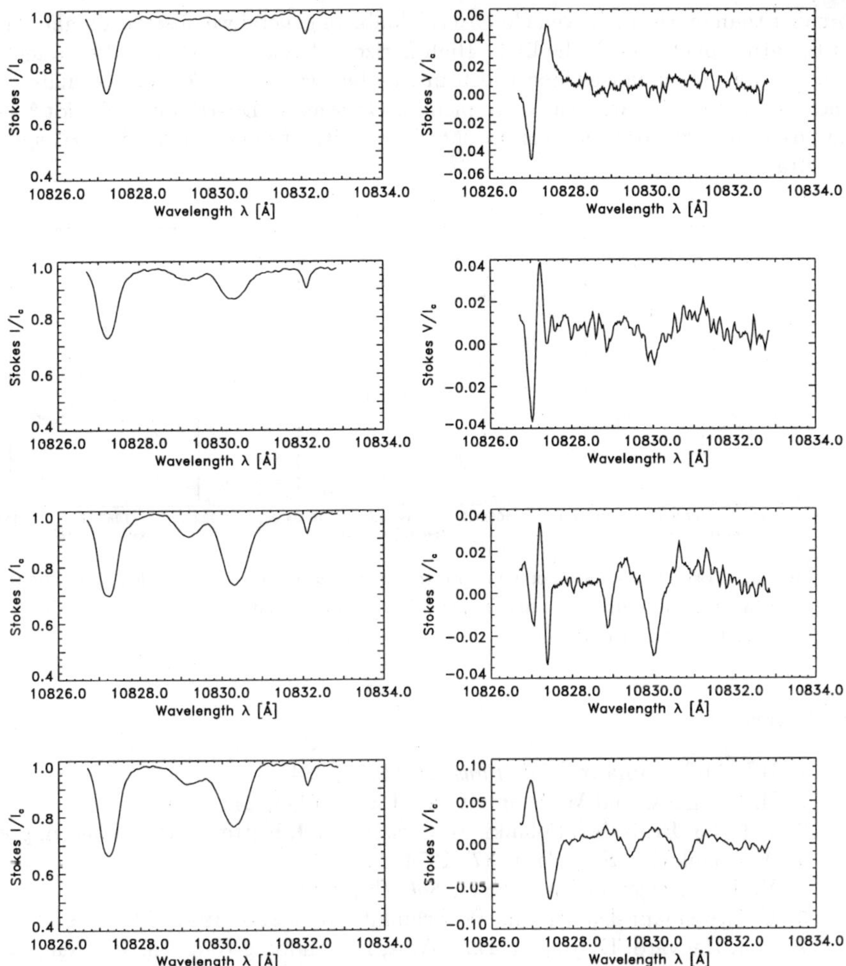

Fig. 2: Stokes I (left) and corresponding Stokes V (right) profiles observed above an active region at the solar limb. The distance to the limb decreases from top to bottom.

component of the triplet, with the smallest oscillator strength is almost as strong as the red component. The middle and right panels of Fig. 3 were observed over a quiet region (the middle panel having been observed closer to the disk than the right panel). The line may be somewhat saturated in the middle panel, but appears to be optically thin in the right-hand panel. It should be noted that the same spectral range is plotted in this figure as in Fig. 2! Above the limb the line is considerably broader than on the disk even close to the limb. In general we observe a higher layer of the atmosphere above the limb. High horizontal velocities have to be present to account for the enhanced line width in this higher layer. We can speculate that these high velocities are related to spicules. Grossmann-Doerth and Schmidt [6] have also derived large horizontal chromospheric velocities from the analysis of H_α spicule spectra.

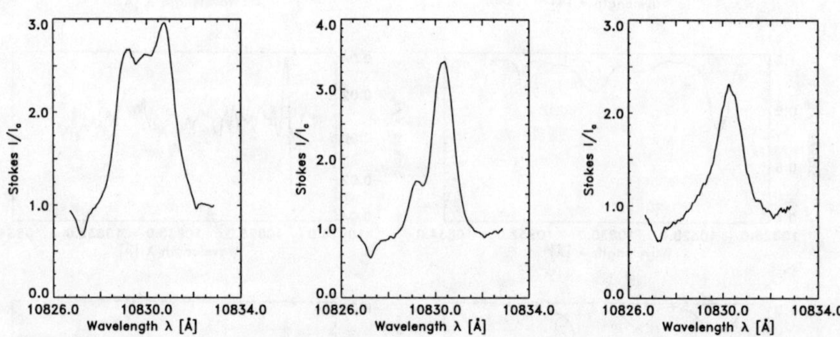

Fig. 3: Stokes I profiles observed above the limb over an active region (left) and above the quiet sun (middle and right). The middle panel has been observed on the same CCD exposure as the right panel, but closer to the disk.

5. References

1. H.I. Abdussamatov, *Sol. Phys.* **16** (1971) 384.
2. H. Balthasar and W. Schmidt, *A&A* **279** (1993) 243.
3. J.H.M.J. Bruls, S.K Solanki, M. Carlsson, R.J. Rutten, *A&A* (1994) in press.
4. A.A. Golovko, *Sol. Phys.* **37** (1974) 113.
5. V.M. Grigorjev and J.M Katz, *Sol. Phys.* **22** (1972) 119.
6. U. Grossmann-Doerth and W. Schmidt, *A&A* **264** (1992) 236.
7. M.J. Hagyard, D. Teuber, E.A. West, E. Tandberg-Hanssen, W. Henze, J.M Beckers, M. Bruner, C.L. Hyder, B.E. Woodgate, *Sol. Phys.* **84** (1983) 13.
8. W. Henze, E. Tandberg-Hanssen , M.J. Hagyard, B.E. Woodgate, R.A. Shine, J.M. Beckers, M. Bruner, J.B. Gurman, C.L. Hyder, E.A. West, *Sol. Phys.* **81** (1982) 231.
9. K. Jahn, *A&A* **222** (1989) 264.
10. J.W. Lee, D.E. Gary, G.J. Hurford, *Sol. Phys.* **144** (1993) 45.
11. V.J. Pizzo, *ApJ* **302** (1986) 785.
12. I. Rüedi, S.K. Solanki, W. Livingston, *A&A* (1994) in press.
13. J. Sánchez Almeida, B.W. Lites, *ApJ* **398** (1992) 359.
14. S.K. Solanki, C.A.P. Montavon, *A&A* **275** (1993) 283.
15. A.D. Wittmann, *Sol. Phys.* **36** (1974) 29.

WHAT HAVE WE LEARNED ABOUT CHROMOSPHERIC OSCILLATIONS FROM HE I 10830Å?

B. FLECK
Space Science Department of ESA, ESTEC
P.O. Box 299, NL-2200 AG Noordijk, The Netherlands

and

F.-L. Deubner, J. Hofmann
Institut für Astronomie und Astrophysik der Universiät Würzburg
Am Hubland, D-97074 Würzburg, Germany

ABSTRACT

Time series of He I 10830 spectra taken simultaneously with spectra of other chromospheric lines (Ca II 8542 Å and Ca II K) are analyzed to study wave propagation in the solar chromosphere. Rather than providing conclusive answers to some of the long standing questions concerning chromospheric oscillations, the new results derived from the He 10830 line raise new puzzling questions. The spatio-temporal wave pattern deduced from the Doppler displacements of the He line differs significantly from those of the two Ca II lines, while the phase difference between the Doppler displacements of He 10830 and Ca K_3 stays close to zero in the whole frequency range observed. This is difficult to reconcile with the low oscillation amplitude observed in the He 10830 line (RMS≈1100 m/s), which is less than half the velocity amplitude derived from the core displacement of Ca K_3. Another surprising result is that the Ca K "bright point" events are only barely visible in the Doppler displacement of the He 10830 line.

1. Introduction

Several fundamental issues concerning chromospheric dynamics are still far from being satisfactorily understood — despite the accumulation of a quite extensive observational data base and despite considerable theoretical efforts made in the past decades. Among the open issues are for instance: What mechanism drives spicules? What is the relevance of the Ca K "bright point" phenomenon for the energy balance of the chromosphere (cf. Carlsson and Stein[1]), and what role might weak intranetwork fields play in that context? What are the wave propagation characteristics in the chromosphere? So far, all attempts to identify running acoustic waves, and in particular running acoustic shock waves have failed.

One of the reasons for this somewhat slow progress in our understanding of chromospheric dynamics might be the fact that, in the "visible", there are only few lines which can be used as a diagnostic to probe the chromosphere. Basically, there are only two lines: Ca II H and K, and one line of the Ca II infrared triplet at 8542 Å. Others might include Hα.

Observing the optically thin He 10830 line we were hoping to have another velocity diagnostics available to probe chromospheric velocity fields, and in addition, as the He 10830 line is generally believed to form high in the chromosphere just below the transition region (cf. e.g. Avrett et al.[10]), to extend the previously observed height range to greater altitudes. For previous studies addressing chromospheric oscillations using the He 10830 line we refer the reader to Lites[2], Lites et al.[3], Fleck et al.[4,5], Hofmann et al.[6], and Bocchialini et al.[7,8,9].

2. Observations

Here we report results from two time series of He 10830 spectra. One was obtained at the Vacuum Tower Telescope at Observatorio del Teide, Tenerife, using the Echelle Spectrograph and one of the "Wright" CCD cameras. Simultaneous spectra of the He line and the adjacent Si line at 10827 Å were recorded for nearly 110 minutes with a cadence of 24 sec. The field of view was increased to 345" by using a focal reducer attached to the exit of the spectrograph. The other series was obtained with the Horizontal Spectrograph and the MDA camera system (RCA CCDs) at the VTT at Sacramento Peak Observatory* in September 1993. It includes simultaneous spectra of He I 10830, Ca II K, Ca II 8542, Si I 10827, Mg b_1 5183, Fe I 5184, and Ni I 5184. The duration of this series is approximately 5 3/4 hours. The spectra were taken at intervals of 7 sec with an exposure time of 3 sec. Effective slit length and pixel size were 43" and 0.12", respectively. Both series were obtained in a quiet region near disk center.

3. Results and Discusson

In Fig. 1 we have displayed the spatio-temporal wave pattern derived from the Doppler displacements of the He 10830 line of the Tenerife data set. As the He line is believed to be optically thin, this should reflect the local velocity field at some (varying) height in the chromosphere. The most striking feature in this diagram which does not resemble any previously seen wave pattern derived from other chromospheric lines like Ca K or Ca 8542, is the particular texture of the wave pattern: Regions 40 to 50 arcsec wide (the typical size of a supergranulation cell) oscillate in a kind of organized, structured fashion for rather extended periods (more than 1 hour), while adjacent regions (cells?) reveal a completely different pattern (but again highly coherent within themselves). For instance, in the region $60'' \leq x \leq 100''$ along the slit one can see a fairly small scale (both in x and t) structure, which changes rather abruptly at $x \approx 115''$ into a much larger scale pattern which also persists over the whole length of the data set. At $x \approx 205''$ another small scale pattern follows which is dominated by fairly high frequency oscillations. In the structure near $x \approx 275''$ the slit crossed an enhanced network element. There, long period oscillations dominate and one can also observe a tendency for downdrafts.

*Operated by the Association of Universities for Research in Astronomy, Inc., under cooperative agreement with the National Science Foundation.

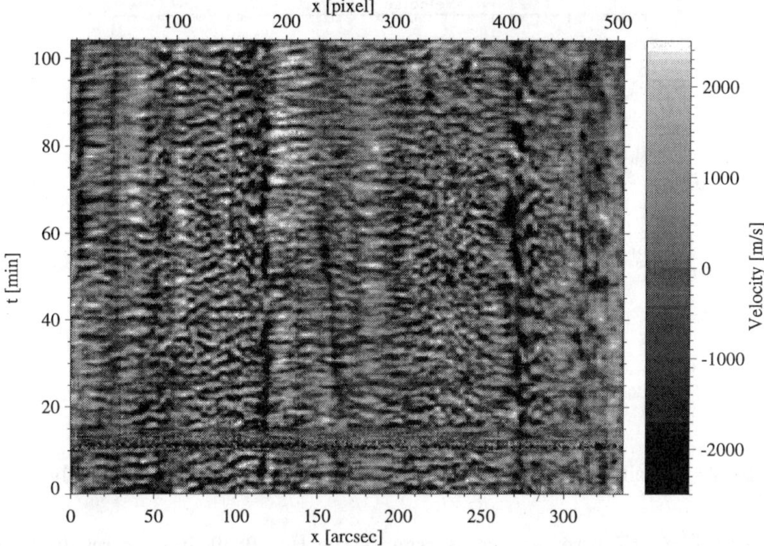

Figure 1: "Velocity" oscillations derived from the Doppler displacements of the He I 10830 line.

The wave field derived from the He line (Fig. 1) is significantly different from what is observed in other chromospheric lines. After all, this finding itself might not be too surprising since the chromosphere is believed to be highly dynamic and dominated by shock waves, the formation process of this line differs markedly from the other chromospheric lines, and since He 10830 likely forms high in the chromosphere just below the transition region, one would quite naturally expect that the He wave pattern differs from that observed in the Ca II lines. However, there are two findings which are difficult to reconcile with this picture, namely the small amplitudes of the He oscillations and the vanishing V–V phase difference between He and Ca K_3 oscillations. We will discuss this further down.

In Fig. 2, we have displayed the velocity power spectra of H3 10830, Ca K_3, Ca 8542, and Mg b_1 (Sac Peak data set). These spectra reveal many of the well known characteristics of solar oscillations: they are dominated by a peak near 3.3 mHz (≈ 3 minutes) in the middle to high chromosphere (CaK_3). The only line which does not fit into this frame is the He line. We found it rather surprising how weak the oscillatory signal of this line is in the 5 to 3 minute band. It is even weaker than in the Mg line which is formed just above the temperature minimum region! The rather high power level of the He spectrum at high frequencies which remains constant beyond 20 mHz is certainly not due to a solar signal enhanced in the He line but is just due to noise (it is highly unlikely that the sun would produce a nearly perfect white spectrum).

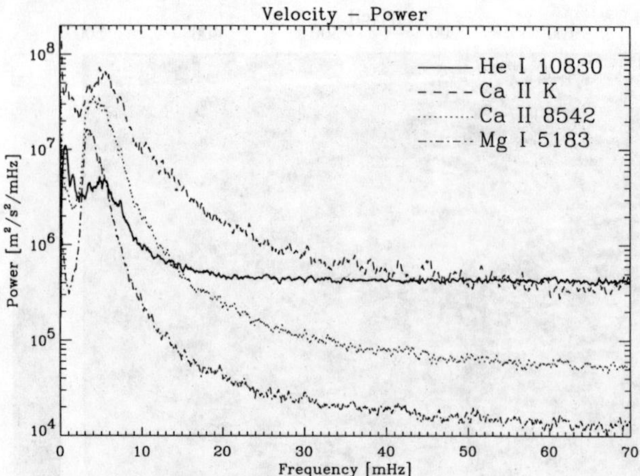

Figure 2: Velocity power spectra of He 10830, Ca K_3, Ca 8542, and Mg b_1

The most problematic error source as regards the He 10830 line are residual time-dependent flat field errors which affect this weak and broad line much more than they affect the other strong and narrow lines, in particular as the RAC CCD chips seem to have a rather unreliable flat field behaviour at this long wavelength. This interpretation of the enhanced high frequency signal as noise is collaborated by the phase and coherence spectra with the other chromospheric lines.

Integration of the He power spectrum displayed in Fig. 2 results in an RMS of the He 10830 velocity oscillations of merely 1100 m/s (including the high frequency noise). This is not even half the amplitude observed in Ca K_3 (RMS $\approx$ 2300 m/s) and even less than the amplitude observed in the Ca II infrared lines (RMS $\approx$ 1300 m/s). The close inspection of the whole time series (see Fig. 3 as an example) also exhibits only little or even no signatures of the Ca K "bright points" in the He 10830 Doppler signal. This finding is the more surprising as the "bright points" driving mechanism - highly steepend shock waves - is one of the most striking dynamic phenomena dominating the whole chromosphere (cf. Carlsson and Stein[1]). How could one explain that a line formed that high in the atmosphere close to the highly dynamic transition region behaves so quiet? Is the He 10830 line possibly formed much lower in the chromosphere? Or are the oscillations simply smeared out because the signal is formed over a very extended height interval? In a recent review, Deubner[11] proposes a simple model of a chromospheric cavity that might account for an evanescent decrease of the velocity amplitudes near its upper reflecting layer. We would not like to speculate more on these questions now. Velocity contribution functions of the He 10830 line calculated dynamically as those derived for the Ca II K line by Carlsson and Stein[1] are urgently needed.

The intensity power spectrum of the He line does not reveal any prominent os-

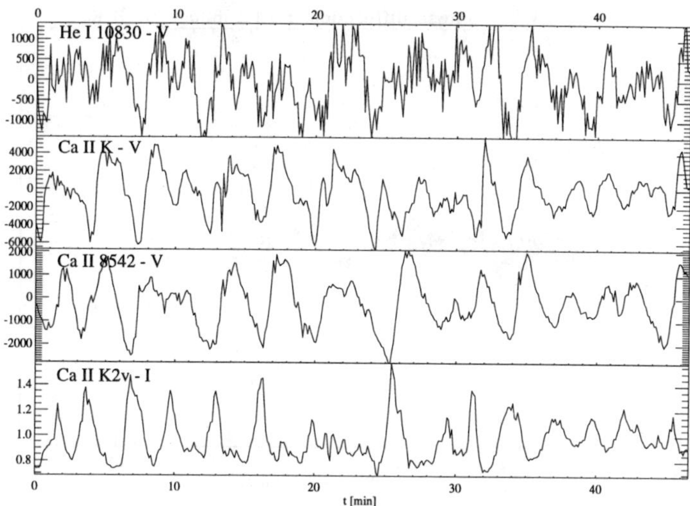

Figure 3: Temporal variations of the intensity of the Ca K_{2v} emission maximum and the Doppler displacements [m/s] of Ca 8542, Ca K_3, and He 10830.

cillatory signature but falls off continuously with increasing frequency. Not even the strong chromospheric 3-min oscillations which dominate the Ca K power spectra leave their imprint in the He 10830 intensity signal. This can be explained by the particular line formation mechanism of the He line (population of the lower level of the He 10830 line by UV back-radiation from the corona; cf. Avrett et al.[10]).

Fig. 4 shows the phase (diamonds) and coherence (dots) spectra between the Doppler displacements of He 10830 and Ca K_3. The coherence is fairly high up to about 9 mHz where it drops rapidly to noise levels. The phase difference between the two signals stays close to 0° in the whole frequency range observed. This can be explained on two ways: either the two signals are formed at very similar heights in the atmosphere, or the phase speed is very high. Neither the particular texture of the He wave pattern (cf. Fig. 1) nor the low RMS found in this line support the idea of a common formation height of the two lines. How then could one argue for high phase speed? Is this another indication of a non-propagating component dominating the chromospheric wave field (cf. e.g. Fleck and Deubner[12])? Or does the magnetic canopy play a significant role? Again, we hesitate to speculate on these questions before we have some state of the art calculations about the velocity contribution function of the He line.

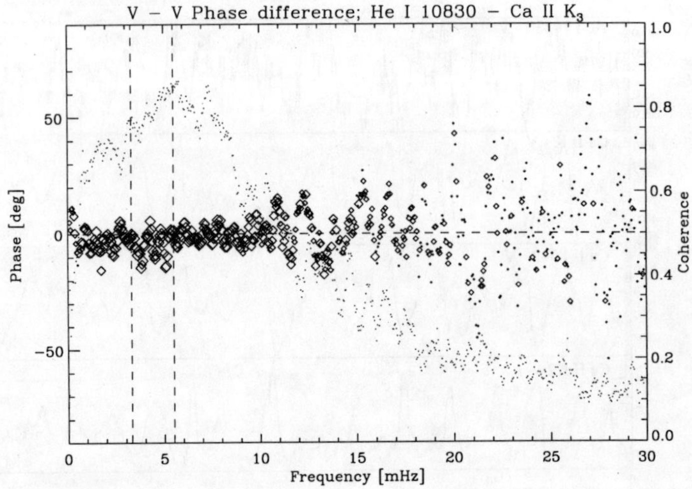

Figure 4: Phase (diamonds) and coherence (dots) between the Doppler displacements of He 10830 and Ca K_3.

4. Acknowledgments

FLD and JH gratefully acknowledge the support and hospitality of the Sac Peak staff and the efficient support by the observers' team at the VTT in particular. Financial support of the Deutsche Forschungsgemeinschaft (DFG) is also gratefully acknowledged.

5. References

1. M. Carlsson and B. Stein, *Proc. Mini-Workshop on Chromospheric Dynamics*, ed. M. Carlsson (University of Oslo, 1994), p. 47.
2. B. Lites, *ApJ* **301** (1986) 1005.
3. B. Lites, R. Rutten, J.H. Thomas, *Solar Surface Magnetism*, eds. R.J. Rutten and C.J. Schrijver (NATO ASI C433, 1994), p. 159.
4. B. Fleck, F.-L. Deubner, D. Maier, W. Schmidt, *Infrared Solar Physics*, eds. D.M.Rabin et al. (IAU Symp. 154, 1994), p. 65.
5. B. Fleck, F.-L. Deubner, J. Hofmann, S. Steffens, *Proc. Mini-Workshop on Chromospheric Dynamics*, ed. M. Carlsson (University of Oslo, 1994), p. 103.
6. J. Hofmann, F.-L. Deubner, B. Fleck, *GONG 1994: Helio- and Asteroseismology from Earth and Space*, eds. R. Ulrich and E. Rhodes (ASP Conference Series, 1995), in press.
7. K. Bocchialini, J.-C. Vial, S. Koutchmy, *ApJ* **423** (1994), L67.

Observing with a Large Coronagraph

The workshop generated much discussion concerning exactly how such a coronagraph would work and what new scientific problems it would address. To focus this we distributed "Large Reflecting Coronagraph Observing Proposals" and asked all of the participants to submit their observing requests and telescope requirements. We received 23 responses for telescope projects ranging from solar coronal and photospheric, to solar system and extragalactic observations.

A sample of these requests and the particular requirements on the telescope and the instrumentation advocated by the investigator follows:

1. Solar Coronal:

- "Observations of Magnetic Topologies and Their Change Leading to a CME" (low polarization and high angular resolution)
- "High Frequency Oscillations in the Corona" (optimize for 1μ and high angular resolution)
- "Formation Mechanism of Spicules and their Influence on the Corona" (emphasize 1-2μ region, high spatial resolution)
- "Quiescent/Explosive non-thermal Kernels Above Active Regions" (IR capability with high spatial resolution)
- "IR Coronal Emission Line Measurements" (spectroscopy to 35μ)
- "Evolution of CME's" (high spatial resolution, IR spectroscopy)
- "Prominence Magnetic Field Studies" (IR spectroscopy, high spatial resolution)
- "Plasmoid Studies" (4m aperture, f/8 to f/12, integral spectropolarimetry instrumentation)

2. Solar Disk:

- "Shocked Granulation" (Low polarization, narrow-band tunable filters)
- "Imaging the Cool Solar Atmosphere" (High resolution, IR capability to 5μ)
- "Sunspot Thermodynamics" (low scattered-light, many pixel, wide field)
- "Spicule Acceleration Mechanisms" (IR and visible, tunable narrow-band filters, high resolution)

- "Studies of Eruptive Phenomena"

3. Extragalactic:

- "Photometry of Very Faint Distant Galaxies in the Close Vicinity of Bright Radio Sources" (3-4m aperture, many-pixel detectors in visible and 1-5μ wavelength region)
- "Faint Star Halos Around cD Galaxies" (low scattered light, many-pixel detectors in visible and 1-2μ IR)

4. Planetary:

- "Studies of the Mercurian Atmosphere, Corona and Magnetosphere" (high resolution, need sensitivity out to 12μ, tunable narrow-band filters)
- "Studies of Near-Earth Orbital Material Density Evolution" (IR, many-pixel, fast readout detectors)

Of course this list is a reflection of the particular interests of the workshop participants. Nevertheless, the diversity in this list illustrates the potential such an instrument could have for a broad range of scientific problems. It is fair to say that most of the participants concluded that a detailed engineering study was needed to understand what competing requirments for aperture, field-of-view, and final effective spatial resolution were sensible. Questions related to superpolishing a large aperture (2-4m) to 0.3nm rms, optical coatings and environmental control could not be resolved in these discussions.

PRELIMINARY LRC CHARACTERISTICS
(PLEASE MAKE CHANGES, SUGGESTIONS, IMPROVEMENTS)

- **Aperture: 2-4 m**

- **Primary Mirror:** f/3.75, off-axis parabola, thermally controlled, superpolished to 3 Å rms; IR emissivity 0.007; no central obstruction.

- **Secondary Mirror:** At prime focus, used for inverse occulting, actively cooled.

- **Adaptive Optics:** Object wavefront sensing for solar and bright nighttime sources; beacon wavefront sensing for faint objects and corona.

- **Detectors:** Visible and IR array detectors up to 4k×4k pixels, $\lambda\lambda$ 0.3 to 25 μ.

- **Cleanliness:** Controlled airflow and electrostatic dust control.

- **Optical performance:** Image sizes (arcsec) @ off axis distance (arcsec): .011 @ 2, .13 @ 23, 5 @ 900. Strehl ratio $1 - (4\pi\sigma/\lambda)^2$ with $\sigma = 3$ Å rms: $(4\pi\sigma/\lambda)^2 = 5.7\times10^{-5}$ @ .5 μ, or 1.4×10^{-7} @ 10 μ.

- **Instrumentally-scattered light:** $\leq 5\times10^{-6}\, F_\lambda$ @ $R/R_\odot = 1.15$

PRELIMINARY CHARACTERISTICS
(PLEASE MAKE CHANGES, SUGGESTIONS, IMPROVEMENTS)

Aperture: 2-4 m

- Primary Mirror: f/2.75, off-axis parabola, thermally controlled, superpolished to 3 Å rms. IR emissivity 0.007, no central obstruction.

- Secondary Mirror: Airplane focus, used for inverse occulting, actively cooled.

- Adaptive Optics: Direct wave front sensing for solar and bright nighttime sources; beacon wavefront sensing for faint objects and corona.

- Detectors: Visible and IR array detectors up to 4k×4k pixels, at 0.3 to 25 μ.

- Cleanliness: Controlled airflow and electrostatic dust control.

- Optical performance: image sizes (arcsec) @ off axis distance (arcsec): 0" @ 2, 15" @ 1.5, ? @ 90". Strehl ratio $1-(4\pi\sigma/\lambda)^2$ with σ = 3 Å rms; $(4\pi\sigma/\lambda)^2 \approx 5.7\times10^{-4}$ @ .5 μ or 1.4×10^{-6} @ 10 μ.

- Instrumentally-scattered light: $< 5\times10^{-7}$ P. @ $R/R_{\odot} = 1.15$

Panel Session

ABSTRACT

Panel members included Ftaclas, Beckers, Jewitt, Livingston, and Schwenn. Each panel member made a short presentation, which was followed by a brief discussion. After the panel member presentations, a general discussion was opened. Session moderation by Zirker.

1. Panel Discussion

Ftaclas: I have several "program points" for the proposed instrument which we have heard about here.

1. It is good to emphasize dual-use technology. There must be a sincere effort from the beginning to incorporate both daytime and nighttime interests into the telescope.

2. Scientists should be bold in identifying the problems to be addressed by this instrument. Problems should be defined with a clear vision and a determined effort should be made to solve them.

3. Start with the science drivers and design a real telescope. Regarding the science drivers: propose to solve the whole problem. Columbus didn't propose to cross half of the Atlantic.

4. Don't do the engineering; instead focus on the role of the science to be done with the telescope. Be sure you can get the science job done.

Now is a good time to get excited about new ideas and to dream about making breakthroughs.

Beckers: I would propose the name "CLEAR" for the telescope: Coronagraphic and Low-Emissivity Astronomical Reflector. I think we should remember that within the Astronomical Division of the NSF, the $20M GONG project is the only project funded in the past 18 years. And GONG is successful because it has broad appeal and widespread support; it is a unique astronomical project.

Regarding CLEAR, I think that 1/2 the proposals for the telescope will be for coronal and solar science, perhaps only 10% of the day use will be for coronal science.

We need to get presentations out to the funding and astronomy communities. The engineering study should be done in 1995, and we should have a real science proposal done in 1997.

Conventional wisdom has shown that a vacuum telescope is needed for high resolution studies. But new studies of mirror seeing show that it is very important for night-time astronomy...

(Ftaclas: Dome and mirror seeing have been proven to be the two most important factors on MKO telescopes.)

Now the seeing at the McMath telescope has been calculated, and a 2 degree temperature rise in the mirror temperature has been shown to degrade the telescope seeing.

(Koutchmy: Remember that the best solar telescope is at Pic du Midi. It is not a vacuum telescope; it was developed by Lyot for granulation studies. Lyot took extensive pains to limit the heating of the interior of the telescope and to control the heating near the entrance aperture. His work implies that a vacuum is not needed.)

(Dunn: I'm not sure that an open air telescope will give you 0.1 arcsec seeing. Ultimately we agree that a vacuum telescope is best?)

(Beckers: The price you pay is that you lose access to the far-IR since it is blocked by the vacuum windows. We don't know whether a proper open-air design can equal a vacuum design.)

Jewitt: I have an external perspective on this discussion, and perhaps a more global view. I'm not sure if there are enough solar scientific goals discussed here to make this project rational. Not having strong goals would be a "kiss-of-death" for any project. You need solar scientific goals that are 100% justifiable, and then support these with night-time and debris science drivers that are also 100% justifiable. What are the science drivers? In the proposals high spatial resolution was stressed. Is the main driver high spatial resolution or coronagraphic potential? Raising money for a project involves psychological techniques. You need support from a large number of people and they have to include the right people.

It seems like solar astronomy has an image problem with the public and with astronomers. With the public the bottom-line is ignorance. The public is interested by fundamental questions. With extragalactic astronomy the fundamental question is "What is the origin of the universe?", and public interest has been focussed there. With planetary astronomy the fundamental question is "What is the origin of the Earth and of life on the Earth?". Again public interest is focussed there and has supported missions to other planets. What is the fundamental question about the Sun? The biggest question used to be "Why does the Sun burn?", but that was answered in the 1930's. A fundamental question must be developed to focus public interest on the Sun. Within the scientific community, education is again the issue. Other astronomers do not read solar papers, so the word must be spread through the astronomy community. Getting monetary support requires education of both communities.

Finally it is important to resolve the technical issues. What is the size of the telescope? Will it use adaptive optics? Is high-resolution or coronagraph capability most important? Where will the site be? Where is the best site, and how do you measure it? The drivers for these issues must be science. Space debris and space weather questions will not distinguish solar astronomy, although they may be interesting additions.

(Ftaclas: There is $1M in EOS without any solar component. Political activism is needed.)

Livingston: The person on the street is not interested in cosmology, but knows that the Sun heats his house and has other direct effects on his life. It seems like this source would support unlimited funding.

(Beckers: But there are 17 new telescopes of 8-10m sze that have been justified on cosmology issues; and they don't do anything that directly helps the guy on the street.)

Well, for the general public, eclipses are intersting: the Sun actually goes away.

Advocates for the McMath upgrade have the following perspective: there is definitely a need for a 4m telescope. There is not much interest in a 1m class aperture, why? Well, magnetic field measurements using Zeeman splitting need a certain signal to noise ratio, and this implies a large telescope, or long exposures. Long exposures are not the way to go. There is also a lot of stress for 12 micron observations with high spatial resolution. Also, instrumentation like the FTS are starved for photons.

An important point is that with an 8m telescope gives 2-3 nights of observing time. Observing runs can be scheduled for longer periods on a 4m telescope.

We must take into account image seeing. Temperature controls on the mirrors are required, and mirror emission for night-time problems is important. Is there new material that can help? SiC is extremely stiff with a high temperature conductivity, and there are a lot of other good things about it. The group in St. Petersburg is working with it now.

Tapping the space debris group is good for potential funding. This seems like an important thing to do. Good advocacy for this telescope among several groups would be an important issue.

Schwenn: From a "space-data" viewpoint, ground-based coronagraphs are a good idea. I would propose the name "LUMICOS" for Large Universal MIrror COronagraph System.

The basic science drivers for this instrument are broad:

1. the solar corona above the limb must be studied, including the structure, dynamics and magnetic topology, of the solar corona and the solar wind; reaching all the way to the Earth.

2. the solar disk can be studied well, esp. B, waves and oscillations.

3. Planetary atmospheres (including moons and comets) are a good target, ionic compositions, dynamical processes, etc.

4. Stellar systems, nebula, extra-solar planets, and gravitational lenses are good targets and a survey is feasible with this telescope. These topics can excite the public interest.

5. There are environmental issues too: space debris, and perhaps atmospheric pollution monitoring. These are not exciting from our scientific perspective, but are nice for side issues.

(Ftaclas: We must decide what is the main problem? Does it require a dedicated telescope?)

(Schwenn: If space debris is the main problem, then would the Air Force pay for it?)

(Beckers: The catalog of 1cm and larger particles is the AF's priority, since those particles cause catastrophic collisions. A second priority is the statistics of the debris field; the time evolution of the debris cloud, which the CLEAR telescope could best address. Since this is a secondary priority, there is not much money available.)

(Ftaclas: So this telescope would not be a patrol instrument.)

When a coronal person shows a single image, there are 20 unsolved questions in that image. This must be communicated to the man on the street. The solar-terrestrial relationship is important. There should be a front-page article in the NY Times saying for example, "A coronal-mass ejection will hit the Earth at 7pm".

Outside 1.1 solar radii the space instruments will provide perhaps the best data. The lower coronal region is where this instrument should be focussed. Are there plasmoids in this region, or jets? The low-laying corona must be studied. Small scales need a big telescope. In the visible 1 arcsec resolution needs only a 10cm telescope; this can be done from orbit. In the IR, or for resolutions of 0.1 arcsec then a meter class instrument is needed, and this must be addressed from the ground. For disk studies at 1.5 microns; to resolve the magnetic structures we need 1-2 meter telescope. If we want to work at 12 microns, then we need at least a 4m.

(Knoelker: Theorists aren't interested so much in the 0.2 arcsec limit. Can we achieve milli-arcsecond resolution with no seeing with this instrument?)

(Vial: We need to look at shorter wavelengths with space-based instruments for better spatial resolution.)

(Beckers: The unique thing about this telescope would be the ability to explore the coronal magnetic fields. With disk magnetic fields the telescope would compete with other instruments.)

It is important to start understanding the trade-offs. For night-time studies you want more photons (bigger telescope) and high spatial resolution.

Let's put this all together then, and find out what are the real science drivers for coronal, disk and astronomy applications. From that we can work on the trade-offs, and what the design of the telescope must be.

I have several caveats for design and planning:

1. "too big" might be bad without proper scientific justification and technical feasibility.

2. "too small" might be as bad since funding agencies are irrational about judging some things.

3. define what is the best telescope choice and go after it.

4. make sure an appropriate site exists; good-seeing and coronal skies might be mutually exclusive.

5. pay attention to the competition in space a planetary programs, and design complimentary programs.

(Livingston: I'm worried about starting a long site survey program.)

2. General Discussion:

Koutchmy: I didn't want to make political comments, but the comment about using the telescope for coronal physics for only 10% of the time seemed incorrect.

GONG was used as an example of success. It was successful because the scientific proposal was to do helioseismology, and people feel that this is worth doing. The prime reason was seismology, but the project was also international and gained support that way. And like all good projects it is one that the man-on-the-street can like. Ultimately that is our duty.

Unfortunately the money is controlled by others, not the men on the street. To get scientific agreement, we need a driver that is different from the coronal heating or the extra-solar planet ideas. The topic might be the solar magnetic connection between the Sun and the Earth, and the physical disturbances caused by the Sun upon the Earth. This way the connection is with the Sun, not with climate or weather. Magnetism is something that everyone feels? This topic might attract the attention of the "deciders". The science driver must be the measurement of the solar B off the disk, since there are no measurements there.

Beckers: Right now we don't know if we can do that.

Knoelker: A political remark – solar physics is not making its case. I think we can make a strong case, for example solar irradiance measurements have a direct impact on climate changes. Such studies include a lot of solar physics.

We must address solar science issues not in contradiction with the low stray light of the coronagraph idea. A question would be "What about the cycles of other stars?" Maybe this is a dumb question for a large telescope. I don't see that the open-air design or the vacuum design questions are so critical, since AO is progressing and has a lot of promise.

Regarding site selection, I suggest that we not waste any more money on this issue, since the decision process is political and has nothing to do with real science. In the case of CME, there are reasons for measuring the magnetic structures directly and not just using density measurements. The general case for CME measurements is very important. Finally I would point out that ground-based projects are one order of magnitude cheaper than space missions, and that many useful things can be done from the ground.

Vial: I have two points. First, this project shouldn't compete with space programs; the design should at least understand what is happening in the space projects. Complementary programs are essential, and simultaneous observing is important. SOHO is a big program and will be ready to observe in 15 months. Collecting simultaneous IR observations could prove the usefulness of such data and be used to justify a larger ground-based instrument.

Secondly, I fully support the idea that having good science goals is the first step. It is important to have rational and realistically defined objectives; the technical issues shouldn't be the first step, but rather the scientific issues.

Ruzmaikin: This project could open a new window for IR coronal work. Also, the space debris idea could be made more scientific, perhaps even focussing on cleaning up the environment by removing the debris? I think we should be open with regard to scientific programs. Remember that Yohkoh has 5 arcsec pixels and TRACE will have 1 arcsec. There is a lot of coronal physics that can be done with that resolution. Scientists at JPL might be interested in collaborations, for instance with the investigation of the solar wind. This project could be done from the ground with the proposed telescope.

Morgan: Two points. First I think that solar seems to be missing a list of subjects and a global argument for its science. In planetary astronomy we get that support through Sagan; what about the solar group?

Secondly, this new project will need joint support, firstly from all solar physics, and secondly from another group. In this way, when another group can't get funding, they would say that " ...if we couldn't get our money, then this solar telescope is a good project for funding." The orbital debris group is a large community, larger than the solar physics group. No one is here from that community, and it seems like there should be direct communication with that group. They should be involved; and a general presentation to that group should be made.

Ftaclas: I have a sense that some of the problems described here aren't hard enough, or important or demanding enough. The biggest NSF project is LIGO, which will try to make a measurement at 10^{-23}. I suggest that we find the LIGO proposal and read it as an example of how to make a successful new start.

Doinidis: This seems like it would be a 24 hour telescope. We need to have input from the other users, although the solar science should be the main driver; the telescope must meet all the solar requirements.

Krishan: There are exciting physics problems which can be addressed with this instrument. General convection problems (ie photospheric observations) are essential to many types of science, and turbulence data (from many solar sources) are useful. Magnetic reconnection issues (which can be addressed with coronal measurements) are important for stellar formation and quasar theories. The Sun can be a powerful astrophysics teacher.

Zayer: Three comments. The man on the street sees us in an ivory tower, without providing much relevance to his concerns. The budget makers need to be re-elected so they care about the concerns of the man on the street. Ultimately we need to educate the general public so that our funding continues. Second, I don't agree that a site with good seeing will not be coronal. They are not mutally exclusive. Finally, I think that dreaming and going after the ultimate telescope is not very practical.

Neidig: The message from Congress is "Science must be strategic". Projects get no money unless it betters the life of the man on the street or the economy. But, Congress would turn us down if we argued for the practical applications of this telescope, so the science issues must come first. We must juggle the science and the pratical applications of this telescope. I've heard a lot about the selling of the telescope, but only a little about the science.

Lindsey: I think we should resist the impulse to generate a project as large as say Gemini. But we shouldn't act like martyrs since we can't get funding; maybe we should put some thought into our connections with the taxpayers. It's a difficult balance between playing chess and having to build the chess-board.

Zayer: Mistakes we make are to emphasize the practical aspects... the guy on the street wants to dream, and they can sense if we are insincere about the practical applications.

Ftaclas: I think that funding agencies fund good science, and that should be stressed.

Kuhn: Seems like the "solar" members of the committee are discussing general items, while the "outside" members are asking for a prioritized list of science questions. We should make a list of questions and sell them.

Jewitt: I think you should have a ranked list of scientific subjects to provide some convergence for this discussion.

Zirker: We could at least start a list of science drivers.

1. Coronal heating.
2. The origin of the solar wind. (Koutchmy: How about "mass loss of the Sun" as a more general topic?)
3. The solar cycle.

Lindsey: I think we should resist the impulse to generate a project as large as [illegible]. But we shouldn't adopt the attitude, since we can't get funding, maybe we should put some thought into our [illegible] with the proposal. It's a difficult balance between playing chess and having to make the chess-board.

Zavert: Mistakes we make are to emphasize the practical aspects, [illegible] value is [illegible] and they can see [illegible] we are [illegible] about the practical applications.

[illegible]: I think that [illegible] should be [illegible].

Kuhn: Seems like the "inside" members of the committee are discussing general [illegible] while the "outside" members are asking for a prioritized list of science questions. We should make a list of questions and sell them.

Levitt: I think you should have a ranked list of scientific subjects to provide some coverage for the discussion.

[illegible]: We could at least make a list of science drivers.

1. Coronal heating

2. The origin of the solar wind. [illegible] How about [illegible] of the [illegible] as a [illegible]?

3. The solar cycle.

WORKSHOP PARTICIPANTS

Vladimir Airapetian, National Solar Observatory, Sunspot, NM

Richard Altrock, USAF Phillips Laborabory/GPSS, Sunspot, NM

Eugene H. Avrett, Harvard-Smithsonian Center for Astrophysics, Cambridge, MA

Thomas R. Ayres, University of Colorado, Boulder, CO

K.S. Balasubramaniam, National Solar Observatory, Sunspot, NM

Tim Bastian, National Radio Astronomy/VLA, Socorro, NM

Jacques Beckers, National Solar Observatory, Tucson, AZ

Karine Bocchialini, Institut d'Astrophysique, CNRS, Paris, France

Peter Cargill, Naval Research Laboratory, Washington, DC

Edward S. Chang, University of Massachusetts, Amherst, MA

Alan Clark, University of Calgary, Calgary, Alberta, Canada

Sidney D'Silva, National Solar Observatory, Tucson, AZ

Steve Doinidis, National Solar Observatory, Sunspot, NM

Alexander Epple, Max-Planck Institute fur Aeronomie, Germany

Bernhard Fleck, European Space Agency, Paris, France

Bernard Fort, Observatoire Midi-Pyrenees, Toulose, France

Al Fowler, National Optical Astronomy Observatories, Tucson, AZ

Christ Ftaclas, Hughes Danbury, Danbury, CT

Dan Gezari, NASA/Goddard Space Flight Center, Greenbelt, MD

Allen Hoffman, SBRC, Galega, CA

Bernard V. Jackson, University of California, La Jolla, CA

Dave Jewitt, Institute for Astronomy, Honolulu, HI

Harrison P. Jones, National Solar Observatory/NASA SW Solar Station, Tucson, AZ

G.C. Joshi, Uttar Pradesh State Observatory, India

Sidney O. Kastner, Math Science Consultants, Greenbelt, MD

Pierre Kaufmann, Lab. Apl. Espaciais, Escola Politecnica, Sao Paulo, Brazil

Stephen L. Keil, USAF Phillips Laboratory/GPSS, Sunspot, NM

Christoph Keller, National Solar Observatory, Tucson, AZ

Iraida S. Kim, Moscow University, Moscow, Russia

Michael Knoelker, Kiepenheuer Institut fur Sonnenphysik, Freiburg, Germany

Serge Koutchmy, Institut d'Astrophysique, CNRS, Paris, France

Vinod Krishan, Indian Institute of Astrophysics, Bangalore, India

Jeff R. Kuhn, National Solar Observatory, Sunspot, NM

Bob Kurtz, IR Lab, Inc., Tucson, AZ

Jing Li, DASOP, Observatoire de Meudon, France

Haosheng Lin, California Institute of Technology, Pasadena, CA

Charles Lindsey, National Solar Observatory, Tucson, AZ

William C. Livingston, National Solar Observatory, Tucson, AZ

Thomas Moran, NASA/Goddard Space Flight Center, Greenbelt, MD

Tom Morgan, SW Research Institute, San Antonio, TX

Karin Muglach, Institute for Astronomy, Austria

Donald F. Neidig, USAF Phillips Laboratory/GPSS, Sunspot, NM

Larry November, National Solar Observatory, Sunspot, NM

Matthew J. Penn, National Solar Observatory, Sunspot, NM

Valentin Pillet, High Altitude Observatory, Boulder, CO

Douglas Rabin, National Solar Observatory, Tucson, AZ

Richard R. Radick, USAF Phillips Laboratory/GPSS, Sunspot, NM

Thomas Rimmele, National Solar Observatory, Sunspot, NM

Thierry Roudier, Univ. Paul Sabatier, France

Isabelle Ruedi, Institute of Astronomy, Zurich, Switzerland

Alexander Ruzmaikin, San Fernando Observatory, Sylmar, CA

Wolfgang Schmidt, Kiepenheuer Institut fur Sonnenphysik, Freiburg, Germany

Rainer Schwenn, Max-Planck Institute fur Aeronomie, Germany

David Sime, High Altitude Observatory, Boulder, CO

Doug Simons, Gemini Project Office, National Solar Observatory, Tucson, AZ

Raymond N. Smartt, National Solar Observatory, Sunspot, NM

Sami K. Solanki, Institut fur Astronomie, Zurich Switzerland

Robert Stencel, University of Denver, Denver, CO

Yoshinori Suematsu, National Astronomical Observatory, Tokyo, Japan

Tokio Tsubaki, Shiga University, Shiga, Japan

Jean-Claude Vial, University De Paris XI, Paris, France

Kadri Vural, Rockwell Science Center, Thousand Oaks, CA

Richard White, High Altitude Observatory, Boulder, CO

Pete Worden, HQ USAF/TA, Washington, DC

Igor Zayer, Lockheed Missiles, Palo Alto, CA

Jack Zirker, National Solar Observatory, Sunspot, NM

Infrared Tools for Solar Astrophysics: What's Next?
15th NSO/Sacramento Peak workshop, Sept. 19-23, 1994
AGENDA

ABSTRACT

A meeting to sample new results derived from extended-object IR observations, and to explore the potential capabilities of large-aperture, low-scattered-light instrumentation

1. Monday, Sept. 19

8:00 Registration (community center building)
9:00 Meeting Introduction: J. Beckers, J. Kuhn (LOC)
9:25 Observing Proposal Gedanken Experiment, Neidig
9:30 Scientific Introduction: Koutchmy
10:30 Participant Introductions(a.) (2min/person, who, where, why, other news), Smartt
12:00 Lunch

1.1. Some important coronal physics questions? (Beckers)

2:00 Zirker, Coronal heating mechanisms (I)
2:40 Cargill, Some observational consequences of coronal heating models (II)
3:20 Airapetian, Optical Diagnostics of Coronal Loop Interactions
3:40 November, 1991 eclipse coronal fine structure
3:50 Poster Reports:
Altrock, Association of Solar Coronal Temperature and Structure from ...
4:00 Break

1.2. New diagnostic possibilities from coronal IR spectroscopy? (Solanki)

4:20 Chang, IR coronal lines from EUV data
5:00 Kastner, Some Questions of Spectroscopic Interest: O, D, Fe(ionized)
5:30 Penn, New observations of IR coronal emission lines
6:30 Hors d'oeuvres at the director's residence

2. Tuesday, Sept. 20

8:30 Participant Introductions (continued), Penn

2.1. What's new, IR detectors? (Kuhn)

9:30 Fowler, IR Detectors in the 1-5 micron Range
10:10 Cooper, 1Kx1K Short Wave IR Arrays for Astronomy

10:50 Break
11:10 Hoffman, SBRC IR Detector Developments
11:50 Group Photograph (in front of community center building)
12:20 Lunch

2.2. What's new, IR instrumentation? (Sime)

2:00 Rabin, Near IR Magnetographs
2:20 Kuhn, Near IR Coronal Spectroscopy
2:40 Break
3:00 Simons, CFHT's Imaging FFT Spectrometer
3:20 Jones, The 1083nm narrow-band filter system
3:40 Kurtz, IR Camera Instrumentation
4:00 Poster report:
Zayer, Tunable Field-widened Etalons for the Visible and Infrared

2.3. Longer wave diagnostics (mm to radio) (Lindsey)

4:10 Bastian
4:50 Kaufman, Solar Burst Emissions in submm/far IR continuum
5:05 Clark, Eclipse measurements of CO and Pfund beta emission distribution above the solar limb using Amber InSb array at the McMath-Pierce
5:15 Poster report:
Clark, Detection of HI n=20-19 submm line in emission at the solar limb using a polarizing FFT spectrometer
5:30 Observatory tour
6:30 BBQ at Community Center
8:30 Don Neidig (our own prize winning local artist) will talk about "Visual Approaches in Art: A Personal View of 19th and 20th Century American Realism" in the community center

3. Wednesday, Sept. 21

8:45 Strawman Telescope Specifications, Penn

3.1. What new IR/coronagraphic capabilities do next generation telescopes offer? (Dunn)

9:00 Beckers, New telescopes, prospects and review
9:50 Smartt, Reflecting coronagraphs: Prospects
10:20 Livingston, Upgrade of McMath-Pierce for better IR use
10:40 Break
10:50 Ftaclas, Faint Sources, Coronagraphs and the Diffraction Limit

11:30 Roudier, Themis and its IR Applications
11:45 Jackson, Stray Light Studies of Coronagraphs
12:00 Lunch
2:00 Keller, IR capabilities of LEST
2:20 Vial, Solar Physics from ground-based IR and space observations
2:40 Poster report:
Schwenn, Pico – A Mirror Coronagraph on Pic Du Midi
Kim, Mirror Coronagraphic Devices and Possible Applications

3.2. New Questions addressible from such Coronagraphic/IR capabilities: (Kim)

3:00 Neidig, Near IR coronagraphic detection of Space Debris
3:20 Jewitt, Coronagraphic observations of near-stellar environments
4:05 Break
4:15 Fort, Coronagraphic Observations of Gravitational Lenses
4:55 Morgan, Coronagraphic Observations of Solar System Objects
5:35 Poster reports:
Tsubaki, Needs for a Large Solar Telescope Viewed from the Trend of Studies Published During the Last 20 Years
6:30 Banquet at the Lodge in Cloudcroft
8:00 Eidenbach, "Prehistoric Solar Observatories in the Southwest"

4. Thursday, Sept. 22

4.1. New results from IR chromospheric and photospheric spectroscopy (Vial)

9:00 Ayres, Thermal Bifurcation Revisited
9:50 Avrett, Two-component Modeling of Solar IR CO Lines
10:40 Lindsey, Theory of Radiative Transfer in Inhomogeneous Media
11:20 Poster reports:
Stein, Is there a Non-magnetic Chromosphere?
Krishan, A Turbulent Model of the Solar Granulation
12:00 Lunch

4.2. IR Magnetic Diagnostics and Photometry (Rabin)

2:00 Solanki, Solar Magnetic Features: An Overview of the Observations
2:40 Lin, IR Photometry of Magnetic Elements
3:00 Schmidt, High Resolution Spectroscopy of Sunspots
3:20 Rimmele, IR Observations of a Disappearing Flux Element
3:40 Balasubramaniam, Magneto-optic Effects in the 1.5 micron Iron Line
4:00 Ruzmaikin, Scaling of Solar Magnetic Fields, Why IR?

4:20 Poster reports:
Knoelker, A Study on the Detectibility of Small-scale Magnetic Fields
Li, Removing the 180deg Ambiguity in Vector Field Measurements
Muglach, Oscillations in Active Plage Regions from 1.56micron Observations
Penn,Kuhn, Measuring Flare Magnetic Fields using HeI 1083nm
Lin, The Magnetic Field Distribution of the Sun
4:30 Break

4.3. Near IR (1083) diagnostics (Keil)

4:40 Poster reports:
Suematsu, He 1083nm observations and chromospheric and coronal activity
Joshi, Molecular Spectroscopic Diagnostics
Schmidt, Limb Observations of He I 1083nm
4:55 Kim, New Prominence Observations Using 1083nm
5:10 Bocchialini, Wave properties of the chromosphere in 1083nm
5:25 Ruedi, Chromospheric magnetic field measurements in HeI 1083
5:40 Fleck, What we've learned about chromospheric oscillations from 1083nm
7:00 Reception at Hubbard Education Facility followed by OmniMax theatre screening of "The Discoverer's" at 8:00pm

5. Friday, Sept. 23

5.1. Observing proposal, telescope specification discussion, summary

9:00 Smartt, Meeting highlights and introduction to panel discussion of Large Mirror Telescope Projects
9:30 Panel Discussion, Telescope capabilities and unique scientific capabilities (Ftaclas, Beckers, Jewitt, Livingston, Schwenn)
12:00 Lunch

NATIONAL SOLAR OBSERVATORY/SACRAMENTO PEAK
FIFTEENTH SUMMER WORKSHOP
19 - 22 SEPTEMBER 1994

1. Bernard V. Jackson
2. Rainer Schwenn
3. Steve Doinidis
4. Bernhard Fleck
5. Eugene H. Avrett
6. Richard White
7. Thomas R. Ayres
8. Wolfgang Schmidt
9. Jing Li
10. David Sime
11. Bernhard Fort
12. Al Fowler
13. Isabella Ruedi
14. Christ Ftaclas
15. Dave Jewitt
16. Alan Clark
17. Iraida Kim
18. Raymond N. Smartt
19. Charles Lindsey
20. Haosheng Lin
21. Vinod Krishan
22. Sidney Kastner
23. Alexander Epple
24. Peter Cargill
25. Larry November
26. Alexander Ruzmaikin
27. Yoshinori Suematsu
28. K.S. Balasubramaniam
29. Karine Bocchailini
30. Igor Zayer
31. Karin Muglach
32. G.C. Joshi
33. Michael Knoelker
34. Tokio Tsubaki
35. Matthew J. Penn
36. Edward S. Chang
37. Christoph Keller
38. Richard R. Radick
39. Vladimir Airapetian
40. Valentin Pillet
41. Thierry Roudier
42. Sami Solanki
43. Phil Wiborg
44. George W. Simon
45. Jeff R. Kuhn
46. William Livingston
47. Harrison P. Jones
48. Jean Claude Vial
49. Serge Koutchmy
50. Donald F. Neidig
51. Bob Kurtz
52. Thomas Morgan
53. Stephen L. Keil
54. Richard B. Dunn
55. Pierre Kaufmann
56. Sidney d'Silva

NSO/SP SUMMER WORKSHOP SERIES SUMMARY

Large-Scale Motions of the Sun: Summer Workshop, Sunspot, N.M., 1-2 September, 1977. S.A. Musman, chairman, 45 participants (NSO/SP Summer Workshop no. **1**, unpublished).

Summary: This first meeting of the NSO/SP Summer Series focused on both the theory and observation of solar oscillations, and on the application of oscillation data to the determination of differential interior rotation. A model of solar rotation was also presented. The application of solar p-modes as a tracer of large-scale horizontal motions was discussed, together with studies of the dynamics of gas and magnetic fields in the outer solar layers.

Small-Scale Magnetic Fields in the Solar Atmosphere: Summer Workshop, Sunspot, N.M., 16-20 July, 1979. L.E. Cram, chairman, 50 participants. (NSO/SP Summer Workshop no. **2**, unpublished).

Summary: Observed properties and theoretical consequences of the structure and intensity of solar magnetic fields were major areas of focus. Observations show that almost all the solar magnetic flux is concentrated in 1-2 kG bundles which occupy only a small fraction of the surface area. Processes responsible for this concentration, as well as the implication to solar cycle dynamo theories, were examined. The need to study in detail the physical processes occurring in single small flux tubes was demonstrated. Such studies require sensitive velocity and magnetic field measurements over relatively long periods, with excellent spatial resolution.

Solar Instrumentation: What's Next? Conference Proceedings, Sunspot, N.M., 14-17 October, 1980. R.B. Dunn, ed., 618 pp, 56 participants. (NSO/SP Summer Workshop no. **3**).

Summary: This workshop, the first to be published in the NSO/SP Summer Series, explored the observational needs of solar and solar-stellar astronomers, and the required instrumentation to fulfill those needs. Sessions were dedicated to LEST, which is in its initial planning stages, and to image stabilization techniques. The use of interferometry for solar observations, as well as the importance of the solar-stellar connection to further the understanding of the Sun, were also discussed.

The Physics of Sunspots: Conference Proceedings, Sunspot, N.M., 14-17 July, 1981. L.E. Cram and J.H. Thomas, eds., 495 pp, 57 participants. (NSO/SP Summer Workshop no.**4**).

Summary: The workshop began with the discussion of relevant observations on the birth, evolution, and emergence of sunspots, as well as the analysis of magnetic features and vector magnetic fields of active regions; and continued with empirical and theoretical models of sunspots. A study of starspots was presented, as were studies of solar dynamical phenomena, such as umbral flares, Evershed effect, oscillations, and the influence of sunspots on the output of solar radiation.

Small-Scale Dynamical Processes in Quiet Stellar Atmospheres: Conference Proceedings, Sunspot, N.M., 25-29 July, 1983. S.L Keil, ed., 469 pp, 66 participants. (NSO/SP Summer Workshop no **5**).

Summary: The study of small-scale dynamical processes in the solar atmosphere was the major focus, including the implication of this work to the interpretation of unresolved observations of stellar atmospheres in general. Sessions were dedicated to high-resolution observations of small-scale solar phenomena and their interpretation; to the modeling of small-scale processes, such as granulation; to the interaction of convection and magnetic fields; and to low-resolution observations, bearing on small-scale dynamical processes such as line asymmetries and wavelength shifts.

Chromospheric Diagnostics and Modeling: Workshop Proceedings, Sunspot, N.M., 13-16 August, 1984. B.W. Lites, ed., 314 pp, 69 participants. (NSO/SP Summer workshop no. **6**).

Summary: The OSO-8 observations revealed that chromospheres (and coronae) are the rule, rather than the exception, for most classes of stars. The aim of the meeting was to enhance the understanding of chromospheres. Sessions dedicated to the observations of chromospheric structures and magnetic fields were followed by talks on the calculation of chromospheric models and heating processes. Working groups were organized on Line Synthesis and Atmospheric Modeling; Magnetic Fields; Space and Ground-Based Collaborative Observations; and Chromospheric Heating and Energy Balance.

The Lower Atmosphere of Solar Flares: Relationships Between Low Temperature Plasmas and High Energy Emissions. Proceedings of the National Solar Observatory/Solar Maximum Mission Symposium, Sunspot, N.M., 20-24 August, 1985. Don Neidig, ed., 503 pp. 58 participants. (NSO/SP Summer Workshop no. **7**).

Summary: The meeting analyzed flare phenomena in the chromosphere (impulsive phase, shocks and condensations, observations of flares in helium lines) and included white light flares, which had, until recently, been often considered a separate flare class. High energy emissions in solar and stellar flares were also reviewed. The meeting ended with sessions on the origin of flare energy, its transport, and on the evidence for and against electron beams as the agent generating hard X-ray bursts in solar flares.

The Internal Solar Angular Velocity: Theory, Observations and Relationship to Solar Magnetic Fields. Symposium Proceedings, Sunspot, N.M., 11-14 August, 1986. B.R. Durney and S. Sofia, eds., 374 pp, 60 participants. (NSO/SP Summer Workshop no. **8**).

Summary: The meeting aimed at the study of the internal angular velocity, both in the radiative zone (where the angular momentum loss due to the solar wind and magnetic fields must play an essential role) and in the convection zone, where the interaction of rotation with convection generates differential rotation - an essential link in the generation of solar magnetic fields. Sessions were also dedicated to observations of surface phenomena as well as of normal modes of oscillations of the Sun, which eventually will provide accurate information of the internal solar angular velocity.

Solar and Stellar Coronal Structure and Dynamics: A Fetschrift in Honor of Dr. John W. Evans, Sunspot, N.M., 17-21 August, 1987. R.C. Altrock, ed., 577 pp, 60 participants. (NSO/SP Summer Workshop no. **9**).

Summary: The festschrift was dedicated to honoring the leadership and accomplishments of John W. Evans, the first director of Sacramento Peak Observatory. The primary focus was on physical processes that result in observable phenomena in coronae of the Sun and stars. The meeting began with talks on stellar coronae, and continued with the solar corona, including processes such as coronal holes and loops, and theories concerning the mechanism for coronal heating. Sections were dedicated to the solar-cycle variations of the corona, to the observations of magnetic fields and their numerical modeling, and to solar mass ejections, transients and flares.

High Spatial Resolution Solar Observations: Proceedings of the Tenth Sacramento Peak Summer Workshop, Sunspot, N.M., 22-26 August, 1988. O. von der Lühe ed., 573 pp. 56 participants. (NSO/SP Summer Workshop no. **10**).

Summary: This workshop studied a number of modern concepts to advance high angular resolution observations of the Sun. It was the intent of the workshop to bring together those who develop high resolution instrumentation with those who intend to use it in their experiments. The topics ranged from planned and newly commissioned solar telescopes, active and adaptive optics, interferometry, and data analysis methods, to observational results of spectroscopy with high spatial resolution, polarimetry, velocity field measurements, and coronal observations.

Solar Polarimetry: Proceedings of the Eleventh National Solar Observatory/Sacramento Peak Summer Workshop, Sunspot, N.M., 27-31 August, 1990. L.J. November, ed., 501 pp, 70 participants. (NSO/SP Summer Workshop no. **11**).

Summary: Due to the inherent anisotropy of the electromagnetic process, its optical manifestations contain polarization information as well as spectral information. Precise polarimetry has been difficult in astronomical observations because the polarization transformation properties of instruments can be quite large and difficult to calibrate. With the advent of improved instrumentation and detector technology, a number of efforts ate underway around the world to obtain precise polarization measurements. This workshop considered technical issues (both facilities and methods); polarimetric observations of solar fine structure, flares, coronal features, and solar/stellar phenomena; the state of the art in solar infrared instrumentation and observations; and interpretive issues.

The Solar Cycle: Proceedings of the National Solar Observatory/Sacramento Peak 12th Summer Workshop, Sunspot, N.M., 15-18 October, 1991. K.L. Harvey, ed., 526 pp, 76 participants. (NSO/SP Summer Workshop no. **12**).

Summary: The workshop focused on six basic questions: 1) What does surface magnetic flux tell us about subsurface magnetic fields? 2) Is the flux we see at the surface causually involved in the solar cycle? How can we use stellar cycles to understand phenomena of the solar cycle? 3) Do the observed large-scale velocity patterns provide information concerning solar cycle mechanisms? 4) The "extended" solar cycle: what does it mean? 5) How does solar luminosity change during a solar cycle? What can we infer about the cycle mechanisms from stellar luminosity variations? 6) Does any existing model, either phenomenological or theoretical, of either solar or stellar cycles, deserve to be a paradigm?

Solar Active Region Evolution – Comparing Models with Observations, Sunspot, N.M. Aug. 31 - Sept. 3, 1993. K.S. Balasubramaniam and G.W. Simon, ed., 3000 pp. 2000 participants (NSO/SP Summer Workshop no. 14).

Summary: The workshop concentrated on comparing observations and models of active region evolution. Topics included: (1) large scale structure and evolution of solar activity on time scales of a solar cycle, (2) kinematic and dynamic models of solar convective processes that drive active region formation and evolution, (3) magnetoconvection, (4) flux tube physics, (5) electric currents, current sheets and magnetic reconnection driven by active region evolution, (6) vector magnetic fields and resulting coronal fields, and (7) observations of physical parameters in active regions and their relationship to models from the photosphere through the corona. The workshop was attended by over 75 participants drawn from 13 countries.